HRW

Chapters 4-6

ADVANCED ALGEBRA

TEACHING RESOURCES

explore

communicate

APPLY

$g(\theta)=15\sin(200\theta)$

HOLT, RINEHART AND WINSTON
Harcourt Brace & Company
Austin • New York • Orlando • Atlanta • San Francisco • Boston • Dallas • Toronto • London

TO THE TEACHER

HRW Advanced Algebra Teaching Resources contains blackline masters that complement regular classroom use of *HRW Advanced Algebra*. They are especially helpful in accommodating students of varying interests, learning styles, and ability levels. The blackline masters are conveniently packaged in four separate booklets organized by chapter content. Each master is referenced to the related lesson and is cross-referenced in the *Teacher's Edition*.

- **Practice Masters** (one per lesson) provide additional practice of the skills and concepts taught in each lesson.
- **Enrichment Masters** (one per lesson) provide stimulating problems, projects, games, and puzzles that extend and/or enrich the lesson material.
- **Technology Masters** (one per lesson) provide computer and calculator activities that offer additional practice and/or alternative technology to that provided in *HRW Advanced Algebra*.
- **Lesson Activity Masters** (one per lesson) connect mathematics to other disciplines, provide family involvement, and address "hot topics" in mathematics education.
- **Chapter Assessment** (one multiple-choice test per chapter and one free response test per chapter)
- **Mid-Chapter Assessment** (one per chapter)
- **Assessing Prior Knowledge and Quiz** (One Assessing Prior Knowledge per lesson and one Quiz per lesson)
- **Alternative Assessment** (two per chapter) is available in two forms, one which entails concepts found in the first half of the chapter and the other which entails concepts found in the second half of the chapter.

Developmental assistance by B&B Communications West, Inc.

Printed in the United States of America

ISBN 0-03-095391-X

2 3 4 5 6 7 066 99 98 97

TABLE OF CONTENTS

NAME ______________________ CLASS __________ DATE __________

Practice & Apply

4.1 Using Matrices to Represent Data

Let $A = \begin{bmatrix} 5 & -2 \\ -1 & 0 \\ 4 & -3 \end{bmatrix}$, $B = \begin{bmatrix} 6 & -1 \\ 7 & 2 \\ 1 & -2 \end{bmatrix}$, $C = \begin{bmatrix} 1 & 0 \\ -4 & 6 \end{bmatrix}$, and $D = \begin{bmatrix} -1 & 7 \\ -2 & 3 \end{bmatrix}$.

1. Give the dimensions of matrix A. ______________________

2. Give the dimensions of matrix D. ______________________

3. Find $A + B$. ______________________

4. Find $A - B$. ______________________

5. Find $C + D$. ______________________

6. Find $D - C$. ______________________

Use a matrix to represent each set of linear equations.

7. $2x - 3y = 6$ and $-2x + 5y = 11$

8. $-3x + 5y = 8$, $2x + y = -4$, and $6x - y = 1$

Joe, Al, and Jose are salespeople in a car dealership that sells subcompact, compact, and luxury cars. The gross dollar sales for the month of October are given in the following matrix.

October Sales

	Subcompact	Compact	Luxury	
Joe	$30,000	$42,000	$40,000	
Al	$40,500	$54,000	$111,000	= A
Jose	$12,000	$68,000	$70,000	

9. What are the dimensions of matrix A? ______________________

10. Describe entry a_{31}. ______________________

11. Describe entry a_{12}. ______________________

12. Describe entry a_{23}. ______________________

NAME ______________________ CLASS ____________ DATE ____________

Practice & Apply

4.2 Matrix Multiplication

Let $A = \begin{bmatrix} 1 & 4 & -2 \\ -5 & 2 & 0 \\ -1 & 5 & -3 \end{bmatrix}$, $B = \begin{bmatrix} 4 \\ 1 \\ 3 \end{bmatrix}$, and $C = [2 \quad 5 \quad -1 \quad -4]$.

Find the indicated element of each product.

1. ab_{31} ____________ **2.** bc_{23} ____________ **3.** ab_{21} ____________

Indicate whether it is possible to find the product.

4. AB ____________ **5.** CA ____________ **6.** BC ____________

Find the product.

7. $\begin{bmatrix} 2 & 4 \\ 8 & -4 \\ -2 & 6 \end{bmatrix} \begin{bmatrix} 7 & 0 \\ 5 & 3 \end{bmatrix}$

8. $\begin{bmatrix} -1 & 2 & -6 \\ 3 & -1 & 4 \end{bmatrix} \begin{bmatrix} -6 & 4 \\ -2 & 3 \\ 1 & -4 \end{bmatrix}$

____________ ____________

9. $\begin{bmatrix} 2 & 0 & 3 \\ -1 & 5 & 1 \end{bmatrix} \begin{bmatrix} 1 \\ -3 \\ 4 \end{bmatrix}$

10. $\begin{bmatrix} -2 \\ -3 \\ 1 \end{bmatrix} [1 \quad 2 \quad 3]$

____________ ____________

Use your graphics calculator to find the product.

11. $\begin{bmatrix} 2 & -1 & 4 \\ -3 & 5 & 7 \\ 4 & 6 & -8 \end{bmatrix} \begin{bmatrix} -4 & 2 & 3 \\ 7 & -1 & -5 \\ 6 & 3 & 1 \end{bmatrix}$

12. $\begin{bmatrix} 3 & -4 & 1 \\ 2 & 6 & -5 \end{bmatrix} \begin{bmatrix} -1 & 0.5 & 0.2 \\ 3 & 0.1 & 0.4 \\ -2 & -0.4 & 0.9 \end{bmatrix}$

____________ ____________

NAME ______________________ CLASS __________ DATE __________

Practice & Apply

4.3 Systems of Two Linear Equations

Indicate whether each of the following systems is inconsistent, dependent, or independent.

1. $2x + y = 3$
$-x + y = -3$ __________

2. $2x - 4y = 4$
$-x + 2y = 2$ __________

3. $4x - 2y = 16$
$2x - y = 8$ __________

4. $5x - y = 4$
$-5x + y = -3$ __________

5. $2x + 3y = 4$
$-6x - 9y = -12$ __________

6. $4x - 3y = -7$
$x + 2y = 1$ __________

Solve each system using elimination by addition.

7. $x + y = 12$
$x - y = 2$ __________

8. $3x + 2y = -7$
$3x + 5y = -4$ __________

9. $y = -9 - 2x$
$3x - y = -1$ __________

10. $x - 2y = 7$
$3x + 4y = 6$ __________

11. $6y = 14 - 5x$
$4x - y = 17$ __________

12. $4x - 3y = -15$
$2x + 4y = 9$ __________

13. $2x + 3y = 10$
$3x + 5y = 19$ __________

14. $5x + 9y = -6$
$3x + 4y = 2$ __________

15. $4x = 7y + 76$
$7x + 4y = -34$ __________

Leona spent \$9.25 at the craft store buying 23¢ and 29¢ colored pencils. She purchased 35 pencils in all.

16. Write the system of equations that would give the number of each type of pencils.

____

17. How many pencils of each type did she buy?

____

The perimeter of a rectangular garden is 180 ft. Four times the length is equal to 5 times the width.

18. Write a system of equations that can be used to find the dimensions of the garden.

____

19. Find the length and the width of the garden. __________________________

20. Find the area of the garden. __________________________

NAME ______________________ CLASS ____________ DATE ____________

Practice & Apply

4.4 Using Matrix Row Operations

Write an augmented matrix for each system of equations.

1. $\begin{cases} 2x + y = 2 \\ 5x - 3y = -17 \end{cases}$ ____________

2. $\begin{cases} 3x - 2y = 4 \\ x - 4y = 15 \end{cases}$ ____________

3. $\begin{cases} 2x + y + z = 1 \\ x + 3y - 4z = 19 \\ 4x - 2y + 3z = -9 \end{cases}$ ____________

4. $\begin{cases} 3x - y + 2z = 3 \\ 2x + 5y - 3z = -12 \\ x - 3y + 4z = 8 \end{cases}$ ____________

Use back substitution to solve each system of equations.

5. $\begin{cases} 5x - 4y = -22 \\ 3y = 9 \end{cases}$ ____________

6. $\begin{cases} 7x - 6y = 24 \\ 4x = 16 \end{cases}$ ____________

7. $\begin{cases} -x + 5y - 2z = -15 \\ 2y - 4z = -18 \\ 3z = 12 \end{cases}$ ____________

8. $\begin{cases} 3x - 5y - 6z = 27 \\ -7y + 4z = -8 \\ -2z = 4 \end{cases}$ ____________

Solve each system of equations using the row reduction method and back substitution.

9. $\begin{cases} x + y = 5 \\ 3x - 2y = 10 \end{cases}$ ____________

10. $\begin{cases} 5x - 2y = -24 \\ x + 3y = 19 \end{cases}$ ____________

11. $\begin{cases} -2x + y = 4 \\ 8x - 3y = 5 \end{cases}$ ____________

12. $\begin{cases} x + 2y - 3z = -12 \\ 3x - 2y + 4z = 8 \\ -4x - 3y + 5z = 25 \end{cases}$ ____________

13. $\begin{cases} 3x + y = 13 \\ y - 4z = 24 \\ -2x - 3y - z = -13 \end{cases}$ ____________

14. $\begin{cases} x - 3y + 5z = 43 \\ 3x - y + z = 15 \\ -x - 5z = -37 \end{cases}$ ____________

Linda made a fruit salad using one apple, two junior bananas, and three small oranges for a total of 185 calories. Jason made a salad using two apples, one junior banana, and two small oranges for a total of 150 calories. Sylvia made her fruit salad using three apples, two junior bananas, and one small orange for a total of 195 calories.

15. Write a system of three linear equations in three variables to represent this situation.

____

16. Solve the system of linear equations using the row reduction method and back substitution to determine the number of calories in each fruit.

____

NAME ______________________ CLASS __________ DATE __________

Practice & Apply

4.5 The Inverse of a Matrix

Determine whether each pair of matrices are inverses.

1. $\begin{bmatrix} 2 & 3 \\ 1 & 2 \end{bmatrix} \begin{bmatrix} 2 & -3 \\ -1 & 2 \end{bmatrix}$ ______________

2. $\begin{bmatrix} 2 & 3 \\ 6 & 9 \end{bmatrix} \begin{bmatrix} -3 & 6 \\ 2 & -4 \end{bmatrix}$ ______________

3. $\begin{bmatrix} -2 & 1 \\ 4 & -3 \end{bmatrix} \begin{bmatrix} 4 & -5 \\ -1 & 2 \end{bmatrix}$ ______________

4. $\begin{bmatrix} 1 & 3 \\ 2 & 7 \end{bmatrix} \begin{bmatrix} 7 & -3 \\ -2 & 1 \end{bmatrix}$ ______________

Find the inverse of each matrix, if it exists. Round numbers to the nearest hundredth. If the inverse matrix does not exist, write does not exist.

5. $\begin{bmatrix} -3 & -2 \\ -2 & 1 \end{bmatrix}$ ______________

6. $\begin{bmatrix} -9 & 5 \\ 6 & -3 \end{bmatrix}$ ______________

7. $\begin{bmatrix} 5 & 3 \\ 2 & 1 \end{bmatrix}$ ______________

8. $\begin{bmatrix} 3 & -2 \\ 4 & 1 \end{bmatrix}$ ______________

9. $\begin{bmatrix} -2 & -3 \\ 6 & 9 \end{bmatrix}$ ______________

10. $\begin{bmatrix} -2 & 5 \\ 1 & -4 \end{bmatrix}$ ______________

A company offers three stock options to its employees. One employee has one share of stock A, two shares of stock B, and three shares of stock C. The cost of all the stock is $125. A second employee has two shares of stock A, three shares of stock B, and four shares of stock C. The cost of the stocks is $195. A third employee has one share of stock A, two shares of stock B, and one share of stock C. The cost of the stocks is $95.

11. Write the system of equations that you would solve to find the price per share of each stock.

12. Write the coefficient matrix.

13. Use your graphics calculator to find the inverse of the coefficient matrix.

NAME ________ CLASS ________ DATE ________

Practice & Apply

4.6 Using Matrix Algebra

Write the matrix equation, $AX = B$, that represents each system.

1. $\begin{cases} x - 3y = 3 \\ 2x + 9y = 11 \end{cases}$

2. $\begin{cases} 3x + y = 13 \\ 2x - y = 2 \end{cases}$

3. $\begin{cases} 3x + 2y - z = 7 \\ x - 4y + z = -9 \\ 5x - 3z = -1 \end{cases}$

4. $\begin{cases} 2x - y - z = -8 \\ 5x + 2y + 3z = 2 \\ x - 3y - 7z = -30 \end{cases}$

Use matrix algebra to solve each system.

5. $\begin{cases} 5x + 3y = -16 \\ 3x + 2y = -9 \end{cases}$ ________

6. $\begin{cases} -4x + 7y = -29 \\ 3x - 5y = 21 \end{cases}$ ________

7. $\begin{cases} 6x - 3y = 15 \\ 4x + 5y = 3 \end{cases}$ ________

8. $\begin{cases} -3x - 4y + 5z = 4 \\ 2x + 5y + 3z = -1 \\ -4x - y + 2z = -8 \end{cases}$ ________

9. $\begin{cases} x + y + z = 5 \\ 2x + z = 8 \\ 3y + 2z = 5 \end{cases}$ ________

10. $\begin{cases} 5x + y - z = 9 \\ x - 2y - z = -6 \\ 3x + 4y + 5z = 0 \end{cases}$ ________

For each system of equations, (a) write the matrix equation that represents it, and (b) use your graphics calculator and matrix algebra to solve it.

11. $\begin{cases} x + y = -1 \\ 2y - z = -3 \\ 3x - 2z = 5 \end{cases}$

12. $\begin{cases} x - 2z = 0 \\ 3y + z = 2 \\ -2x + y = 5 \end{cases}$

13. $\begin{cases} 2x - 3y + z - w = 3 \\ 4x + 3y + 2z + w = -6 \\ x + 2y - 3z - w = 4 \\ x - y - z + w = 8 \end{cases}$

14. $\begin{cases} 2x + y + 2z = -5 \\ x - 2y + 3z - 4w = -13 \\ 3x + y - z + 2w = 2 \\ -x - 3y + 3w = 1 \end{cases}$

NAME ______________________ CLASS ______________ DATE ______________

Practice & Apply

4.7 Exploring Transformations Using Matrices

Let $A = \begin{bmatrix} -1 & -6 & -3 & 2 \\ 4 & 4 & -2 & -2 \end{bmatrix}$ and $B = \begin{bmatrix} 3 & -3 & 0 \\ 4 & 1 & -2 \end{bmatrix}$.

1. Plot the vertices stored in matrix A. Use a straightedge to connect the vertices. What is the shape of the object represented by matrix A?

   ____

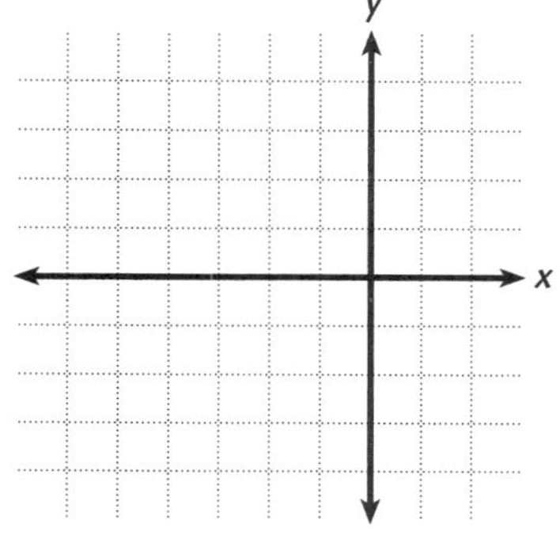

2. Plot the vertices stored in matrix B. Use a straightedge to connect the vertices. What is the shape of the object represented by matrix B?

   ____

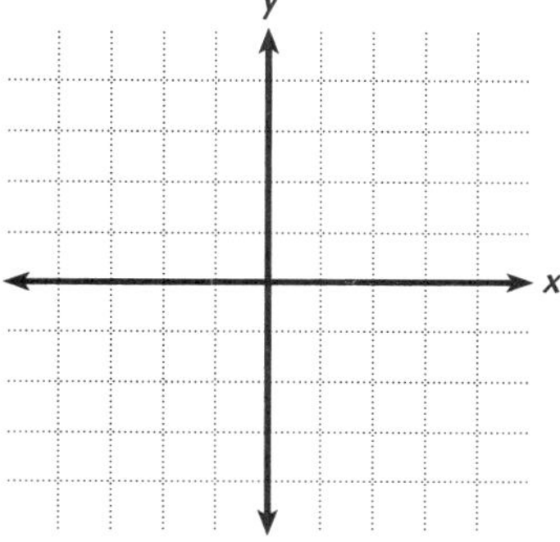

Use your graphics calculator to transform each matrix. Write the transformation matrix you used.

3. Rotate matrix A 180° clockwise. ______________________________

4. Rotate matrix B 90° counterclockwise. ______________________________

5. Enlarge matrix A 3 times. ______________________________

6. Shrink matrix B to $\frac{1}{2}$ its size. ______________________________

Use triangle *ABC* with vertices *A*(−1,5), *B*(4,1), and *C*(2,−3) in Exercises 7–8.

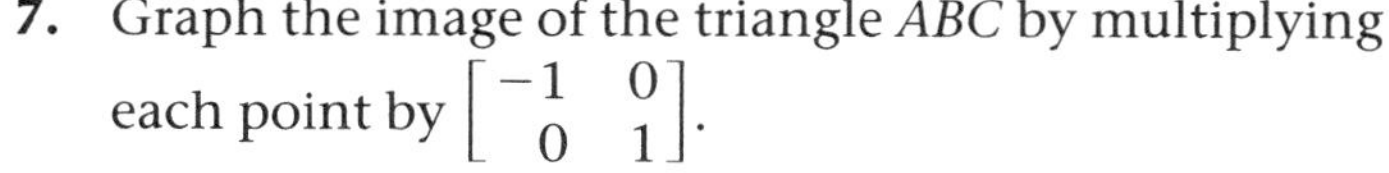

7. Graph the image of the triangle ABC by multiplying each point by $\begin{bmatrix} -1 & 0 \\ 0 & 1 \end{bmatrix}$.

8. What is the resulting image? ______________________

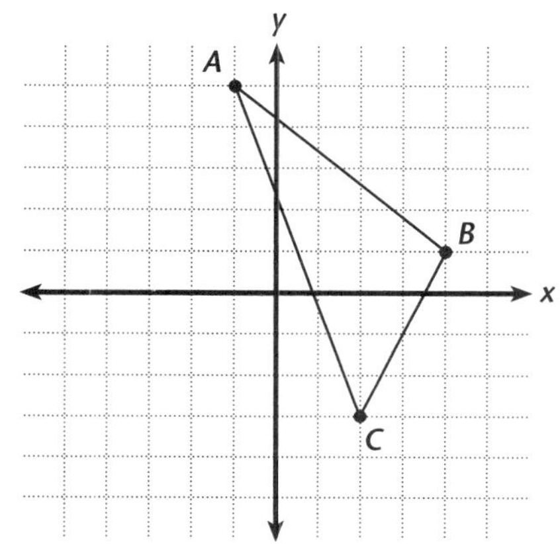

Practice & Apply

4.8 Exploring Systems of Linear Inequalities

Graph the solution to the following systems of linear inequalities.

1. $\begin{cases} y > 2x - 4 \\ y \leq 5 - x \end{cases}$

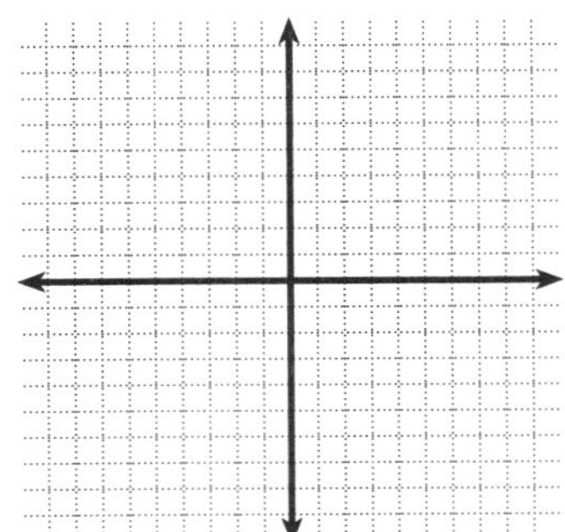

2. $\begin{cases} x + y > 1 \\ x - y < 4 \end{cases}$

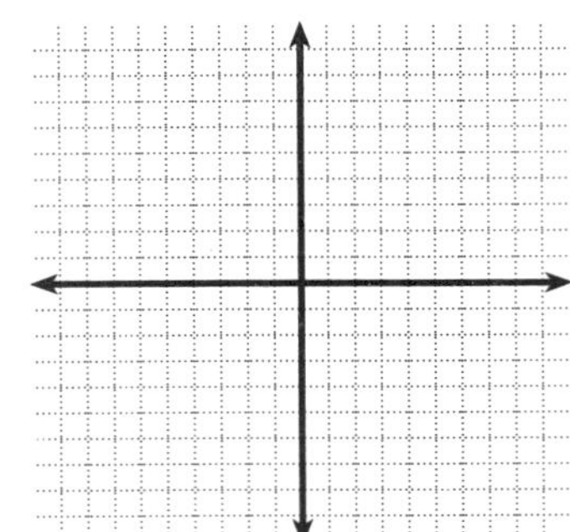

3. $\begin{cases} y \leq 3x - 4 \\ 2x + y > -2 \end{cases}$

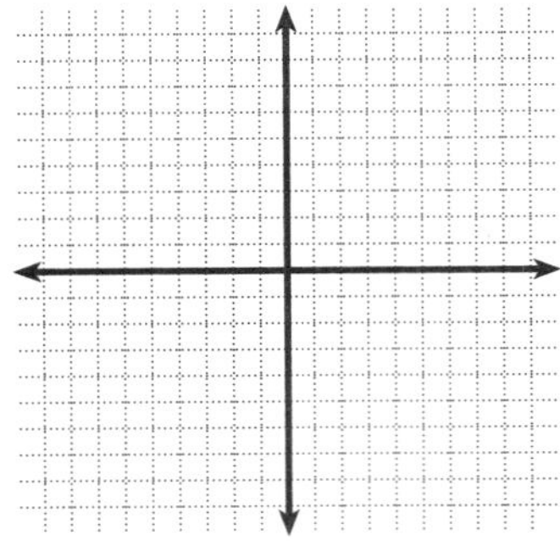

4. $\begin{cases} 3x + y \geq -4 \\ x - y \leq -3 \end{cases}$

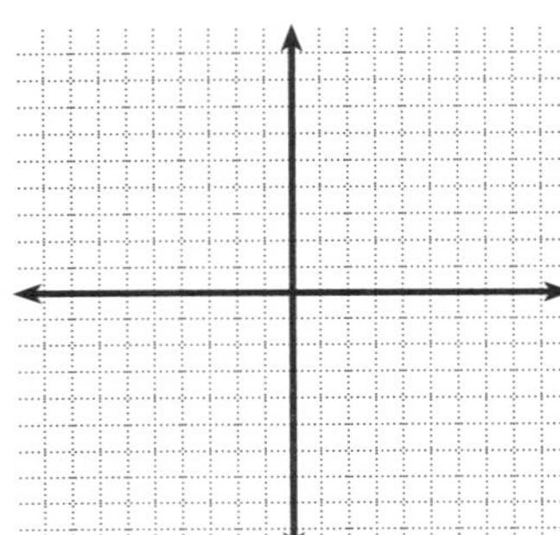

5. $\begin{cases} x \leq 1 \\ x + y > 0 \end{cases}$

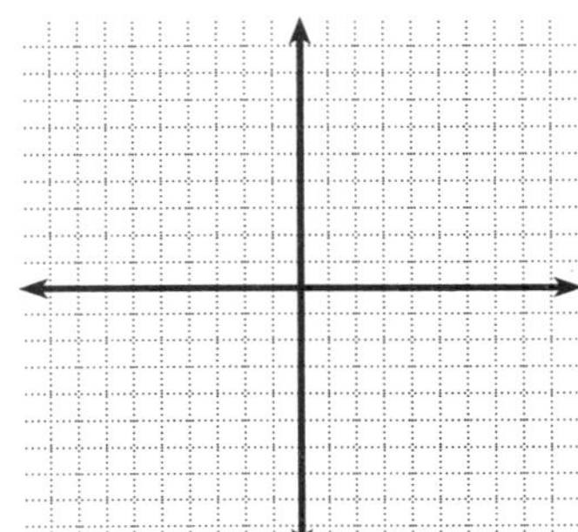

6. $\begin{cases} y > -2 \\ 4x - y > -1 \end{cases}$

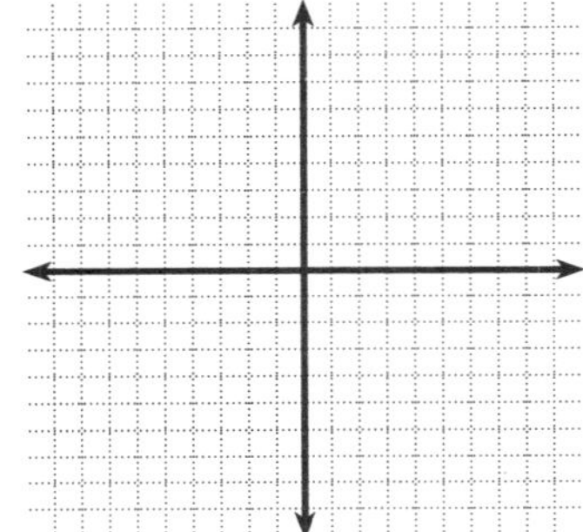

7. $\begin{cases} x + 2y \geq 4 \\ 4y \geq x - 8 \\ y \leq 1 \end{cases}$

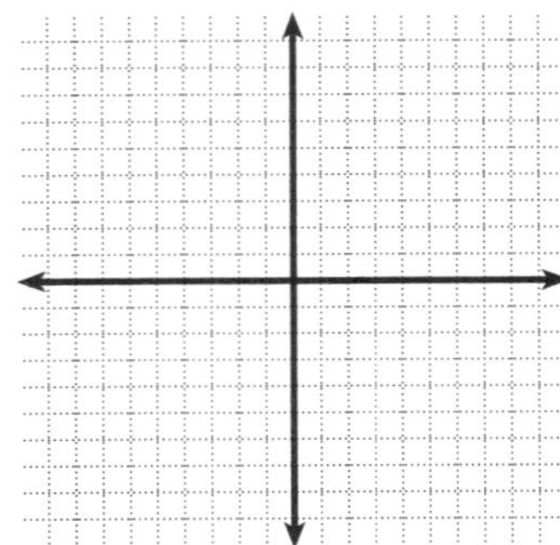

8. $\begin{cases} y \leq 3 \\ x - y < 2 \\ 4x + 2y \leq 5 \end{cases}$

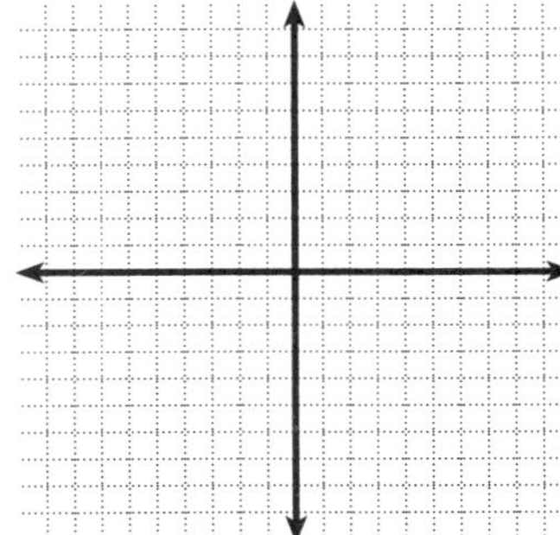

9. $\begin{cases} 3x - y < -1 \\ x + 2y \leq 4 \\ 3y > x \end{cases}$

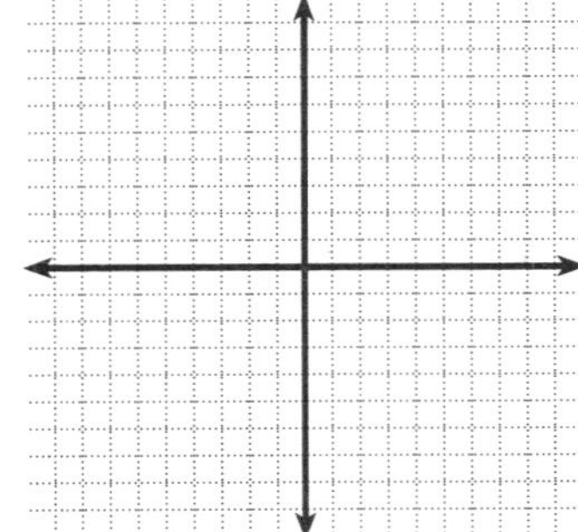

NAME ______________________ CLASS __________ DATE __________

Practice & Apply

4.9 Introduction to Linear Programming

Graph the feasible region for each set of constraints. Then determine any four points in the feasible region and find a value for the objective function $P = 2x + 3y$.

1. $\begin{cases} 4y \geq x - 2 \\ y \leq -x + 5 \\ x \geq 0 \\ y \geq 0 \end{cases}$

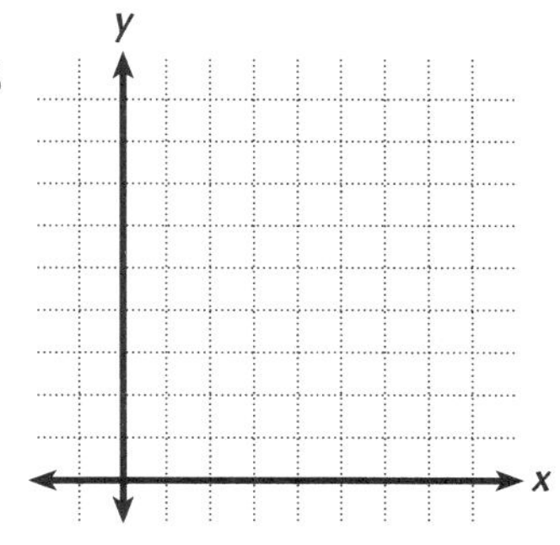

2. $\begin{cases} x + y \leq 7 \\ x - y \leq 3 \\ x \geq 0 \\ y \geq 0 \end{cases}$

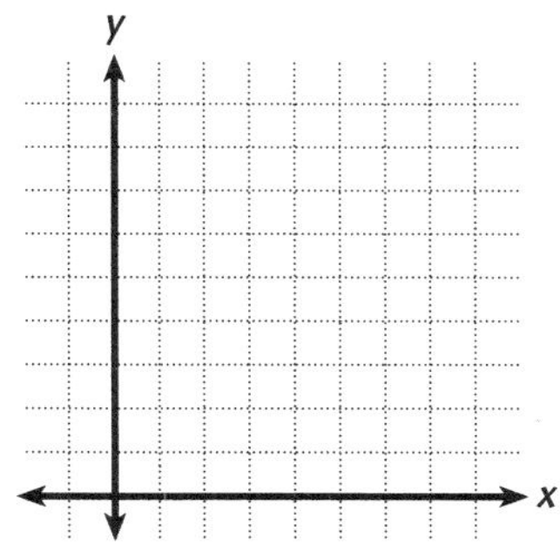

3. $\begin{cases} y \geq 2x - 4 \\ x + 3y \leq 9 \\ x \geq 0 \\ y \geq 0 \end{cases}$

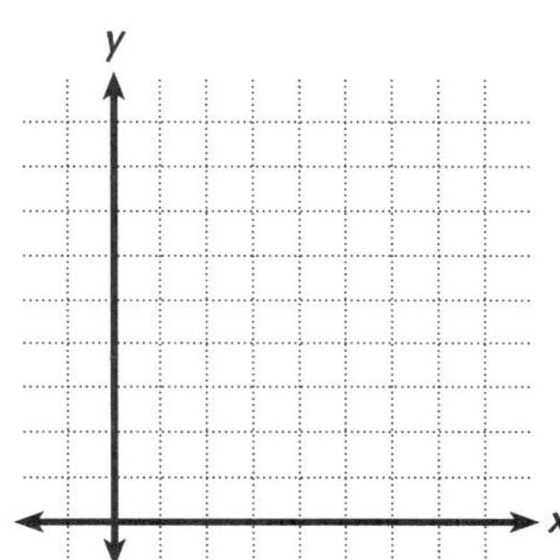

4. $\begin{cases} x + 2y \leq 11 \\ 4x + 3y \leq 24 \\ x \geq 0 \\ y \geq 0 \end{cases}$

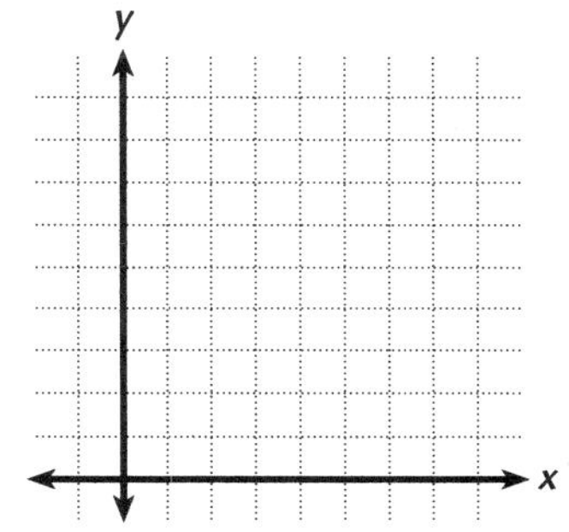

A manufacturer makes two types of calculators, a scientific model and a graphics model. Each model is assembled in two stages. The time required for the scientific model in the first stage is 2 hours, and 5 hours are required in the second stage. The graphics model requires 4 hours in the first stage and 2 hours in stage two. The maximum work hours available per week is 80 hours. The manufacturer makes a profit of $25 for each scientific model and $40 for each graphics model.

5. Create a table and use it to describe this situation with inequalities and an objective function related to cost.

6. On the grid provided, graph the constraints for the number of calculators that the manufacturer can produce.

7. Determine the value of the objective function for four points with integer coordinates in the feasible region in the previous problem.

NAME ______________________ CLASS ____________ DATE ____________

Practice & Apply

4.10 Linear Programming in Two Variables

For each objective function under the given constraints, (a) graph the constraints, (b) shade the feasible region, (c) find the coordinates of the vertices of the feasible region, (d) find the minimum value of the object function, and (e) find the maximum value of the object function.

1. $P = x + 5y$

Constraints:

$$\begin{cases} x + 3y \leq 18 \\ x - y \leq 2 \\ x \geq 0 \\ y \geq 0 \end{cases}$$

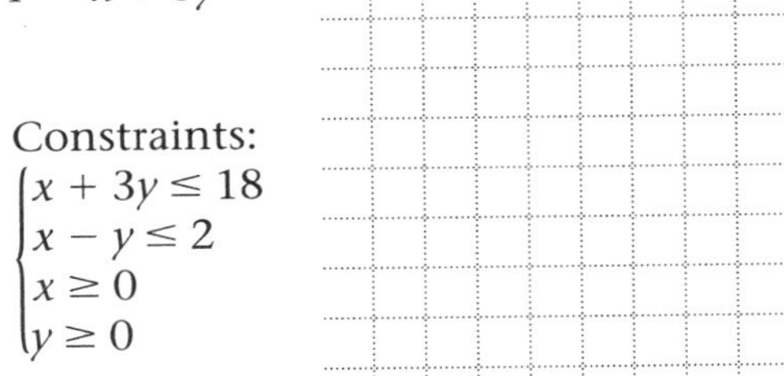

2. $C = 4x - y$

Constraints:

$$\begin{cases} 2x - 3y \geq -12 \\ 2x + y \leq 12 \\ x \geq 0 \\ y \geq 0 \end{cases}$$

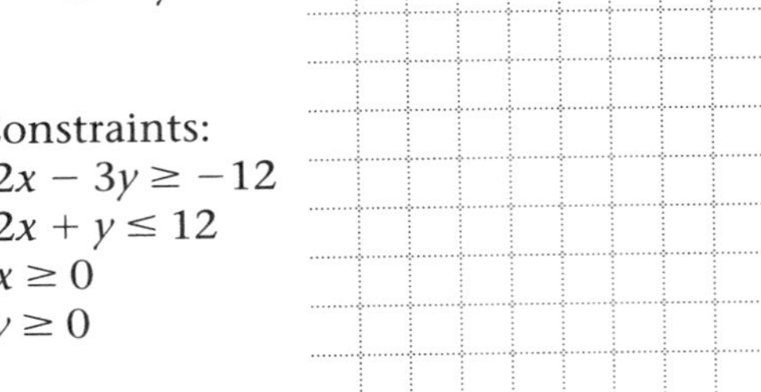

3. $F = 2x + 3y$

Constraints:

$$\begin{cases} x + 3y \leq 15 \\ 4x + 3y \leq 24 \\ x \geq 0 \\ y \geq 0 \end{cases}$$

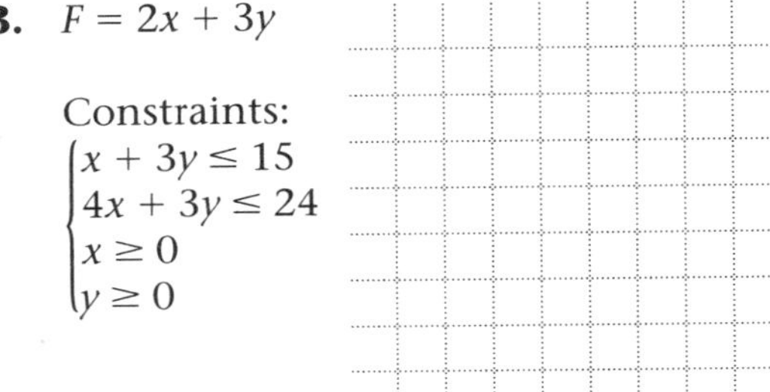

4. $R = 6x - 2y$

Constraints:

$$\begin{cases} 4x - y \geq -2 \\ x + y \leq 7 \\ x \geq 0 \\ y \geq 0 \end{cases}$$

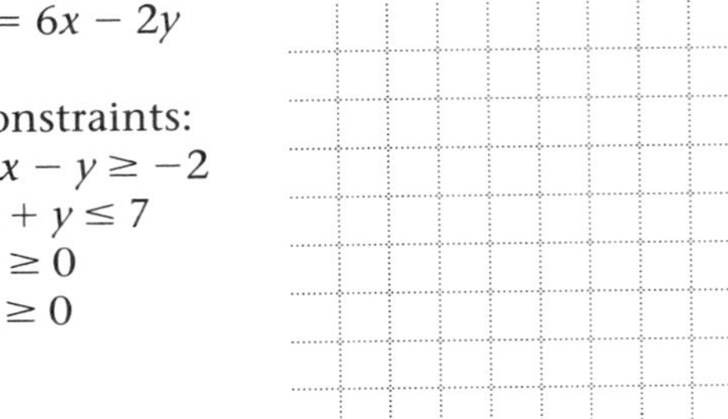

A company manufactures earrings and rings. It takes 10 g of metal to make an earring and 15 g to make a ring. The company has 600 g of the metal and 17.5 hours to work. It takes 15 min to make an earring and 30 min to make a ring. The profit on each earring is \$3, and the profit on each ring is \$5.

5. Graph the feasible region for this situation.

6. Determine the objective function. ______________________

7. Find the vertices of the feasible region.

____

8. Find the maximum for the objective function that satisfies the constraints.

____

NAME ______________________ CLASS __________ DATE __________

Enrichment

4.1 Adding Matrices Without a Calculator

Two matrices can be added if their dimensions are the same. To find the sum, add the corresponding elements. For example if

$$A = \begin{bmatrix} a_{11} & a_{12} & a_{13} \\ a_{21} & a_{22} & a_{23} \end{bmatrix} \text{ and } B = \begin{bmatrix} b_{11} & b_{12} & b_{13} \\ b_{21} & b_{22} & b_{23} \end{bmatrix} \text{ then}$$

$$A + B = \begin{bmatrix} a_{11} + b_{11} & a_{12} + b_{12} & a_{13} + b_{13} \\ a_{21} + b_{21} & a_{22} + b_{22} & a_{23} + b_{23} \end{bmatrix}$$

Find the sum of the following matrices without using a calculator. If the matrices cannot be added, state the reason.

1. $\begin{bmatrix} 8 & 15 \\ 21 & 19 \end{bmatrix} + \begin{bmatrix} 23 & 57 \\ 14 & 36 \end{bmatrix}$

____

2. $\begin{bmatrix} -19 & 26 \\ 53 & 78 \\ 92 & -31 \end{bmatrix} + \begin{bmatrix} 25 & 36 & 73 \\ 18 & -56 & 29 \end{bmatrix}$

____

3. $\begin{bmatrix} 81 & -19 & 27 \\ 15 & 82 & -9 \end{bmatrix} + \begin{bmatrix} 23 & -17 & 42 \\ -31 & 48 & 53 \end{bmatrix}$

____

4. $\begin{bmatrix} 5.32 & 4.17 & 8.25 \\ 6.01 & 9.38 & 4.32 \\ 6.75 & 8.19 & 3.71 \end{bmatrix} + \begin{bmatrix} 2.34 & 5.62 & 3.14 \\ 5.27 & 1.09 & 3.78 \\ 6.92 & 8.14 & 7.63 \end{bmatrix}$

____

5. $\begin{bmatrix} 4\frac{1}{2} & 6\frac{1}{3} \\ 7\frac{2}{5} & 8\frac{1}{4} \end{bmatrix} + \begin{bmatrix} 5\frac{2}{3} & 7\frac{1}{4} \\ 3\frac{5}{6} & 4\frac{9}{10} \end{bmatrix}$

____

6. $\begin{bmatrix} 8\frac{1}{7} & -2\frac{1}{2} \\ -5\frac{6}{7} & 3\frac{1}{6} \\ 2\frac{1}{3} & -4\frac{1}{8} \end{bmatrix} + \begin{bmatrix} 6\frac{1}{3} & 4\frac{1}{10} \\ 3\frac{4}{5} & -2\frac{1}{8} \\ 5\frac{7}{8} & -1\frac{1}{6} \end{bmatrix}$

____

7. $\begin{bmatrix} 1.16 & 2.37 & 5.84 \\ 3.21 & 4.85 & 6.27 \end{bmatrix} + \begin{bmatrix} 5.34 & 4.71 & 8.93 \\ 4.12 & 5.27 & 3.89 \end{bmatrix}$

____

8. $\begin{bmatrix} 112 & -156 & 129 \\ 346 & 427 & 181 \\ -147 & 206 & -119 \end{bmatrix} + \begin{bmatrix} 153 & 227 & -118 \\ -301 & 463 & 529 \\ 412 & 849 & -208 \end{bmatrix}$

____

NAME ______________________ CLASS ____________ DATE ____________

Enrichment
4.2 Miscalculations

Each matrix multiplication contains one incorrect entry in the product matrix. Write this number on the answer blank that corresponds to the problem number. When you are finished, a number will result that has special meaning in math.

1. $\begin{bmatrix} 3 \\ 2 \\ 5 \\ 1 \end{bmatrix} \begin{bmatrix} 5 & 6 \end{bmatrix} = \begin{bmatrix} 15 & 18 \\ 3 & 12 \\ 25 & 30 \\ 5 & 6 \end{bmatrix}$

2. $\begin{bmatrix} 0 & 1 \end{bmatrix} \begin{bmatrix} 2 \\ 3 \end{bmatrix} = \begin{bmatrix} 1 \end{bmatrix}$

3. $\begin{bmatrix} 2 & 1 \\ -1 & 3 \end{bmatrix} \begin{bmatrix} 6 & 2 \\ 5 & -2 \end{bmatrix} = \begin{bmatrix} 17 & 2 \\ 4 & -8 \end{bmatrix}$

4. $\begin{bmatrix} -3 & 2 & 1 \\ -1 & 4 & 5 \\ -2 & 1 & 0 \end{bmatrix} \begin{bmatrix} 2 & -1 \\ 4 & 3 \\ 2 & -1 \end{bmatrix} = \begin{bmatrix} 1 & 8 \\ 24 & 8 \\ 0 & 5 \end{bmatrix}$

5. $\begin{bmatrix} 6 & -1 \\ 3 & 1 \\ 2 & 3 \\ 1 & 5 \\ 4 & -3 \end{bmatrix} \begin{bmatrix} -1 & 2 \\ 3 & -2 \end{bmatrix} = \begin{bmatrix} -9 & 14 \\ 0 & 4 \\ 7 & -2 \\ 14 & -8 \\ 5 & 14 \end{bmatrix}$

6. $\begin{bmatrix} 2 & 1 & 3 \\ 6 & -4 & -2 \end{bmatrix} \begin{bmatrix} 5 & 1 \\ -1 & 2 \\ 3 & -2 \end{bmatrix} = \begin{bmatrix} 9 & -2 \\ 28 & 2 \end{bmatrix}$

7. $\begin{bmatrix} -1 & 1 & 3 \\ 2 & -2 & 1 \\ 4 & -3 & 0 \end{bmatrix} \begin{bmatrix} 2 & -1 & 2 \\ 0 & 1 & -2 \\ 3 & 2 & 2 \end{bmatrix} = \begin{bmatrix} 7 & 8 & 3 \\ 7 & 2 & 8 \\ 8 & -7 & 10 \end{bmatrix}$

8. $\begin{bmatrix} 5 & -1 & 2 \\ 3 & 1 & 4 \\ -2 & 2 & 3 \end{bmatrix} \begin{bmatrix} 2 \\ 1 \\ -4 \end{bmatrix} = \begin{bmatrix} 6 \\ -9 \\ -14 \end{bmatrix}$

9. $\begin{bmatrix} 3 & 2 \\ 2 & -1 \\ 3 & 4 \\ 5 & -6 \\ 1 & 3 \end{bmatrix} \begin{bmatrix} 3 & 1 & 7 \\ -2 & 4 & 1 \end{bmatrix} = \begin{bmatrix} 5 & 11 & 23 \\ 8 & -2 & 13 \\ 5 & 19 & 25 \\ 27 & -19 & 29 \\ -3 & 13 & 10 \end{bmatrix}$

10. $\begin{bmatrix} 1 & -3 & 2 \\ 4 & -1 & 3 \end{bmatrix} \begin{bmatrix} 2 \\ -1 \\ 4 \end{bmatrix} = \begin{bmatrix} 13 \\ 4 \end{bmatrix}$

____ . ____ ____ ____ ____ ____ ____ ____ ____ ____

1 . 2 3 4 5 6 7 8 9 10

11. What is special about the resulting number?

____

NAME ______________________ CLASS ____________ DATE ____________

Enrichment

4.3 Solving Three Equations in Three Variables

Systems of linear equations in three variables can also be solved by the elimination method. For example, to solve

$$\text{①}\quad x + y + z = 3$$
$$\text{②}\quad x - y + 2z = 11$$
$$\text{③}\quad x + 2y - z = -7$$

use equations 2 to eliminate the variable y from equations 1 and 3.

$$\begin{array}{r} x + y + z = 3 \\ x - y + 2z = 11 \\ \hline \text{④}\quad 2x + 3z = 14 \end{array} \qquad \begin{array}{r} 2(x - y + 2z = 11) \\ x + 2y - z = -7 \\ \hline \text{⑤}\quad 3x + 3z = 15 \end{array}$$

Solve this new system for x or z.

$$\begin{array}{rr} \text{④} & 2x + 3z = 14 \\ \text{⑤} & 3x + 3z = 15 \\ \hline & -x = -1 \\ & x = 1 \end{array}$$

Use equation ④ or ⑤ to find z. $\quad 2(1) + 3z = 14$, $z = 4$

Use equation ①, ②, or ③ to find y. $\quad 1 + y + 4 = 3$, $y = -2$

$(x, y, z) = (1, -2, 4)$

Solve each system for (x, y, z).

1. $\begin{cases} x + y - z = 2 \\ 2x - y + z = -3 \\ x - y - z = 0 \end{cases}$ ______________

2. $\begin{cases} 2x + y + z = 5 \\ x - y + 3z = -11 \\ 3x + 2y - 4z = 26 \end{cases}$ ______________

3. $\begin{cases} x - 2y + 3z = -6 \\ 2x + y - 4z = -7 \\ 5x + 3y - 2z = 10 \end{cases}$ ______________

4. $\begin{cases} 4x - 3y + z = 9 \\ 2x + y - 3z = -7 \\ 3x + 2y + z = 12 \end{cases}$ ______________

5. $\begin{cases} x + y - z = 15 \\ 2x - y + z = 0 \\ 3x + 2y - 3z = 38 \end{cases}$ ______________

6. $\begin{cases} 2x - 3y + z = 35 \\ x + y - 3z = -29 \\ 2x - 4y + 5z = 77 \end{cases}$ ______________

7. $\begin{cases} x + y - 3z = -8.25 \\ 4x - y + z = 15.25 \\ 2x + y - 4z = -10 \end{cases}$ ______________

8. $\begin{cases} x + y + z = 0.2 \\ x - y - z = 0.4 \\ 2x + y - z = 0.9 \end{cases}$ ______________

9. $\begin{cases} x - y + z = 3.1 \\ -x + y + z = 6.1 \\ x + y - z = 0.3 \end{cases}$ ______________

NAME ______________________ CLASS ____________ DATE ____________

Enrichment

4.4 Echelon Form

The form $\left[\begin{array}{ccc|c} 1 & 0 & 0 & a \\ 0 & 1 & 0 & b \\ 0 & 0 & 1 & c \end{array}\right]$ is known as echelon form. If you can write an augmented matrix in echelon form, the solutions can be read directly from the matrix, without using back substitution, where $x = a$, $y = b$, and $z = c$.

Write each system as an augmented matrix. Then use row reduction to obtain a matrix in echelon form and give the solution in the form (x, y, z).

1. $\begin{cases} 2x + y - z = -1 \\ x - y + 3z = 8 \\ x + y + z = 2 \end{cases}$ ______________

2. $\begin{cases} x + y - z = 1 \\ 2x + 3y + z = 7 \\ 2x - y + 4z = 17 \end{cases}$ ______________

3. $\begin{cases} x - y + z = 5 \\ 2x + y + z = 4 \\ x - 2y + 3z = 12 \end{cases}$ ______________

4. $\begin{cases} 2x - y + 3z = -7 \\ x + 4y - 2z = 17 \\ 3x + y + 2z = 2 \end{cases}$ ______________

5. $\begin{cases} x + 5y - 3z = 14 \\ 2x + y - z = 10 \\ x - 2y + z = 0 \end{cases}$ ______________

6. $\begin{cases} x - 3y + 2z = -10 \\ 2x + y + z = 5 \\ 3x - 2y + 3z = -5 \end{cases}$ ______________

7. $\begin{cases} x + y - 4z = -2 \\ 2x - y + z = 2 \\ 3x + 2y - 2z = 3 \end{cases}$ ______________

8. $\begin{cases} x - 2y + 3z = 13 \\ 2x + 5y - 3z = -19 \\ x + 4y - 5z = -21 \end{cases}$ ______________

9. $\begin{cases} x + y - 3z = 8 \\ 2x - 3y + z = -6 \\ x - y + z = -2 \end{cases}$ ______________

10. $\begin{cases} x + y + z = 6 \\ 2x - 3y + 5z = -11 \\ x + 3y - 4z = 19 \end{cases}$ ______________

11. $\begin{cases} 3x - 2y + z = 16 \\ x + 3y + 4z = 9 \\ 2x - y + 3z = 15 \end{cases}$ ______________

12. $\begin{cases} 2x - 4y + 3z = -8 \\ x + 3y - 2z = 9 \\ 3x + 2y + z = 13 \end{cases}$ ______________

13. $\begin{cases} x + y - 3z = -21 \\ 2x - y + z = 12 \\ 3x + 2y + 2z = 7 \end{cases}$ ______________

14. $\begin{cases} 2x + 5y - 3z = -11 \\ 3x - 2y + 4z = 7 \\ 2x + 3y - 2z = -10 \end{cases}$ ______________

15. $\begin{cases} 3x + 6y - 4z = -42 \\ 2x + 2y + 3z = 14 \\ 4x + 3y - 5z = -34 \end{cases}$ ______________

NAME ______________________ CLASS ____________ DATE ____________

Enrichment

4.5 An Alternate Method

Another way you can find the inverse of a matrix $A = \begin{bmatrix} a & b \\ c & d \end{bmatrix}$ is using row operations on the augmented matrix $\left[\begin{array}{cc:cc} a & b & 1 & 0 \\ c & d & 0 & 1 \end{array}\right]$ until it is of the form $\left[\begin{array}{cc:cc} 1 & 0 & e & f \\ 0 & 1 & g & h \end{array}\right]$. Then $A^{-1} = \begin{bmatrix} e & f \\ g & h \end{bmatrix}$.

For example, to find A^{-1} for $A = \begin{bmatrix} 1 & 2 \\ -1 & 3 \end{bmatrix}$ follow these steps:

$$\left[\begin{array}{cc:cc} 1 & 2 & 1 & 0 \\ -1 & 3 & 0 & 1 \end{array}\right] \xrightarrow{R_1 + R_2 \to R_2} \left[\begin{array}{cc:cc} 1 & 2 & 1 & 0 \\ 0 & 5 & 1 & 1 \end{array}\right] \xrightarrow{\frac{1}{5}R_2 \to R_2}$$

$$\left[\begin{array}{cc:cc} 1 & 2 & 1 & 0 \\ 0 & 1 & \frac{1}{5} & \frac{1}{5} \end{array}\right] \xrightarrow{R_1 - 2R_2 \to R_1} \left[\begin{array}{cc:cc} 1 & 0 & \frac{3}{5} & -\frac{2}{5} \\ 0 & 1 & \frac{1}{5} & \frac{1}{5} \end{array}\right]$$

$$A^{-1} = \begin{bmatrix} \frac{3}{5} & -\frac{2}{5} \\ \frac{1}{5} & \frac{1}{5} \end{bmatrix}.$$

Find the inverse of each matrix by using row operations. Verify that $A^{-1}A = I$.

1. $\begin{bmatrix} 1 & -1 \\ 2 & 3 \end{bmatrix}$ ______________

2. $\begin{bmatrix} 2 & 5 \\ 1 & 3 \end{bmatrix}$ ______________

3. $\begin{bmatrix} 1 & -2 \\ 2 & 5 \end{bmatrix}$ ______________

4. $\begin{bmatrix} 5 & -1 \\ 4 & 2 \end{bmatrix}$ ______________

5. $\begin{bmatrix} -3 & 2 \\ 5 & 1 \end{bmatrix}$ ______________

6. $\begin{bmatrix} 6 & 5 \\ 2 & 1 \end{bmatrix}$ ______________

7. $\begin{bmatrix} 1 & 0 \\ -3 & 4 \end{bmatrix}$ ______________

8. $\begin{bmatrix} 5 & -5 \\ 2 & 4 \end{bmatrix}$ ______________

9. $\begin{bmatrix} -1 & 6 \\ 1 & 0 \end{bmatrix}$ ______________

NAME ______________________ CLASS __________ DATE __________

Enrichment

4.6 Missing Entries

Multiply A^{-1} by A to determine missing entries.

1. $A = \begin{bmatrix} 1 & b \\ -1 & a \end{bmatrix}, A^{-1} = \begin{bmatrix} \frac{1}{3} & -\frac{2}{3} \\ \frac{1}{3} & \frac{1}{3} \end{bmatrix}$

____

2. $A = \begin{bmatrix} a & 2 \\ -1 & b \end{bmatrix}, A^{-1} = \begin{bmatrix} \frac{1}{2} & -1 \\ \frac{1}{2} & 0 \end{bmatrix}$

____

3. $A = \begin{bmatrix} a & b \\ 3 & 1 \end{bmatrix}, A^{-1} = \begin{bmatrix} -\frac{1}{4} & \frac{1}{4} \\ \frac{3}{4} & \frac{1}{4} \end{bmatrix}$

____

4. $A = \begin{bmatrix} 1 & a \\ 2 & b \end{bmatrix}, A^{-1} = \begin{bmatrix} \frac{1}{5} & \frac{2}{5} \\ \frac{2}{5} & -\frac{1}{5} \end{bmatrix}$

____

5. $A = \begin{bmatrix} b & 0 \\ 4 & a \end{bmatrix}, A^{-1} = \begin{bmatrix} -\frac{1}{2} & 0 \\ 2 & 1 \end{bmatrix}$

____

6. $A = \begin{bmatrix} 3 & a \\ b & 5 \end{bmatrix}, A^{-1} = \begin{bmatrix} \frac{5}{14} & -\frac{1}{14} \\ -\frac{1}{14} & \frac{3}{14} \end{bmatrix}$

____

7. $A = \begin{bmatrix} 0 & a & 4 \\ 1 & 0 & b \\ -4 & 4 & c \end{bmatrix}, A^{-1} = \begin{bmatrix} \frac{1}{4} & 0 & -\frac{1}{4} \\ \frac{1}{8} & \frac{1}{2} & \frac{1}{8} \\ \frac{1}{8} & -\frac{1}{2} & -\frac{1}{8} \end{bmatrix}$

____

8. $A = \begin{bmatrix} 2 & 0 & b \\ a & 1 & -1 \\ 0 & c & 2 \end{bmatrix}, A^{-1} = \begin{bmatrix} 1 & \frac{1}{2} & -\frac{1}{4} \\ 1 & 1 & 0 \\ -1 & -1 & \frac{1}{2} \end{bmatrix}$

____

9. $A = \begin{bmatrix} 2 & -1 & b \\ a & 1 & 0 \\ 4 & c & -2 \end{bmatrix}, A^{-1} = \begin{bmatrix} \frac{1}{3} & \frac{1}{3} & 0 \\ -\frac{1}{3} & \frac{2}{3} & 0 \\ \frac{2}{3} & \frac{2}{3} & -\frac{1}{2} \end{bmatrix}$

____

10. $A = \begin{bmatrix} 3 & a & 5 \\ 1 & 2 & b \\ 0 & c & 1 \end{bmatrix}, A^{-1} = \begin{bmatrix} -2 & 7 & -4 \\ -\frac{1}{2} & \frac{3}{2} & -\frac{1}{2} \\ \frac{3}{2} & -\frac{9}{2} & \frac{5}{2} \end{bmatrix}$

____

NAME ______________________ CLASS __________ DATE __________

Enrichment

4.7 Representing Other Transformations

The triangle with vertices $A(1, 3)$, $B(3, -3)$, and $C(-3, 1)$ can be represented by the matrix

$$\begin{bmatrix} 1 & 3 & -3 \\ 3 & -3 & 1 \end{bmatrix}.$$

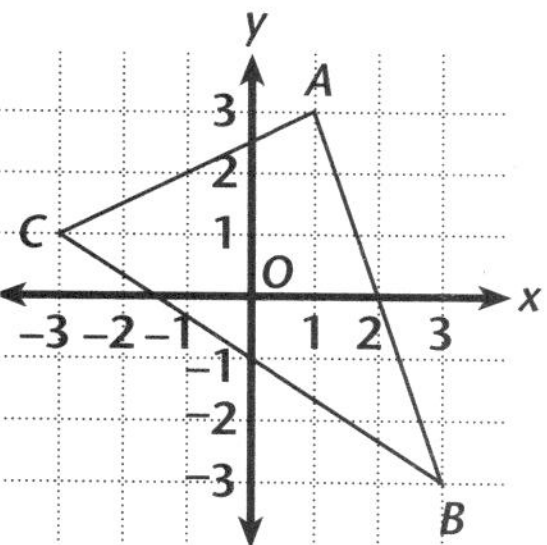

The reflection of ΔABC over the y-axis is given by

$$\begin{bmatrix} -1 & 0 \\ 0 & 1 \end{bmatrix} \begin{bmatrix} 1 & 3 & -3 \\ 3 & -3 & 1 \end{bmatrix}$$

$$= \begin{bmatrix} -1 & -3 & 3 \\ 3 & -3 & 1 \end{bmatrix}$$

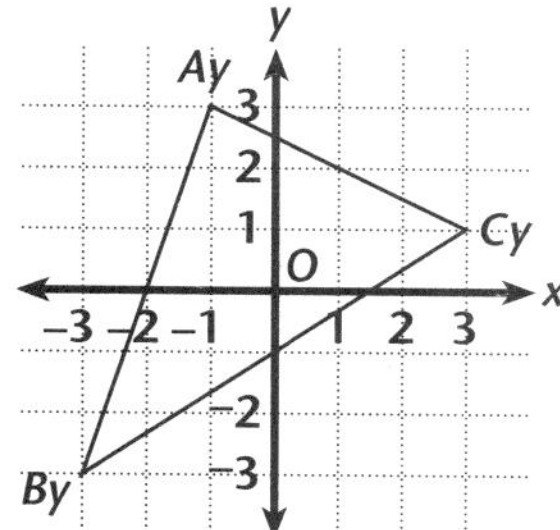

The reflection of ΔABC over the x-axis is given by

$$\begin{bmatrix} 1 & 0 \\ 0 & -1 \end{bmatrix} \begin{bmatrix} 1 & 3 & -3 \\ 3 & -3 & 1 \end{bmatrix}$$

$$= \begin{bmatrix} 1 & 3 & -3 \\ -3 & 3 & -1 \end{bmatrix}$$

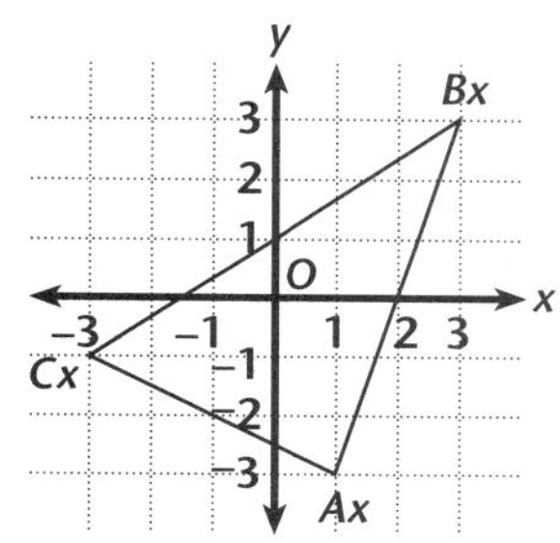

Write the matrix that results when the figure represented by each matrix is (a) reflected over the y-axis, and (b) reflected over the x-axis.

1. $\begin{bmatrix} 2 & -1 & 0 \\ 4 & 3 & 2 \end{bmatrix}$

2. $\begin{bmatrix} 5 & 3 & -1 \\ -1 & 2 & 4 \end{bmatrix}$

3. $\begin{bmatrix} 3 & 1 & 6 \\ 4 & 5 & 2 \end{bmatrix}$

4. $\begin{bmatrix} 5 & 0 & 6 \\ 2 & 3 & -4 \end{bmatrix}$

5. $\begin{bmatrix} 5 & -1 & 2 \\ 7 & -5 & 3 \end{bmatrix}$

6. $\begin{bmatrix} 0 & -1 & 2 \\ 6 & 4 & -5 \end{bmatrix}$

NAME ______________________ CLASS ____________ DATE ____________

Enrichment

4.8 Intersections of Boundary Lines

Graph each system of inequalities on the grid provided. Then find the coordinates of the vertices of the figure formed by the boundary lines.

1. $\begin{cases} 3x + 3y \leq 9 \\ x \geq 1 \\ y \geq 0 \end{cases}$

____

2. $\begin{cases} x + 2y \leq 10 \\ 2x - y \geq 3 \\ x \geq 0 \\ y \geq 0 \end{cases}$

____

3. $\begin{cases} 2x + 3y \leq 18 \\ 3x - 2y \geq 6 \\ x \geq 1 \\ y \geq 1 \end{cases}$

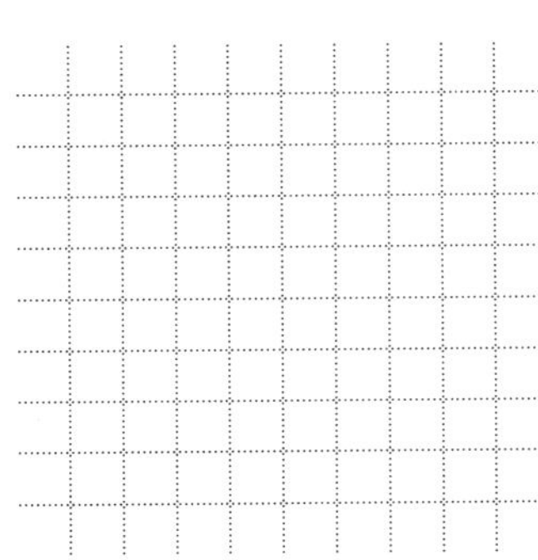

____

4. $\begin{cases} 3x + 4y \leq 24 \\ 2x - y \geq 6 \\ x \geq 4 \\ y \geq -1 \end{cases}$

____

5. $\begin{cases} -2x + 3y \leq 18 \\ x + y \geq 4 \\ x \geq 0 \\ y \geq 0 \\ x \leq 6 \end{cases}$

____

6. $\begin{cases} 2x - 5y \leq 20 \\ x + 2y \geq 6 \\ x \geq 0 \\ x + y \leq 10 \end{cases}$

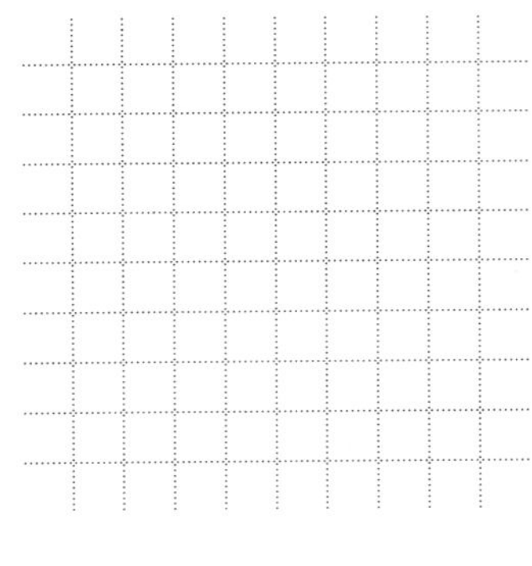

____

NAME ______________________ CLASS ____________ DATE ____________

Enrichment

4.9 Using Vertices of the Feasible Solution to Find Optimal Solutions

If a feasible region has an optimal solution, whether a maximum or a minimum, it will occur at one of the vertices of the feasible region.

For each problem, determine the constraints and the objective function, find the vertices of the feasible region, and test each vertex until you find the optimal solution.

1. ABC Toy Company makes trains and buses. It takes 5 minutes to carve each train, and 2 minutes to paint it. It takes 3 minutes to carve each bus, and 5 minutes to paint each. The carving machine and the painting machine each work only 8 hours a day. Demands for their products mean that at least 30 trains and 50 buses must be completed each day. If ABC Toy Company makes a profit of $10 per train and $8 per bus, how many trains and how many buses should ABC Toy Company make each day in order to maximize their profits?

____

____

____

2. ABC Toy Company also makes dump trucks and motorcycles. Each dump truck requires 10 wheels and each motorcycle requires 2 wheels. Each dump truck requires 5 ounces of plastic and each motorcycle requires 3 ounces of plastic. Each day, 500 wheels and 200 ounces of plastic are available. If the company can make a profit of $5 on a dump truck and $8 on a motorcycle, how many of each should the company make in order to maximize its daily profits?

____

____

____

3. ABC Toy Company also makes tea sets and Tomato-Patch dolls. Each tea set takes 20 minutes to cut out and 5 minutes to assemble. Each doll takes 10 minutes to cut out and 25 minutes to assemble. Each operation is done simultaneously during an 8-hour work day. If at least 10 tea sets and 5 dolls must be made each day, and it costs $5 to make a tea set and $15 to make a doll, what will the minimum cost be to ABC Toy Company?

____

____

____

NAME ______________________ CLASS __________ DATE __________

Enrichment

4.10 Numerous Constraints

Graph each system of constraints on the grid provided. Then find the vertices of the feasible region and the maximum value of each objective function.

1. $P = 2x + 3y$

$$\begin{cases} x + y \le 10 \\ x + 2y \le 12 \\ -x + y \le 2 \\ x \ge 0 \\ y \ge 0 \end{cases}$$

____

2. $P = 2x + 2y$

$$\begin{cases} 5x + 8y \le 80 \\ x + y \le 12 \\ x - y \le 6 \\ x \ge 0 \\ y \ge 0 \end{cases}$$

____

3. $P = 3x + 5y$

$$\begin{cases} x + y \le 12 \\ x + 2y \le 20 \\ y - x \le 6 \\ x \le 8 \\ x \ge 0 \\ y \ge 0 \end{cases}$$

____

4. $P = 4x + 3y$

$$\begin{cases} 2x + y \ge 12 \\ 3x + 5y \ge 30 \\ x + y \le 16 \\ x \le 12 \\ y \le 12 \\ x \ge 0 \\ y \ge 0 \end{cases}$$

____

NAME ______________________ CLASS ____________ DATE ____________

Technology

4.1 Spreadsheets and Matrices

Since a matrix is a rectangular array of data and a spreadsheet is as well, you can use a spreadsheet to hold data. The spreadsheet shown contains the following matrices, each highlighted by an outline.

$$R = \begin{bmatrix} 5 & -4 & 3 \\ 3 & 0 & -2 \\ 0 & 9 & -9 \end{bmatrix} \quad S = \begin{bmatrix} 10 & 0 & 0 \\ 4 & -5 & 1 \\ 6 & 6 & 6 \end{bmatrix} \quad T = \begin{bmatrix} 3 & -3 & 3 & -3 \\ -2 & 2 & -2 & 2 \\ 5 & 5 & 6 & 6 \end{bmatrix} \quad U = \begin{bmatrix} 1 & 3 & 5 & 7 \\ 2 & 4 & 6 & 8 \\ 1 & 0 & 1 & 0 \end{bmatrix}$$

	A	B	C	D	E	F	G	H	I	J	K	L	M	N
1	5	−4	3		10	0	0							
2	3	0	−2		4	−5	1							
3	0	9	−9		6	6	6							
4														
5	3	−3	3	−3		1	3	5	7					
6	−2	2	−2	2		2	4	6	8					
7	5	5	6	6		1	0	1	0					

Use the spreadsheet to answer each exercise.

1. Write formulas in the highlighted box for the sum $R + S$.

2. Write formulas in the highlighted box for the difference $T - U$.

3. Find $R + S$. ______________________

4. Find $T - U$. ______________________

5. Find $R - S$. ______________________

6. Find $T + U$. ______________________

Let $\boldsymbol{W} = \begin{bmatrix} \mathbf{0} & \mathbf{-1} & \mathbf{-2} \\ \mathbf{3} & \mathbf{6} & \mathbf{-8} \\ \mathbf{10} & \mathbf{8} & \mathbf{-4} \end{bmatrix}$.

7. Find $W + W$. ______________________

8. Find $3W$ and $4W$. ______________________

9. Describe how to find $10W$. ______________________

10. Describe how to find $-4W$. ______________________

NAME ______________________ CLASS __________ DATE __________

Technology

4.2 Exploring Questions About Matrix Multiplication

You can explore some interesting questions about matrix multiplication by using a spreadsheet. The spreadsheet shown contains matrices L and M and a place for the product LM.

$$L = \begin{bmatrix} 1 & 2 & 3 \\ 4 & 5 & 6 \end{bmatrix} \qquad M = \begin{bmatrix} 0 & 2 \\ 5 & 3 \\ -1 & 7 \end{bmatrix}$$

	A	B	C	D	E	F	G	H	I
1	1	2	3		0	2			
2	4	5	6		5	3			
3					−1	7			
4									

Use the spreadsheet shown.

1. Write a formula for Cell H1 that contains the element for the first row and first column of LM. ______________________

2. Write the formula for Cell I2 that contains the element for the second row and second column of LM. ______________________

3. Find LM. ______________________

4. Find ML. ______________________

5. Find a matrix Y with at least one nonzero element such that $LY = \begin{bmatrix} 1 & 0 \\ 1 & 0 \end{bmatrix}$. ______________________

6. Find a matrix N with at least one nonzero element such that $NM = \begin{bmatrix} 0 & 1 \\ 0 & 0 \end{bmatrix}$. ______________________

7. Find a matrix Y with at least one nonzero element such that $LY = \begin{bmatrix} 0 & 0 \\ 0 & 0 \end{bmatrix}$. ______________________

8. Find a matrix N with at least one nonzero element such that $NM = \begin{bmatrix} 0 & 0 \\ 0 & 0 \end{bmatrix}$. ______________________

NAME ______________________ CLASS __________ DATE __________

Technology

4.3 Spreadsheets and Linear Systems

Suppose you have \$2000 to invest and place part of the money in an account that pays 9% interest annually and the rest in an account that pays 5% interest annually. If you want to receive \$120 in interest altogether, how much should be placed in each account?

If you let x represent the amount invested at 9% and y represent the amount invested at 5%, you can write the system shown.

$$x + y = 2000$$
$$0.09x + 0.05y = 120$$

Using substitution, you can combine the two equations to get $0.09x + 0.05(2000 - x) = 120$. With a spreadsheet, you can experiment with different values of x to see how close to \$120 you are. In Column A, enter a sequence of x-values starting with a reasonable estimate.

	A	B	C
1	X	Y	INTEREST
2	400	1600	116.00
3	420	1580	116.80
4	440	1560	117.60
5	...		

Cell B2 contains 2000−A2.
Cell C2 contains 0.09*A2+0.05*B2.

The spreadsheet indicates that \$440 placed into the 9% account is not enough to achieve the \$120 interest goal.

Use the spreadsheet shown.

1. Find x and y such that the \$2000 will yield \$120 in interest. ______________

2. Find x and y such that the \$2000 will yield \$100 in interest. ______________

3. Find x and y such that \$3000 will yield \$240 in interest. ______________

4. Find x and y such that \$3000 will yield \$200 in interest. ______________

5. Find x and y such that \$5000 will yield \$350 in interest. ______________

6. Find x and y such that \$4000 will yield \$260 in interest. ______________

7. Find x and y such that \$5000 will yield \$350 in interest. ______________

8. Find x and y such that \$5000 will yield \$262.50 in interest. ______________

9. Refer to Exercise 1. Explain how you know that there is only one plan that will yield \$120 in interest.

____

NAME ______________________ CLASS ____________ DATE ____________

Technology

4.4 A Tutorial on Row Reduction

Rows 2, 3, and 4 of the spreadsheet shown contain the augmented matrix for

$$\begin{cases} x - 2y + 3z = 4 \\ 2x + y - 4z = 3 \\ -3x + 4y - z = -32 \end{cases}$$

To create rows 5, 6, and 7 from rows 2, 3, and 4, let

A5=A2
A6=−2*A2+A3
A7=3*A2+A4

Then highlight rows 5, 6, and 7 and use the FILL RIGHT command.

To create rows 8, 9, and 10 from rows 5, 6, and 7, let

A8=A5
A9=A6/5
A10=A7

	A	B	C	D	E
1	X	Y	Z		
2	1	−2	3		4
3	2	1	−4		3
4	−3	4	−1		−2
5	1	−2	3	0	4
6	0	5	−10	0	−5
7	0	−2	8	0	10
8	1	−2	3	0	4
9	0	1	−2	0	−1
10	0	−2	8	0	10
11	1	0	−1	0	2
12	0	1	−2	0	−1
13	0	0	4	0	8

Then highlight rows 8, 9, and 10 and use the FILL RIGHT command.

To create rows 11, 12, and 13 from rows 8, 9, and 10, let

A11=2*A9+A8
A12=A9
A13=2*A9+A10

Then highlight rows 11, 12, and 13 and use the FILL RIGHT command.

You can complete the remainder of the row reduction in a similar fashion.

Use a spreadsheet to solve each system of equations by row reduction.

1. $\begin{cases} x - 2y - 3z = -1 \\ 2x + y + z = 6 \\ x + 3y - 2z = 13 \end{cases}$ ______________

2. $\begin{cases} 2x - 3y + 2z = -3 \\ -3x + 2y + z = 1 \\ 4x + y - 3z = 4 \end{cases}$ ______________

3. $\begin{cases} 5x + 2y - z = -7 \\ x - 2y + 2z = 0 \\ 3y + z = 17 \end{cases}$ ______________

4. $\begin{cases} x - y + z = 0 \\ 2x - 3z = -1 \\ -x - y + 2z = -1 \end{cases}$ ______________

5. $\begin{cases} x + y + z = 9 \\ x - y - z = -15 \\ x + y - z = -5 \end{cases}$ ______________

6. $\begin{cases} 3x + 2y + 2z = 3 \\ 2x + 4y - z = 8 \\ 2x - 4y + z = 0 \end{cases}$ ______________

NAME ______________________ CLASS __________ DATE __________

Technology

4.5 A Pattern for Inverses

You may be surprised to learn that with some experimentation you can find a patterned relationship between the elements of a certain type of matrix and the elements of its inverse.

In Exercises 1–6, let $M = \begin{bmatrix} a & b \\ c & d \end{bmatrix}$. First calculate $ad - bc$. Then use a graphics calculator to find the inverse of each matrix. Record your results for reference.

1. $M = \begin{bmatrix} 11 & 2 \\ 5 & 1 \end{bmatrix}$

____

2. $M = \begin{bmatrix} 7 & 9 \\ 3 & 4 \end{bmatrix}$

____

3. $M = \begin{bmatrix} -17 & 5 \\ -7 & 2 \end{bmatrix}$

____

4. $M = \begin{bmatrix} 1 & 2 \\ -1 & -1 \end{bmatrix}$

____

5. $M = \begin{bmatrix} 10 & 9 \\ 11 & 10 \end{bmatrix}$

____

6. $M = \begin{bmatrix} -2 & -13 \\ -3 & -20 \end{bmatrix}$

____

7. What is true of $ad - bc$ from Exercises 1–6?

____

8. If $M = \begin{bmatrix} a & b \\ c & d \end{bmatrix}$ and $ad - bc = 1$, describe how to find the inverse of M from a, b, c, and d.

____

9. Find the inverse of $\begin{bmatrix} a & a+1 \\ a-1 & a \end{bmatrix}$. ______________________

10. Find the inverse of $\begin{bmatrix} a+1 & 1 \\ a(a+2) & a+1 \end{bmatrix}$. ______________________

NAME ______________________ CLASS __________ DATE __________

Technology

4.6 Related Linear Systems

One of the advantages to finding the inverse of a matrix is that it allows you to solve systems of linear equations whose coefficient matrices are equal but whose constant matrices are unequal. For example, the two systems shown have equal coefficient matrices but unequal constant matrices.

System 1 $\begin{cases} 2x - 3y = 19 \\ 3x + 5y = 38 \end{cases}$ System 2 $\begin{cases} 2x - 3y = 9.5 \\ 3x + 5y = 57 \end{cases}$

To solve each system, find the inverse M of $\begin{bmatrix} 2 & -3 \\ 3 & 5 \end{bmatrix}$. Then find the products $M\begin{bmatrix} 19 \\ 38 \end{bmatrix}$ and $M\begin{bmatrix} 9.5 \\ 57 \end{bmatrix}$.

$$M\begin{bmatrix} 19 \\ 38 \end{bmatrix} = \begin{bmatrix} 11 \\ 1 \end{bmatrix} \qquad M\begin{bmatrix} 9.5 \\ 57 \end{bmatrix} = \begin{bmatrix} 11.5 \\ 4.5 \end{bmatrix}$$

For system 1: $x = 11$ and $y = 1$. For system 2: $x = 11.5$ and $y = 4.5$.

Use a graphics calculator to solve each system.

1. $\begin{cases} 2x - 3y = 23 \\ 3x + 5y = -13 \end{cases}$ ______________

2. $\begin{cases} 2x - 3y = -25 \\ 3x + 5y = 29 \end{cases}$ ______________

3. $\begin{cases} 2x - 3y = -1 \\ 3x + 5y = 8 \end{cases}$ ______________

4. $\begin{cases} 6x + 3y = 30 \\ 7x - 2y = 13 \end{cases}$ ______________

5. $\begin{cases} 6x + 3y = 39 \\ 7x - 2y = 18 \end{cases}$ ______________

6. $\begin{cases} 6x + 3y = 48 \\ 7x - 2y = 23 \end{cases}$ ______________

7. $\begin{cases} 3x - 3y = 17 \\ 2x - 5y = 19 \end{cases}$ ______________

8. $\begin{cases} 4x - 3y = 16 \\ 2x - 5y = 22 \end{cases}$ ______________

9. $\begin{cases} 4x - 3y = 15 \\ 2x - 5y = 25 \end{cases}$ ______________

10. $\begin{cases} 4x + 5y = 95 \\ 9x - 2y = 68 \end{cases}$ ______________

11. $\begin{cases} 4x + 5y = 93 \\ 9x - 2y = 90 \end{cases}$ ______________

12. $\begin{cases} 4x + 5y = 91 \\ 9x - 2y = 112 \end{cases}$ ______________

NAME ______________________ CLASS __________ DATE __________

Technology

4.7 Finding the Matrix of a Transformation

The diagram shows polygon *ABCD* transformed into polygon *A′B′C′D′*. Is there a matrix $M = \begin{bmatrix} r & s \\ t & u \end{bmatrix}$ that accomplishes the transformation?

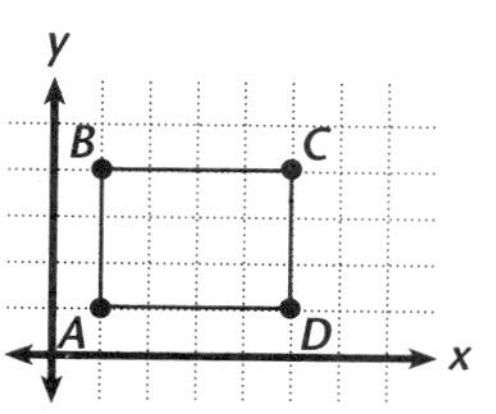

In other words, are there values of *r, s, t,* and *u* such that:

$M\begin{bmatrix} 1 \\ 1 \end{bmatrix} = \begin{bmatrix} 2 \\ 2 \end{bmatrix}$ $\quad M\begin{bmatrix} 1 \\ 4 \end{bmatrix} = \begin{bmatrix} 5 \\ 4 \end{bmatrix}$ $\quad M\begin{bmatrix} 5 \\ 4 \end{bmatrix} = \begin{bmatrix} 9 \\ 4 \end{bmatrix}$ $\quad M\begin{bmatrix} 5 \\ 1 \end{bmatrix} = \begin{bmatrix} 6 \\ 1 \end{bmatrix}$

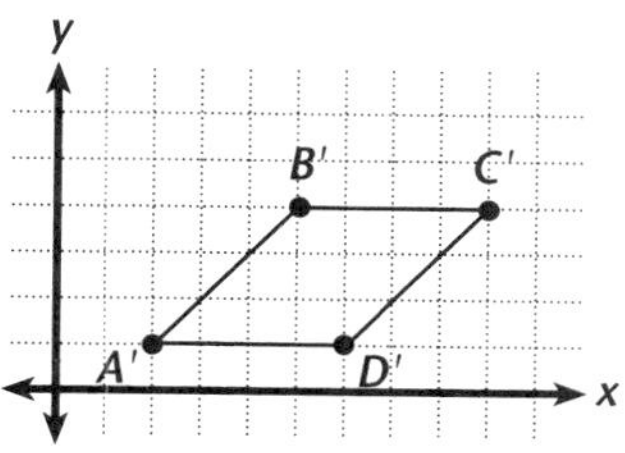

Write a system of equations for each matrix equation.

1. $\begin{bmatrix} r & s \\ t & u \end{bmatrix}\begin{bmatrix} 1 \\ 1 \end{bmatrix} = \begin{bmatrix} 2 \\ 1 \end{bmatrix}$ ______________________

2. $\begin{bmatrix} r & s \\ t & u \end{bmatrix}\begin{bmatrix} 1 \\ 4 \end{bmatrix} = \begin{bmatrix} 5 \\ 4 \end{bmatrix}$ ______________________

3. $\begin{bmatrix} r & s \\ t & u \end{bmatrix}\begin{bmatrix} 5 \\ 4 \end{bmatrix} = \begin{bmatrix} 9 \\ 4 \end{bmatrix}$ ______________________

4. $\begin{bmatrix} r & s \\ t & u \end{bmatrix}\begin{bmatrix} 5 \\ 1 \end{bmatrix} = \begin{bmatrix} 6 \\ 1 \end{bmatrix}$ ______________________

5. From Exercises 1–4, write one system of equations to find *r* and *s*, and one for *t* and *u*.

____

6. Solve the system from Exercise 5. ______________________

Find a matrix $M = \begin{bmatrix} r & s \\ t & u \end{bmatrix}$ such that polygon *ABCD* is transformed into *A′B′C′D′*.

7.

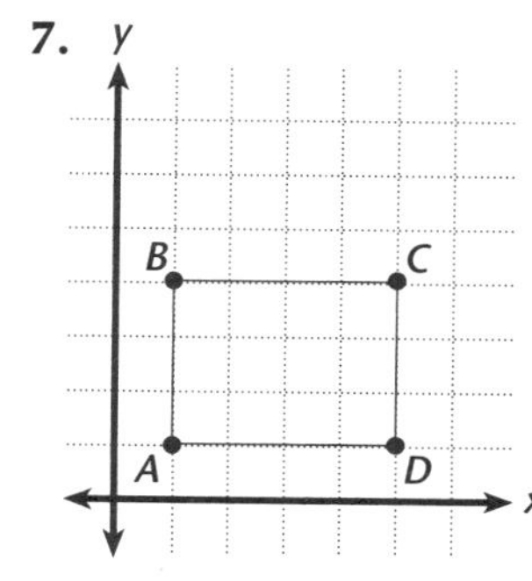

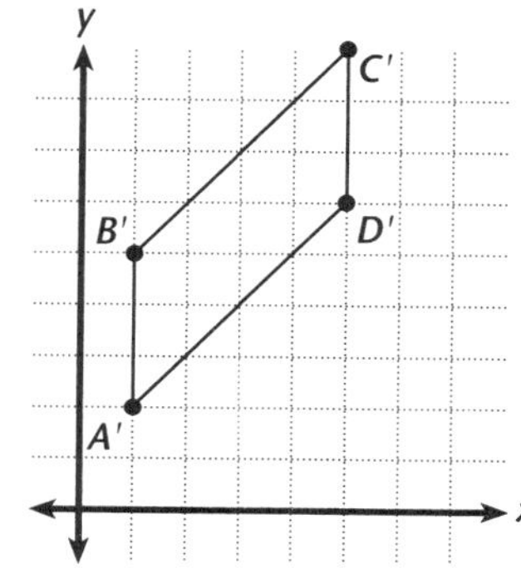

____

8.

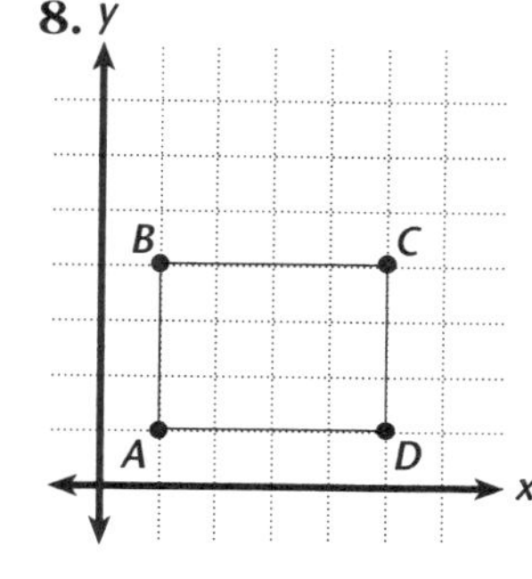

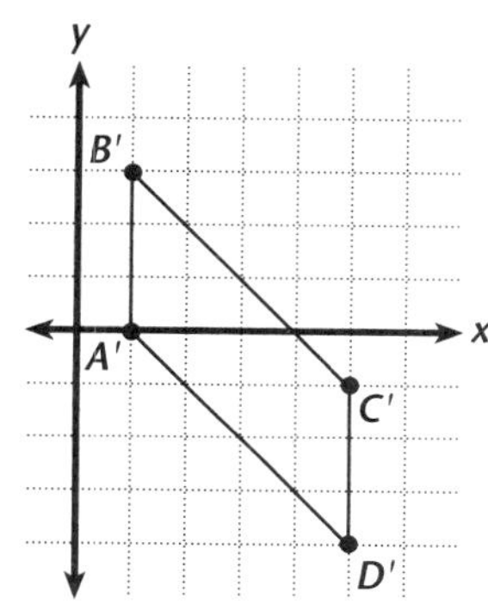

____

NAME ______________________ CLASS __________ DATE __________

Technology

4.8 Bar Graphs and Linear Inequalities

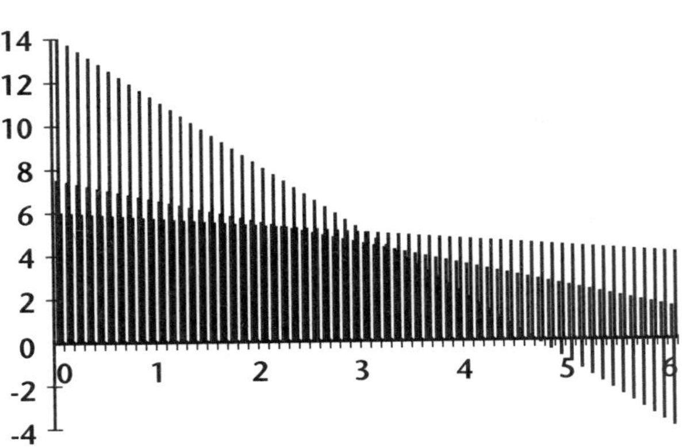

If the number of bars in a bar graph is great enough, you can simulate the intersection of the solution regions of a set of linear inequalities with the use of a bar graph. The diagram shows the solution of the system.

$$\begin{cases} x \geq 0 \\ y \geq 0 \\ x + y \leq 7.5 \\ 3x + y \leq 14 \\ x + 3y \leq 18 \end{cases}$$

To create the graph,

- Write each inequality involving x and y in terms of y.
- To obtain very narrow bars, increment x by 0.1. These entries will be in column A.

Enter

- Cell B2 as 7.5−A2. Cell C2 as −3*A2+14. Cell D2 as −A2/3+6.
- Using the FILL DOWN command, you can complete the data set.

	A	B	C	D
1	X	Y1	Y2	Y3
2	0.0	7.5	14.0	6.00
3	0.1	7.4	13.7	5.97
4	0.2	7.3	13.4	5.93
	...			
60	5.8	1.7	−3.4	4.07
61	5.9	1.6	−3.7	4.03
62	6.0	1.5	−4.0	4.00

Use a spreadsheet to graph the solution region for each set of inequalities.

1. $\begin{cases} x \geq 0 \\ y \geq 0 \\ y \leq x \\ y \leq -x + 6 \end{cases}$

2. $\begin{cases} 0 \leq x \leq 6 \\ y \geq 0 \\ y \leq 5x \\ 3x + 5y \leq 28 \end{cases}$

3. $\begin{cases} x \geq 0 \\ y \geq 0 \\ y \leq -0.5x + 6 \\ y \leq -2x + 12 \end{cases}$

4. $\begin{cases} 0 \leq x \leq 6 \\ y \geq 0 \\ y \leq -0.1x + 12 \\ y \leq 3.6x + 1 \\ y \leq -3.6x + 29 \end{cases}$

5. $\begin{cases} x \geq 0 \\ y \geq 0 \\ y \leq 6 \\ y \leq -1.5x + 9 \end{cases}$

6. $\begin{cases} 0 \leq x \leq 6 \\ 0 \leq y \leq 6 \\ y \leq -0.6x + 13.4 \\ y \leq 1.6x + 2.4 \end{cases}$

7. Explain how to make a bar graph solution region even finer than what is shown on this page.

____

NAME ______________________________ CLASS ______________ DATE ______________

Technology

4.9 Constraint Equations

A linear programming problem is shown. Notice that a constraint equation, $y = 0.25x$ has been added to the set of inequalities. The graphics calculator display shows the feasible region along with the graph of $y = 0.25x$.

Maximize $P = 15x + 8y$ given that

$$\begin{cases} x \geq 0 \\ 0 \leq y \leq 4 \\ y \leq 0.5x + 3 \\ y \leq -x + 12 \\ y \leq -0.33x + 6 \end{cases}$$

and that

$y = 0.25x$

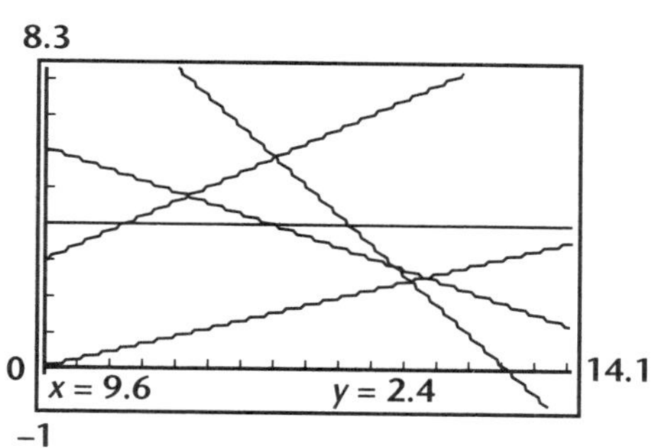

The maximum value of P given both the inequalities and the equation occurs at the point where the graph of $y = 0.25x$ intersects the boundary of the feasible region. You can use the trace feature of the graphics calculator to find those coordinates to be (9.6, 2.4). The maximum value of P is $15(9.6) + 8(2.4) = 163.2$.

Find the maximum value of P given that $x \geq 0$ and $y \geq 0$.

1. $P = 15x + 8y$

$$\begin{cases} y \leq 4 \\ y \leq 0.5x + 3 \\ y \leq -x + 12 \\ y \leq -0.33x + 6 \end{cases}$$

$y = 0.5x$

____

2. $P = 15x + 8y$

$$\begin{cases} y \leq 4 \\ y \leq 0.5x + 3 \\ y \leq -x + 12 \\ y \leq -0.33x + 6 \end{cases}$$

$y = x$

____

3. $P = 15x + 8y$

$$\begin{cases} y \leq 4 \\ y \leq 0.5x + 3 \\ y \leq -x + 12 \\ y \leq -0.33x + 6 \end{cases}$$

$y = 3.44x$

____

4. $P = 10.5x + 7.6y$

$$\begin{cases} x \leq 8 \\ y \leq 2.2x + 3 \\ y \leq 1.4x + 4 \\ y \leq -0.7x + 9 \\ y \leq -0.2x + 7 \end{cases}$$

$y = 0.25x$

____

5. $P = 10.5x + 7.6y$

$$\begin{cases} x \leq 8 \\ y \leq 2.2x + 3 \\ y \leq 1.4x + 4 \\ y \leq -0.7x + 9 \\ y \leq -0.2x + 7 \end{cases}$$

$y = 0.57x$

____

6. $P = 10.5x + 7.6y$

$$\begin{cases} x \leq 8 \\ y \leq 2.2x + 3 \\ y \leq 1.4x + 4 \\ y \leq -0.7x + 9 \\ y \leq -0.2x + 7 \end{cases}$$

$y = 2.56x$

____

7. Suppose $A(r, s)$ is on $y = mx$ $(m > 0)$ and on the boundary of the feasible region. Why does the maximum value of P occur at $A(r, s)$?

____

NAME ______________________ CLASS ____________ DATE ____________

Technology

4.10 Minimums, Maximums, and Fixed Profits

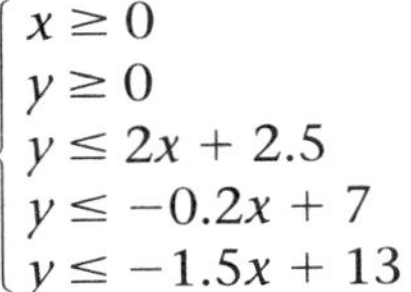

$$\begin{cases} x \geq 0 \\ y \geq 0 \\ y \leq 2x + 2.5 \\ y \leq -0.2x + 7 \\ y \leq -1.5x + 13 \end{cases}$$

Profit level:
$12 = x + 3y$

In some situations, a problem will contain a set of constraints and a profit level that is fixed, such as the situation at the right. What are the least and greatest production levels x that will yield the fixed profit level?

The graphics calculator display shows the feasible region and the profit equation for the given problem. The least and greatest production levels give the indicated profit. These points are represented by the intersection points of the lower horizontal line.

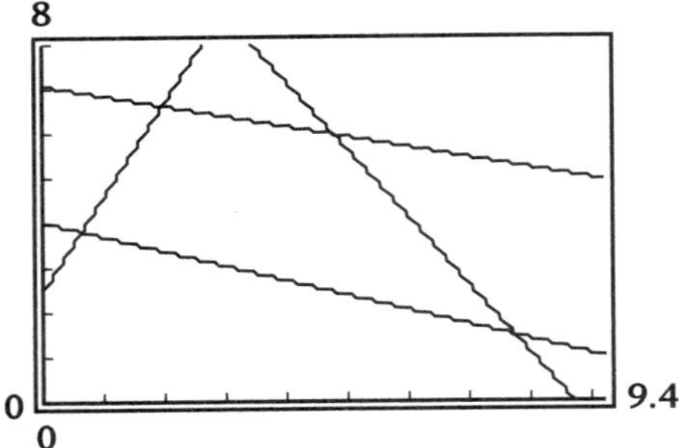

To find the x-coordinates of these points, use the trace feature on your graphics calculator. You will find them to be $x = 0.6$ and $x = 7.7$. If $0.6 \leq x \leq 7.7$, the manufacturer will realize a fixed profit.

If you also want to find the values of y that correspond to the minimum and maximum values of x, you can read them from the display as you use trace key.

Find the minimum and maximum values of x that give the desired profit level P.

1. $\begin{cases} x \geq 0 \\ y \geq 0 \\ y \leq 4 \\ y \leq -1.5x + 8 \\ y \leq -0.75x + 5 \end{cases}$
Profit level:
P: $y = 3$

____

2. $\begin{cases} x \geq 0 \\ y \geq 0 \\ y \leq 4 \\ y \leq -1.5x + 8 \\ y \leq -0.75x + 5 \end{cases}$
Profit level:
P: $12 = 4x + y$

____

3. $\begin{cases} x \geq 0 \\ y \geq 0 \\ y \leq 4 \\ y \leq -1.5x + 8 \\ y \leq -0.75x + 5 \end{cases}$
Profit level:
P: $2 = -x + y$

____

4. $\begin{cases} x \geq 0 \\ y \geq 0 \\ y \leq x + 2 \\ y \leq -0.5x + 8 \\ y \leq -4x + 36 \end{cases}$
Profit level:
P: $36 = 2x + 9y$

____

5. $\begin{cases} x \geq 0 \\ y \geq 0 \\ y \leq x + 2 \\ y \leq -0.5x + 8 \\ y \leq -4x + 36 \end{cases}$
Profit level:
P: $12 = -3x + 4y$

____

6. $\begin{cases} x \geq 0 \\ y \geq 0 \\ y \leq x + 2 \\ y \leq -0.5x + 8 \\ y \leq -4x + 36 \end{cases}$
Profit level:
P: $72 = 2x + 9y$

____

7. Suppose that no fixed profit level is given for Exercises 4–6. Instead you are given only that $P = 2x + 9y$. What would be the minimum and maximum profits?

____

NAME ______________________ CLASS ____________ DATE ____________

Lesson Activity
4.1 Matrices and Assignments

Malik, Adelso, and Paula were hired to work part time at a soon-to-be-opened supermarket. They have taken a job preference test to determine which of three jobs: cashier, deli clerk, or stock clerk they like best. Their scores, which range from 1 to 10, are stored in a rectangular array as shown. The higher the score, the higher the preference for the job.

	Cashier	Deli clerk	Stock clerk
Malik	5	4	8
Adelso	6	2	3
Paula	4	9	2

$$\begin{bmatrix} 5 & 4 & 8 \\ 6 & 2 & 3 \\ 4 & 9 & 2 \end{bmatrix} = \text{Job Preferences}$$

The manager wants to assign one worker to only one job so that each worker's preference is maximized. To represent the data in an organized manner, the manager stores the data in this rectangular array, using a "1" to indicate that the worker was assigned the job and a "0" to indicate that the worker was not assigned the job.

	Cashier	Deli clerk	Stock clerk
Malik	1	0	0
Adelso	0	1	0
Paula	0	0	1

$$\begin{bmatrix} 1 & 0 & 0 \\ 0 & 1 & 0 \\ 0 & 0 & 1 \end{bmatrix} = \text{Worker Assignments}$$

To find the total score of the individual assignments, the manager finds the sum of the preference score for each person assigned to the job. In the given array, Malik is assigned the cashier job; Adelso, the deli clerk job; and Paula the stock clerk job. Therefore, the total score is $5 + 2 + 2 = 9$.

1. How many other assignment arrays are possible? ______________________

2. Use matrix notation to list the other possible assignment matrices.

3. Describe the assignment that yields the maximum score.

____

4. What is that score? ______________________

NAME ______________________ CLASS __________ DATE __________

Lesson Activity

4.2 Matrices and Directed Graphs

A directed graph is a graph consisting of points (vertices) and directed line segments. Between any two points only one directed line segment can be drawn so the number of directed segments that connect one point to another can be placed into a matrix. If two points are connected by a directed segment, the entry is coded "1." If there is no connection, the entry is coded "0." This directed graph is represented by matrix G.

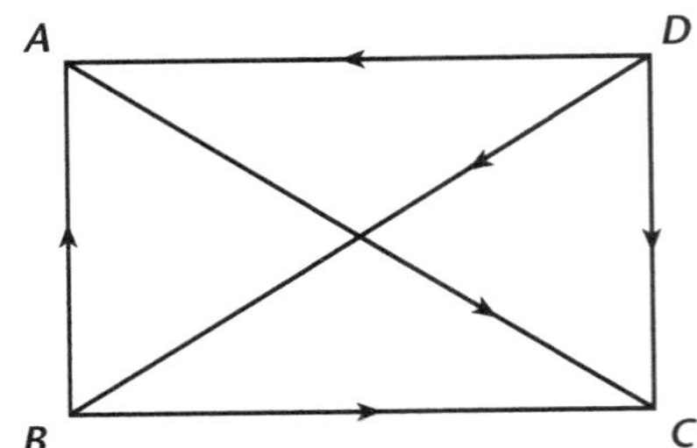

$$\begin{array}{c} \\ \boldsymbol{A} \\ \boldsymbol{B} \\ \boldsymbol{C} \\ \boldsymbol{D} \end{array} \begin{array}{c} \begin{array}{cccc} \boldsymbol{A} & \boldsymbol{B} & \boldsymbol{C} & \boldsymbol{D} \end{array} \\ \begin{bmatrix} 0 & 0 & 1 & 0 \\ 1 & 0 & 1 & 0 \\ 0 & 0 & 0 & 0 \\ 1 & 1 & 1 & 0 \end{bmatrix} \end{array} = G$$

Matrix G shows a relation among the vertices of the directed graph. Vertex D is connected to A, B, and C. Vertex B is connected to A and C, and vertex A is connected to C. the vertices can be arranged to form another matrix, as shown. Notice that all the elements above the main diagonal are 1s and all the elements on or below the main diagonal are 0s.

$$\begin{array}{c} \\ \boldsymbol{D} \\ \boldsymbol{B} \\ \boldsymbol{A} \\ \boldsymbol{C} \end{array} \begin{array}{c} \begin{array}{cccc} \boldsymbol{D} & \boldsymbol{B} & \boldsymbol{A} & \boldsymbol{C} \end{array} \\ \begin{bmatrix} 0 & 1 & 1 & 1 \\ 0 & 0 & 1 & 1 \\ 0 & 0 & 0 & 1 \\ 0 & 0 & 0 & 0 \end{bmatrix} \end{array} = N$$

Use matrix *N* to answer the questions.

1. How many directed segments begin at *D*? ______________________

2. Name these directed segments. ______________________

3. How many direct segments begin at *C*? ______________________

4. How are the vertices arranged in matrix *N*? ______________________

____

5. Represent this directed graph as a matrix using 1s and 0s.

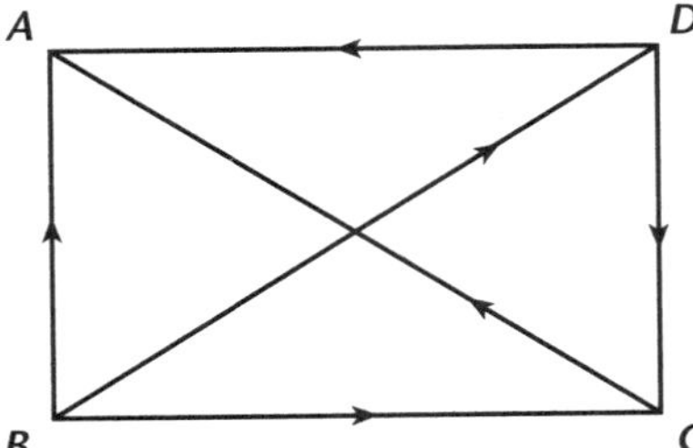

NAME ______________________ CLASS ____________ DATE ____________

Lesson Activity

4.3 Matrices and Tours

Linda wants to visit her aunt, her grandparents, and a friend, each of whom lives in a different city. Linda wants to leave from home, visit each party exactly once, and return home on the same day. In addition, she wants to minimize the total distance she travels. To help plan her trip, Linda arranges the cities and the distances in miles between each city in a matrix. For example, the matrix indicates that the distance from City 1 to City 4 is 15 miles.

$$\begin{array}{c} \\ \mathbf{1} \\ \mathbf{2} \\ \mathbf{3} \\ \mathbf{4} \end{array} \begin{array}{c} \begin{array}{cccc} \mathbf{1} & \mathbf{2} & \mathbf{3} & \mathbf{4} \end{array} \\ \begin{bmatrix} 0 & 17 & 10 & 15 \\ 17 & 0 & 6 & 12 \\ 10 & 6 & 0 & 14 \\ 15 & 12 & 14 & 0 \end{bmatrix} \end{array} = D$$

As Linda leaves her home in City 1, she has a choice of going to any one of the three remaining cities. Suppose Linda decides to go to City 2 first, City 4 second, City 3 third, and then return home. She can write the data in an assignment matrix, where a "1" in any row indicates that Linda travels from that city to another city and a "0" indicates no connection between the cities. Let the ordering of the cities—1, 2, 4, 3, 1—be called a "tour."

$$\begin{array}{c} \\ \mathbf{1} \\ \mathbf{2} \\ \mathbf{3} \\ \mathbf{4} \end{array} \begin{array}{c} \begin{array}{cccc} \mathbf{1} & \mathbf{2} & \mathbf{3} & \mathbf{4} \end{array} \\ \begin{bmatrix} 0 & 1 & 0 & 0 \\ 0 & 0 & 0 & 1 \\ 1 & 0 & 0 & 0 \\ 0 & 0 & 1 & 0 \end{bmatrix} \end{array} = T$$

1. How many tours are possible? ______________________

2. Using matrix notation, list all the possible tours.

3. What are the dimensions of Matrix D and Matrix T? ______________________

4. Find the product matrix TD.

5. Add $td_{11} + td_{22} + td_{33} + td_{44}$. ______________________

6. What does this sum represent? ______________________

7. Find the product of each possible tour matrix and matrix D.

8. Which tour minimizes the distance Linda travels? ______________________

Lesson Activity

4.4 Matrices and Route Problems

The direct highway routes (not through any other city) for several cities are shown on this diagram.

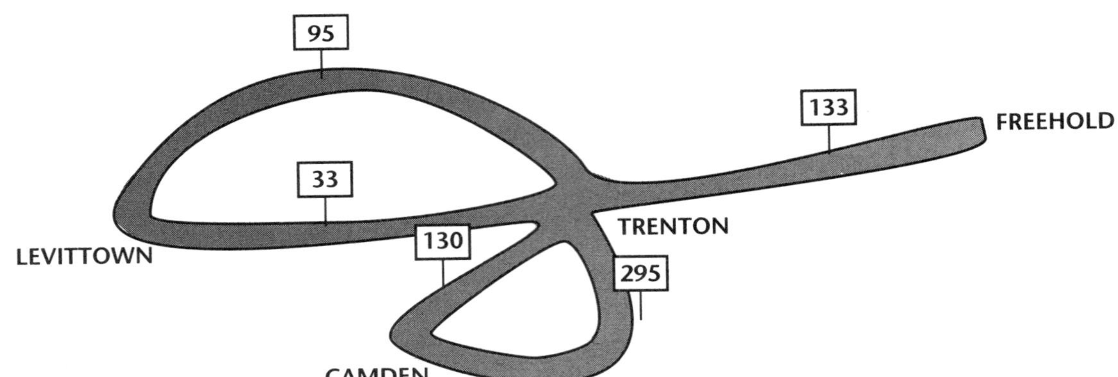

This information can be placed into a matrix showing the total number of direct routes between each pair of cities.

	Trenton	**Levittown**	**Freehold**	**Camden**
Trenton	0	2	1	2
Levittown	2	0	0	0
Freehold	1	0	0	0
Camden	2	0	0	0

$= R$

1. How many direct routes are there from Camden to Trenton? ______________

2. In how many ways can you travel directly from Trenton to one other city? ______________

3. Why are the entries along the main diagonal all 0s?

____

4. Multiply matrix R by itself.

5. What does this matrix product indicate? ____

____

6. How many ways are there of traveling from Levittown to Freehold by traveling through exactly one other city? ______________

7. Suppose matrix R is multiplied by itself three times. What does the product indicate? ____

NAME ______________________ CLASS ____________ DATE ____________

Lesson Activity

4.5 Matrices and Codes

One way to conceal the meanings of words is to substitute a number for a letter. Because of the frequency of letters, it is often easy to break this type of code. The use of matrices provides a means of coding a message in such a way that the code is almost impossible to detect. For example, matrix M codes the word "yes" by using 25 for y, 5 for e, and 19 for s; this code would be easy to break.

$$M = \begin{bmatrix} 25 \\ 5 \\ 19 \end{bmatrix}$$

To make the code more difficult to break, first form matrix A so that it has 3 rows and 3 columns. Placing 1s on the main diagonal and 0s below the main diagonal. Use any other number to fill the remaining entries.

$$A = \begin{bmatrix} 1 & 3 & 5 \\ 0 & 1 & 2 \\ 0 & 0 & 1 \end{bmatrix}$$

Next, form coding matrix C by multiplying matrix A by its transpose. In the transpose of a matrix, the rows and columns are switched.

$$C = \overset{A}{\begin{bmatrix} 1 & 3 & 5 \\ 0 & 1 & 2 \\ 0 & 0 & 1 \end{bmatrix}} \overset{A^t}{\begin{bmatrix} 1 & 0 & 0 \\ 3 & 1 & 0 \\ 5 & 2 & 1 \end{bmatrix}} = \begin{bmatrix} 35 & 13 & 5 \\ 13 & 5 & 2 \\ 5 & 2 & 1 \end{bmatrix}$$

To code the word "yes," find the product matrix CM.

$$CM = \begin{bmatrix} 35 & 13 & 5 \\ 13 & 5 & 2 \\ 5 & 2 & 1 \end{bmatrix} \begin{bmatrix} 25 \\ 5 \\ 19 \end{bmatrix} = \begin{bmatrix} 1035 \\ 388 \\ 154 \end{bmatrix}$$

Now the word "yes" is coded by the message 1035 388 154.

1. How could you decode this message? ______________________

2. Code the word "cat" using matrices.

3. Why are matrix codes so difficult to break?

____

To make the code even more difficult to break, a letter of the alphabet can be used as a filler. Code and decode the following messages by using a 4×4 matrix with x as a filler, if needed.

4. Math is fun.

5. Matrices are great.

NAME ____________________ CLASS __________ DATE __________

Lesson Activity

4.6 Matrices and Game Theory

In many chance situations, matrices can be used to represent outcomes and probabilities, and matrix operations can be used to predict events.

For example, suppose Rachel and Carl play the game of "heads and tails" for $0.10 a toss with Rachel trying to match heads or tails, and Carl winning only if the coins are unmatched. The game of "heads and tails" can be represented as a matrix.

$$\begin{matrix} & \textbf{Heads} & \textbf{Tails} \\ \textbf{Heads} & +10 & -10 \\ \textbf{Tails} & -10 & +10 \end{matrix} \quad \begin{bmatrix} +10 & -10 \\ -10 & +10 \end{bmatrix} = R$$

Matrix R represents Rachel's payoff matrix.

$$\begin{matrix} & \textbf{Heads} & \textbf{Tails} \\ \textbf{Heads} & -10 & +10 \\ \textbf{Tails} & +10 & -10 \end{matrix} \quad \begin{bmatrix} -10 & +10 \\ +10 & -10 \end{bmatrix} = C$$

Matrix C represents Carl's payoff matrix.

Notice that the sum of R and C is the matrix $\begin{bmatrix} 0 & 0 \\ 0 & 0 \end{bmatrix}$. A game with this property is known as a zero-sum game. A zero-sum game is one in which one player must lose whenever the other wins.

In another game, Rachel and Carl each have two coins. They both hold either zero, one, or two coins in a clenched fist and open their fists at the same time. If they are both holding the same number of coins, then Rachel takes the coin or coins Carl is holding. If they are holding a different number of coins, then Carl takes the coins Rachel is holding.

1. What is Rachel's payoff matrix?

2. What is Carl's payoff matrix?

3. Is the game a zero-sum game? Explain.

____

NAME ______________________ CLASS __________ DATE __________

Lesson Activity

4.7 Matrices and Transformations

Matrix R lists the coordinates of the vertices of rhombus $PQRS$:

$$R = \begin{bmatrix} -2 & 0 & 2 & 0 \\ 0 & 4 & 0 & -4 \end{bmatrix}$$

Enter matrix R in a graphics calculator. Find the product matrix and describe what happens to the rhombus when its vertex matrix is operated on by each of the following 2×2 matrices.

1. $A = \begin{bmatrix} 1 & 0 \\ 0 & -1 \end{bmatrix}$ ______________________

2. $B = \begin{bmatrix} -1 & 0 \\ 0 & 1 \end{bmatrix}$ ______________________

3. $C = \begin{bmatrix} 0 & 1 \\ 1 & 0 \end{bmatrix}$ ______________________

4. $D = \begin{bmatrix} 0 & -1 \\ -1 & 0 \end{bmatrix}$ ______________________

5. $E = \begin{bmatrix} 0 & -1 \\ 1 & 0 \end{bmatrix}$ ______________________

6. $F = \begin{bmatrix} -1 & 0 \\ 0 & -1 \end{bmatrix}$ ______________________

7. $G = \begin{bmatrix} 0 & 1 \\ -1 & 0 \end{bmatrix}$ ______________________

8. $H = \begin{bmatrix} 1 & 1 \\ -1 & 1 \end{bmatrix}$ ______________________

9. A transformation that preserves distances is called an isometry. Which of the preceding matrix transformations seem to be isometries?

____

Investigate what happens to the rhombus when its matrix is multiplied by each of the following matrix products.

10. AF ______________________

11. FB ______________________

12. BF ______________________

13. $(AF)^{-1}$ ______________________

NAME ______________________ CLASS __________ DATE __________

Lesson Activity

4.8 Matrices and Production

A company produces chairs and tables. Each chair requires 5 board-feet of mahogany and 10 hours of labor. Each table requires 20 board-feet of mahogany and 15 hours of labor. Each week, 400 board-feet of mahogany and 450 hours of labor may be used. Each chair brings in $45 profit and each table generates $80 profit.

The production manager must determine how many chairs and tables should be produced each week so the company's profits can be maximized using the resources available.

Use matrices to model this situation.

1. Complete the matrix showing the requirements for each chair and each table.

	Chair	**Table**
Mahogany		
Labor		

2. Complete the matrix showing the resources available each week.

	Availability
Mahogany	
Labor	

3. Write two inequalities to represent the data in the matrices.

____

4. Graph the inequalities on the grid provided.

5. Write two inequalities that represent the constraints for the number of tables and the number of chairs.

____

6. Write an expression to represent the profit.

____

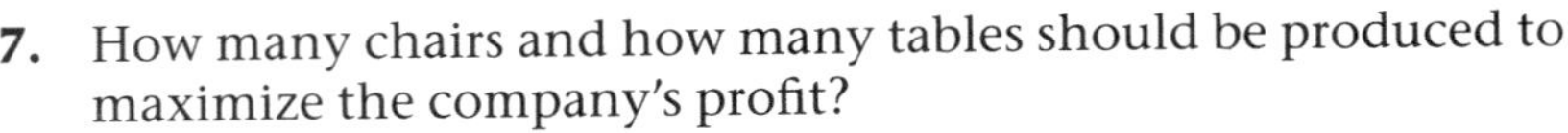

7. How many chairs and how many tables should be produced to maximize the company's profit?

____

Lesson Activity

4.9 Matrices, Linear Programming, and Game Theory

In a two-player game there are optional strategies, one for each player, which assures that player of winning at least a given number of times. Consider the following game.

Rebecca has a dime and a quarter, and Chris has a penny and a dime. Each selects a coin and simultaneously shows it to the other. Rebecca wins \$1 if the coins match. She wins \$2 if a quarter and a penny appear. She loses \$1 if only one dime appears.

Linear algebra can be used to find Rebecca's optimal strategy. First construct Rebecca's payoff matrix.

		Chris	
		penny	**dime**
Rebecca	**dime**	-1	1
	quarter	2	-1

A system of three linear inequalities, with x representing the probability that Rebecca plays the dime, y representing the probability that she plays the quarter, and v representing the value of the game, can be written using Rebecca's payoff matrix.

The system $\begin{cases} -x + 2y \leq v \\ x - y \leq v \\ x + y = 1 \\ x > 0 \\ y > 0 \end{cases}$ models the game for Rebecca.

1. What do the constraints $x > 0$ and $y > 0$ represent? ______________________

2. What does the equation $x + y = 1$ represent? ______________________

3. The ordered triple (x, y, v) is called the optimum point. Use matrices to find the optimum point. ______________________

4. Describe what the optimum point indicates. ______________________

____

5. Let a represent the probability that Chris plays the penny and let b represent the probability that Chris plays the dime. Write a system of equations that models the game for Chris.

____

6. Use matrices to find the optimum point for Chris. ______________________

NAME ______________________ CLASS __________ DATE __________

Lesson Activity
4.10 Matrices and Probabilities

For the past five years, a state has conducted a survey to find out what percent of its residents moved from the cities to the suburbs or moved from the suburbs to the cities. The results of the survey were placed in matrix A.

$$\text{From}\ \begin{matrix} \textbf{Cities} \\ \textbf{Suburbs} \end{matrix} \overset{\text{To}}{\overset{\begin{matrix}\textbf{Cities} & \textbf{Suburbs}\end{matrix}}{\begin{bmatrix} 0.90 & 0.10 \\ 0.05 & 0.95 \end{bmatrix}}} = A$$

Currently, 40% of the population of the state lives in the suburbs and 60% in the cities. The total population of the state has remained constant from year to year. Matrix B represents the percent of the population living in a city or in a suburb.

$$\overset{\begin{matrix}\textbf{City} & \textbf{Suburb}\end{matrix}}{\begin{bmatrix} 0.60 & 0.40 \end{bmatrix}} = B$$

The probability of living in the cities or in the suburbs after one year is found using the product matrix BA.

$$BA = \begin{bmatrix} 0.60 & 0.40 \end{bmatrix} \begin{bmatrix} 0.90 & 0.10 \\ 0.05 & 0.95 \end{bmatrix} = \begin{bmatrix} 0.56 & 0.44 \end{bmatrix}$$

1. Determining the probability of living in the cities or in the suburbs after two years by finding the product BA^2. ____________

2. Find the probability of living in the cities or in the suburbs after three years. ____________

3. Find the probability after five years. ____________

4. As the number of years increase, what appears to happen to the probability of living in the cities or in the suburbs?

____

NAME ______________________ CLASS __________ DATE __________

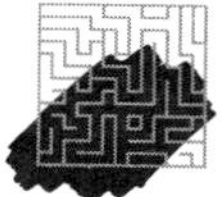

Assessing Prior Knowledge

4.1 Using Matrices to Represent Data

Complete the table.

x	1	2	3
$y = 2x + 6$			
$y = 12x - 18$			

NAME ______________________ CLASS __________ DATE __________

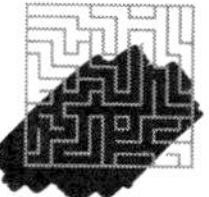

Quiz

4.1 Using Matrices to Represent Data

The matrix shows the high school absentee record for one week.

$$\begin{array}{l} \\ \textbf{Freshmen} \\ \textbf{Sophomores} \\ \textbf{Juniors} \\ \textbf{Seniors} \end{array} \begin{array}{c} \begin{array}{ccccc} \textbf{M} & \textbf{T} & \textbf{W} & \textbf{R} & \textbf{F} \end{array} \\ \begin{bmatrix} 8 & 6 & 8 & 10 & 12 \\ 9 & 8 & 10 & 14 & 10 \\ 6 & 12 & 10 & 16 & 12 \\ 11 & 9 & 12 & 13 & 15 \end{bmatrix} \end{array} = M$$

1. Describe the data in the first row, third column.

____

2. In what location do you find that 13 seniors were absent? ______________________

3. Give the dimensions of matrix M. ______________________

4. Describe entry m_{32}. ______________________

5. Describe entry m_{25}. ______________________

6. In the space provided, represent the set of linear equations using a matrix equation.

$$\begin{cases} -4x + 7y = 12 \\ -3x + 6y = 5 \\ 5x - y = -2 \end{cases}$$

Let $A = \begin{bmatrix} 2 & 3 & -1 \\ 0 & -4 & 5 \end{bmatrix}$ and $B = \begin{bmatrix} 0 & -11 & 9 \\ -4 & 4 & -2 \end{bmatrix}$.

7. Find $A + B$.

8. Find $B - A$.

NAME ______________________ CLASS ____________ DATE ____________

Assessing Prior Knowledge

4.2 Matrix Multiplication

Give the dimensions of each matrix.

1. $M = \begin{bmatrix} -5 & 0 & 7 & -11 \\ 0 & 3 & 7 & -19 \end{bmatrix}$

____

2. $N = \begin{bmatrix} -15 & 7 \\ 10 & 9 \\ 6 & 6 \end{bmatrix}$

____

3. $B = \begin{bmatrix} -7 & 0 & 13 \\ -2 & 4 & -2 \\ 12 & 1 & 10 \end{bmatrix}$

____

4. $A = \begin{bmatrix} 6 & 16 & 5 & 9 & -3 \end{bmatrix}$

____

- -

NAME ______________________ CLASS ____________ DATE ____________

Quiz

4.2 Matrix Multiplication

If possible, find the products of each pair of matrices. If multiplication is not possible, explain why.

1. $\begin{bmatrix} 5 \\ 1 \\ 0 \end{bmatrix} \begin{bmatrix} -1 & 4 & -3 \end{bmatrix}$

____

2. $\begin{bmatrix} -3 & 5 \\ 1 & -2 \end{bmatrix} \begin{bmatrix} 0 & 3 & -2 \end{bmatrix}$

____

3. $\begin{bmatrix} 4 & 1 & 0 & 2 \\ -1 & 5 & 3 & -2 \\ -2 & 3 & 5 & 0 \end{bmatrix} \begin{bmatrix} 1 \\ -2 \\ 3 \\ 0 \end{bmatrix}$

____

4. $\begin{bmatrix} 1 & 0 \\ 0 & 1 \end{bmatrix} \begin{bmatrix} 2 \\ 3 \end{bmatrix}$

____

5. The refreshment stand at Birchwood Beach sells three types of candy: Krazy Kats, Choco Mania, and Wonder Wafers. On Tuesday, 14 Krazy Kats were sold at 35¢ each, 23 Choco Manias at 50¢ each, and 18 Wonder Wafers at 40¢ each. Use matrix multiplication to determine the total revenue from the three types of candy.

NAME ______________________ CLASS __________ DATE __________

Assessing Prior Knowledge

4.3 Systems of Two Linear Equations

Express each equation in slope-intercept form.

1. $3x + 4y = 12$ ______________________

2. $x - \frac{y}{4} + 8 = 0$ ______________________

3. $0.1y + 0.2 = x$ ______________________

4. $\frac{5x}{4y} = 1$ ______________________

NAME ______________________ CLASS __________ DATE __________

Quiz

4.3 Systems of Two Linear Equations

Indicate whether each of the following systems is inconsistent, dependent, or independent.

1. $\begin{cases} 2x - 2y = 4 \\ x + y = 3 \end{cases}$

____

2. $\begin{cases} 2x + 3y = 6 \\ -4x - 6y = -12 \end{cases}$

____

3. $\begin{cases} 4x - y = 7 \\ 4x - y = -2 \end{cases}$

____

4. $\begin{cases} y + x = 15 \\ 2x + 2y = 12 \end{cases}$

____

Solve each system of linear equations using any method.

5. $\begin{cases} 5x - 2y = 3 \\ 5x + 2y = -6 \end{cases}$

____

6. $\begin{cases} \frac{1}{3}x - y = 2 \\ x - 3y = 12 \end{cases}$

____

7. $\begin{cases} -x + 3y = 5 \\ 2x - 3y = 4 \end{cases}$

____

8. $\begin{cases} y = 3x - 5 \\ 2x + 3y = 12 \end{cases}$

____

NAME ________________ CLASS ________ DATE ________

Assessing Prior Knowledge

4.4 Using Matrix Row Operations

Find the value of y in each equation.

1. $y = 2x - z$ if $x = 4$ and $z = -2$ ________

2. $z = 2y - 3x$ if $x = -6$ and $z = -1$ ________

NAME ________________ CLASS ________ DATE ________

Quiz

4.4 Using Matrix Row Operations

Use back substitution to solve each system of equations.

1. $\begin{cases} 5x + 2y - 4z = -6 \\ 7y - z = 56 \\ 2z = 28 \end{cases}$

____

2. $\begin{cases} x - 3y + 4z = -20 \\ 2y + 7z = 12 \\ -z = 2 \end{cases}$

____

3. Write the augmented matrix for the system of equations.

$$\begin{cases} 3x - 2y - z = -24 \\ -2x + 5y + 3z = 15 \\ x + 6y - 4z = -5 \end{cases}$$

____

Solve each system of equations using the row reduction method and back substitution.

4. $\begin{cases} 2x + 3y = 9 \\ y - 4z = -2 \\ -5x - 3y - 4z = 7 \end{cases}$

____

5. $\begin{cases} -x + 3y - 6z = 15 \\ 3x - 2y + 4z = -10 \\ -5x - 3y + 2z = -8 \end{cases}$

____

NAME ______________________ CLASS __________ DATE __________

Assessing Prior Knowledge

4.5 The Inverse of a Matrix

Find the multiplicative inverse of each number.

1. -3 __________ **2.** $-\frac{1}{7}$ __________ **3.** 2.5 __________ **4.** $-\frac{6}{5}$ __________

NAME ______________________ CLASS __________ DATE __________

Quiz

4.5 The Inverse of a Matrix

1. Determine whether the pair of matrices are inverses.

$$\begin{bmatrix} 1 & -3 \\ -2 & 4 \end{bmatrix} \begin{bmatrix} -2 & -\frac{3}{4} \\ -1 & -\frac{1}{2} \end{bmatrix}$$

____

Use row reduction to find the inverse of each matrix.

2. $\begin{bmatrix} 4 & -5 \\ 2 & -7 \end{bmatrix}$

____

3. $\begin{bmatrix} 4 & 3 \\ -6 & 2 \end{bmatrix}$

____

4. $\begin{bmatrix} -5 & 2 \\ 8 & -3 \end{bmatrix}$

____

5. $\begin{bmatrix} 2 & 0 \\ 3 & 5 \end{bmatrix}$

____

Use your graphics calculator to find the inverse matrix, if it exists. If it does not exist, write does not exist.

6. $\begin{bmatrix} 4 & 7 \\ 3 & 5 \end{bmatrix}$

____

7. $\begin{bmatrix} 1 & 2 & 0 \\ 0 & 2 & -2 \\ 1 & 0 & 2 \end{bmatrix}$

____

NAME ______________________ CLASS __________ DATE __________

Mid-Chapter Assessment

Chapter 4 (Lessons 4.1–4.5)

Write the letter that best answers the question or completes the statement.

______ **1.** If matrix $A = \begin{bmatrix} 2 & 4 & 3 & 6 \\ -1 & 0 & -9 & 1 \\ -6 & -2 & -3 & 5 \end{bmatrix}$, the data in location a_{23} is

a. 5 **b.** −2 **c.** 6 **d.** −9

______ **2.** If the product of two matrices is $[-2 \quad 3]$, then they could be

a. $[1 \quad 0 \quad -3]$, $\begin{bmatrix} 2 & 0 & 3 \\ -2 & 1 & -1 \end{bmatrix}$

b. $[-2]$, $[3]$

c. $[3 \quad 0 \quad -1]$, $\begin{bmatrix} 0 & 1 \\ 5 & 4 \\ 2 & 0 \end{bmatrix}$

d. $\begin{bmatrix} 4 & 5 \\ 0 & 2 \\ 1 & 4 \end{bmatrix}$, $\begin{bmatrix} 2 & 3 & -1 \\ -2 & 0 & 1 \end{bmatrix}$

______ **3.** Which is the solution to the system of equations $\begin{cases} 3x + 2y = 8 \\ 5y = 4x - 3? \end{cases}$

a. no solution **b.** (−2, 1) **c.** (2, 1) **d.** (−1, 7)

4. Which system of equations is represented by the augmented matrix?

$$\left[\begin{array}{ccc|c} 2 & 0 & 3 & 2 \\ 0 & 1 & 6 & -3 \\ -2 & 5 & 0 & -1 \end{array}\right]$$

a. $\begin{aligned} 2x + 3y &= 2 \\ y + 6z &= -3 \\ -2x + 5y &= -1 \end{aligned}$

b. $\begin{aligned} 2x + 3z &= 2 \\ y + 6z &= -3 \\ -2y + 5z &= -1 \end{aligned}$

c. $\begin{aligned} 2x + 3z &= 2 \\ y + 6z &= -3 \\ -2x + 5y &= -1 \end{aligned}$

d. $\begin{aligned} 2x + 3z &= 2 \\ x + 6y &= -3 \\ -2x + 5z &= -1 \end{aligned}$

5. Find the product.

$$\begin{bmatrix} 5 & -4 & 3 \\ -2 & 0 & 1 \\ -4 & 3 & -2 \end{bmatrix} \begin{bmatrix} 2 & 8 \\ 0 & 4 \\ -1 & 5 \end{bmatrix}$$ ______

6. Use any method to solve the system of linear equations.

$\begin{cases} 3x + y = 8 \\ -5y = 6 - 8x \end{cases}$ ______

7. Solve the system using row reduction and back substitution.

$\begin{cases} 3x - 2y + 5z = -7 \\ 2x + y - 3z = 12 \\ x - 5z = 17 \end{cases}$ ______

NAME ____________________ CLASS ____________ DATE ____________

Assessing Prior Knowledge

4.6 Using Matrix Algebra

Classify each system as independent, dependent, or inconsistent.

1. $\begin{cases} 3x + 4y = -10 \\ -6x - 8y = 20 \end{cases}$ ________

2. $\begin{cases} 3x + 4y = -8 \\ 6x + 8y = 16 \end{cases}$ ________

3. $\begin{cases} -2x + 5y = 15 \\ 10x + 2y = 6 \end{cases}$ ________

- -

NAME ____________________ CLASS ____________ DATE ____________

Quiz

4.6 Using Matrix Algebra

For each system of equations, write the matrix equation that represents it, use matrix algebra to solve it, and show your check of the solution.

1. $\begin{cases} x + 2y = 11 \\ 2x + 3y = 18 \end{cases}$

____

2. $\begin{cases} 9b + 2c = 14 \\ 3a + 2b + c = 5 \\ a - b = -1 \end{cases}$

____

3. $\begin{cases} -l = -4 - n \\ 2m = n - l \\ l = 6 - m - n \end{cases}$

____

4. $\begin{cases} x + 9y + 2z = 14 \\ 3x + 2y + z = 5 \\ x - y + 2z = -5 \end{cases}$

____

NAME ______________________ CLASS __________ DATE __________

Assessing Prior Knowledge

4.7 Exploring Transformations Using Matrices

Plot and connect each set of points. Identify the resulting polygon.

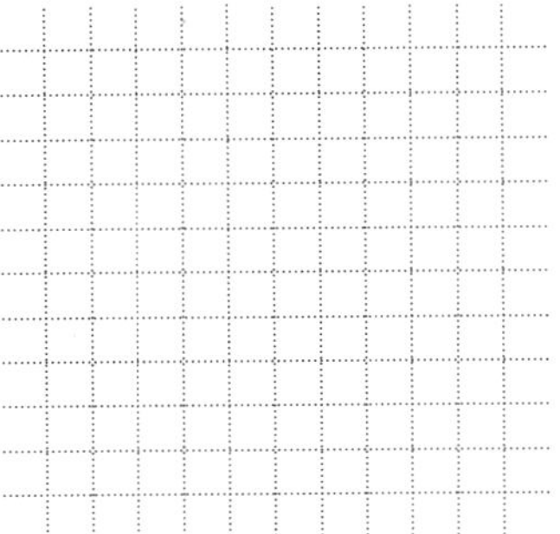

1. $P(3, 0)$, $R(-3, 0)$, $Q(0, 4)$ ______________

2. $A(-3, 3)$, $B(-3, -3)$, $C(3, -3)$, $D(3, 3)$ ______________

3. $H(6, 2)$, $I(-4, 2)$, $J(-4, -3)$, $K(6, -3)$ ______________

4. $X(4, 5)$, $Y(0, -5)$, $Z(-6, -5)$, $W(-2, 5)$ ______________

NAME ______________________ CLASS __________ DATE __________

Quiz

4.7 Exploring Transformations Using Matrices

Let $A = \begin{bmatrix} -4 & -8 & -5 & -1 \\ 1 & 5 & 8 & 4 \end{bmatrix}$, $B = \begin{bmatrix} 6 & 6 & 8 & 8 \\ -3 & 3 & 1 & -1 \end{bmatrix}$, and $C = \begin{bmatrix} 2 & 2 & 4 \\ 2 & 4 & 2 \end{bmatrix}$.

1. Use matrices to give vertices of the object resulting from transforming matrix A through a 180° clockwise rotation.

____

2. Enlarge the object represented by matrix C to 3 times its size and give the vertices of the result in matrix form.

____

3. Rotate the object represented by matrix B through a 90° counterclockwise rotation and give the vertices of the resulting object in matrix form.

____

NAME ______________________ CLASS __________ DATE __________

Assessing Prior Knowledge

4.8 Exploring Systems of Linear Inequalities

Graph each inequality on the coordinate plane.

1. $y < 2x - 1$

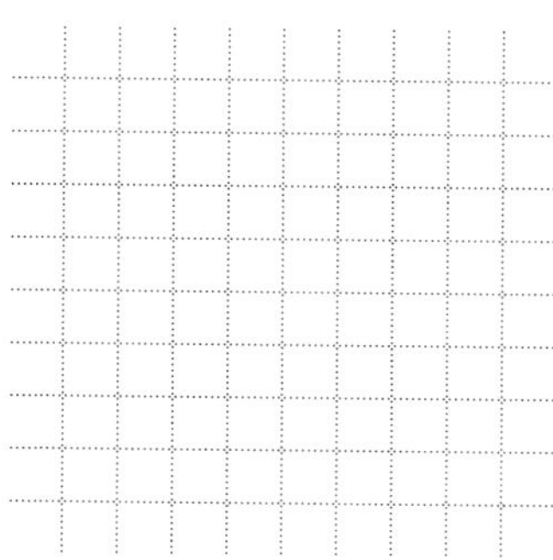

2. $y \geq 3x + 2$

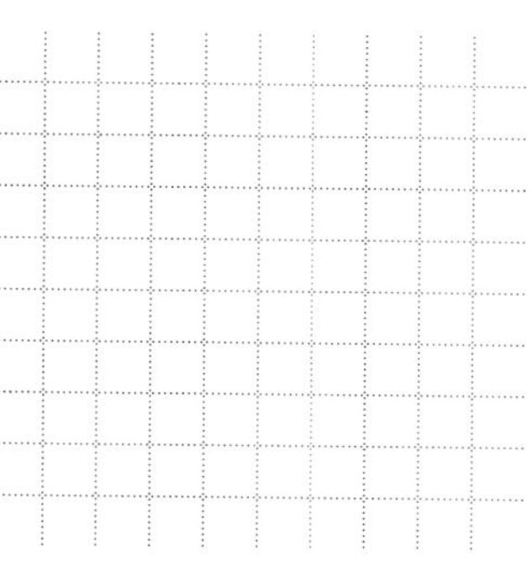

NAME ______________________ CLASS __________ DATE __________

Quiz

4.8 Exploring Systems of Linear Inequalities

Graph the solution to each of the following systems of linear inequalities on the grid provided.

1. $\begin{cases} x + y \leq 5 \\ x - y \geq -2 \end{cases}$

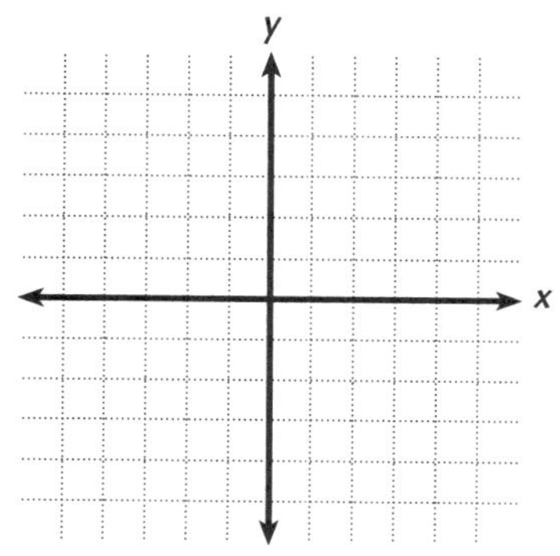

2. $\begin{cases} x \geq 0 \\ y \leq 3 \\ x + y \geq 2 \end{cases}$

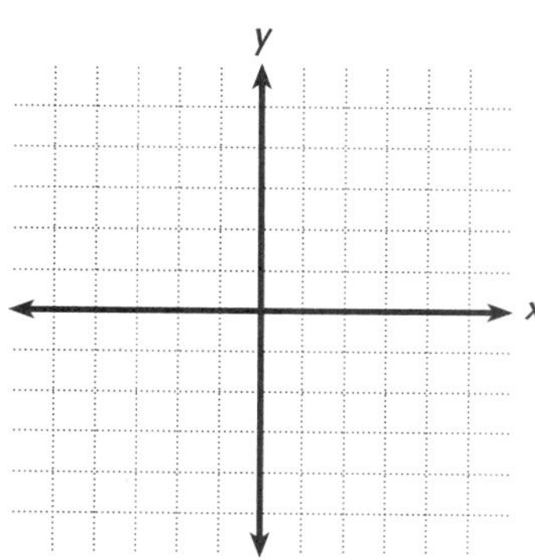

Graph the solution to each of the following systems of linear inequalities under the condition that both *x* and *y* are nonnegative.

3. $\begin{cases} x + y \leq 8 \\ 3x - y \leq 15 \end{cases}$

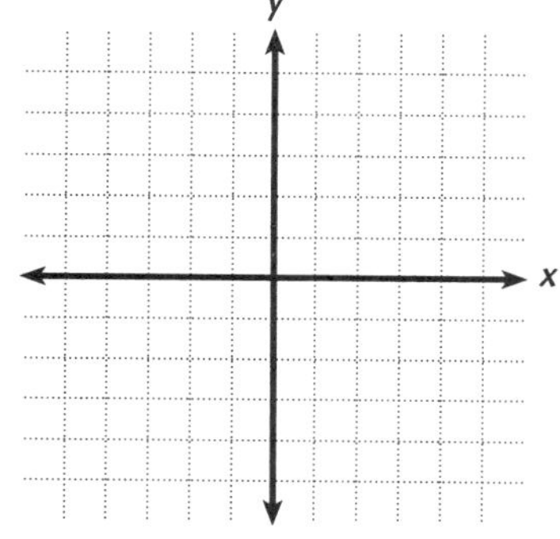

4. $\begin{cases} x - y \leq -4 \\ 2x + 5y \geq 16 \\ x + 2y < 4 \end{cases}$

NAME ______________________ CLASS ____________ DATE ____________

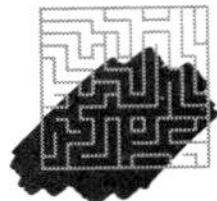

Assessing Prior Knowledge

4.9 Introduction to Linear Programming

Find the slope and y-intercept of the graph of each linear equation.

1. $40x + 15y = 600$ ____________ **2.** $-12x = 3 + 36y$ ____________ **3.** $\frac{7x}{2y} = 30$ ____________

NAME ______________________ CLASS ____________ DATE ____________

Quiz

4.9 Introduction to Linear Programming

Use the grid provided to graph the feasible region for each set of constraints. Determine any three points in the feasible region.

1. $\begin{cases} x + y \leq 4 \\ -x - 2y \geq -8 \\ x \geq 0 \\ y \geq 0 \end{cases}$

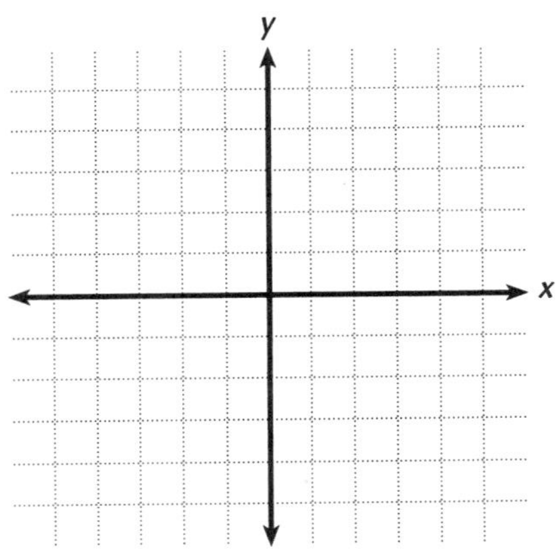

____

2. $\begin{cases} x + y \leq 3 \\ y \leq -2x + 4 \\ x \geq 0 \\ y \geq 0 \end{cases}$

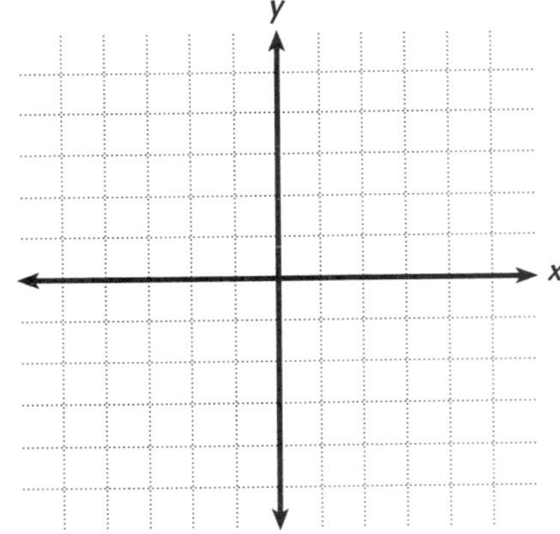

____

3. Find a value of the objective function $P = 4x + 3y$ determined by the feasible region of Exercise 1. If none occurs, write not feasible.

____

4. Find a value of the objective function $T = 3x + y$ determined by the feasible region of Exercise 2. If none occurs, write not feasible.

____

NAME ______________________ CLASS __________ DATE __________

Assessing Prior Knowledge

4.10 Linear Programming in Two Variables

Find the solution to each system of equations.

1. $\begin{cases} x - 2y = -8 \\ 3x + y = 18 \end{cases}$ ________

2. $\begin{cases} 4x + 2y = 2 \\ 5x - y = -22 \end{cases}$ ________

3. $\begin{cases} 5x - 4y = 15 \\ 3x - 2y = 7 \end{cases}$ ________

NAME ______________________ CLASS __________ DATE __________

Quiz

4.10 Linear Programming in Two Variables

1. Maximize $P = 2x + 2y$.

Constraints $\begin{cases} 2x + 3y \leq 15 \\ 2x + y \leq 12 \\ x \geq 0 \\ y \geq 0 \end{cases}$

____

2. Minimize $C = 3x + y$.

Constraints $\begin{cases} x + 2y \leq 8 \\ 2x + y \leq 6 \\ x \geq 0 \\ y \geq 0 \end{cases}$

____

3. Maximize $P = x + 4y$.

Constraints $\begin{cases} x + 3y \leq 6 \\ 3x + 4y \leq 12 \\ x \geq 0 \\ y \geq 0 \end{cases}$

____

4. Maximize $C = 2x + 3y$.

Constraints $\begin{cases} x + 4y \leq 16 \\ -2x + 3y \geq -9 \\ x \geq 0 \\ y \geq 0 \end{cases}$

____

NAME ______________________ CLASS __________ DATE __________

Chapter Assessment

Chapter 4, Form A, page 1

Write the letter that best answers the question or completes the statement.

_____ **1.** The product $\begin{bmatrix} 2 & 3 & 0 \\ 1 & -1 & 2 \end{bmatrix} \begin{bmatrix} -1 & 0 \\ 5 & 2 \\ 8 & 1 \end{bmatrix}$ is:

a. $[21 \quad 0]$

b. $\begin{bmatrix} -2 & 0 \\ 15 & -2 \\ 0 & 2 \end{bmatrix}$

c. $[13 \quad -4]$

d. $\begin{bmatrix} 13 & 6 \\ 10 & 0 \end{bmatrix}$

_____ **2.** If $A = \begin{bmatrix} 2 & 0 & -1 \\ 5 & 8 & -3 \end{bmatrix} \begin{bmatrix} 1 & -4 & 3 \\ 2 & 6 & -1 \\ 0 & 5 & 8 \end{bmatrix}$, the element ab_{23} of the product AB is:

a. -6 **b.** -13 **c.** 2 **d.** -17

_____ **3.** The system of equations $\begin{cases} 2x + 5y = -12 \\ -10y = 24 + 4x \end{cases}$ has

a. no solution.

b. infinitely many solutions.

c. the solution $(-1, -2)$.

d. the solution $(-6, 0)$.

_____ **4.** The inverse of $\begin{bmatrix} 4 & 2 \\ 5 & 3 \end{bmatrix}$

a. is $\begin{bmatrix} 3 & -5 \\ -2 & 4 \end{bmatrix}$.

b. does not exist.

c. is $\begin{bmatrix} 3 & -2 \\ -5 & 4 \end{bmatrix}$.

d. is $\begin{bmatrix} \frac{3}{2} & -1 \\ -\frac{5}{2} & 2 \end{bmatrix}$.

_____ **5.** Which system of equations does the matrix equation

$$\begin{bmatrix} 3 & 0 & 1 \\ 0 & -1 & 2 \\ 4 & -2 & 0 \end{bmatrix} \begin{bmatrix} x \\ y \\ z \end{bmatrix} = \begin{bmatrix} 4 \\ 0 \\ -3 \end{bmatrix}$$ represent?

a. $\begin{cases} 3x + y = 4 \\ -y + 2z = 0 \\ 4x - 2y = -3 \end{cases}$

b. $\begin{cases} 3x + z = 4 \\ -y + 2z = 0 \\ 4y - 2z = -3 \end{cases}$

c. $\begin{cases} 3x + z = 4 \\ -x + 2y = 0 \\ 4x - 2y = -3 \end{cases}$

d. $\begin{cases} 3x + z = 4 \\ -y + 2z = 0 \\ 4x - 2y = -3 \end{cases}$

Chapter Assessment

Chapter 4, Form A, page 2

______ **6.** If a matrix representing the vertices of a polygon is transformed by $\begin{bmatrix} 0 & -1 \\ 1 & 0 \end{bmatrix}$, the polygon is

a. rotated clockwise 90°.
b. enlarged 2 times its size.
c. rotated counterclockwise 90°.
d. rotated counterclockwise 180°.

______ **7.** What is the transformation matrix that enlarges a figure to 3 times its original size?

a. $\begin{bmatrix} 0 & 3 \\ 3 & 0 \end{bmatrix}$
b. $\begin{bmatrix} -1 & 0 \\ 0 & 1 \end{bmatrix}$
c. $\begin{bmatrix} 3 & 0 \\ 0 & 3 \end{bmatrix}$
d. $\begin{bmatrix} \frac{1}{3} & 0 \\ 0 & \frac{1}{3} \end{bmatrix}$

______ **8.** Which of the following ordered pairs maximizes the expression $P = 3x + 2y$?

a. (0, 0) **b.** (−2, 3) **c.** (3, 2) **d.** (2, 3)

9. Which of the following ordered pairs minimizes the expression $C = x + 3y$?

a. (2, 3) **b.** (0, 2) **c.** (1, 0) **d.** (4, 1)

______ **10.** Let $A = \begin{bmatrix} 3 & 2 & 5 \\ 4 & -1 & 0 \\ -2 & 6 & 3 \end{bmatrix}$ and $B = \begin{bmatrix} -2 & 4 & 0 \\ 5 & -1 & 4 \\ 2 & -6 & 3 \end{bmatrix}$. Then $B - A =$

a. $\begin{bmatrix} 1 & 6 & 5 \\ 9 & -2 & 4 \\ 0 & 0 & 6 \end{bmatrix}$
b. $\begin{bmatrix} -5 & 2 & -5 \\ 1 & 0 & 4 \\ 4 & -12 & 0 \end{bmatrix}$
c. does not exist
d. $\begin{bmatrix} 5 & -2 & 5 \\ -1 & 0 & -4 \\ -4 & 12 & 0 \end{bmatrix}$

______ **11.** Which of the following is true about operations on matrices?

I. Matrix addition is commutative.
II. Matrix subtraction is commutative.
III. Matrix multiplication is commutative.

a. I only
b. I and II only
c. I, II, and III
d. none of the above

Chapter Assessment

Chapter 4, Form B, page 1

1. Use a matrix equation to represent the following set of linear equations.

$$\begin{cases} -3x + 4y + z = 5 \\ 2x - 5y - 5z = -7 \\ x - 2y + 5z = 12 \end{cases}$$ ______________________

2. Find the product: $\begin{bmatrix} 4 & 8 \\ 0 & -2 \end{bmatrix} \begin{bmatrix} -1 & 5 & 3 \\ 2 & -4 & 0 \end{bmatrix}$

____

3. Indicate whether the system is inconsistent, dependent, or independent.

$$\begin{cases} 5x - y = 3 \\ 3y = 15x - 6 \end{cases}$$ ______________________

4. Perform the indicated row operation on matrix A.

$$A = \left[\begin{array}{rrr|r} -1 & 3 & -4 & 2 \\ 2 & 5 & -3 & 1 \\ -4 & 3 & 2 & -5 \end{array}\right] \quad -4R_1 + R_3 \longrightarrow R_3$$

____

5. Use row reduction and back substitution to solve.

$$\begin{cases} 4x + y - z = -2 \\ x + 3y - 4z = 1 \\ 2x - y + 3z = 4 \end{cases}$$ ______________________

6. Determine the inverse, if it exists of $\begin{bmatrix} 1 & -2 & 0 \\ 0 & 3 & 1 \\ 2 & -1 & 2 \end{bmatrix}$.

____

7. Write the matrix equation representing the following system.

$$\begin{cases} 2x + y = 3 + z \\ 4x + 4z = y \\ -3y = 6 - 2z \end{cases}$$

8. Homerooms at Stevens High held a week-long contest to collect change for UNICEF. Only pennies, nickels, and dimes could be put into the jar. After the first day, Kate's class counted their change and found that the number of coins was 269, their total value was $10.34, and the number of pennies was twice the number of nickels. Find the number of each coin using a matrix equation.

NAME ______________________________ CLASS ____________ DATE ____________

Chapter Assessment

Chapter 4, Form B, page 2

9. Let $B = \begin{bmatrix} 2 & 4 & 4 & 2 \\ 6 & 6 & -2 & -2 \end{bmatrix}$ be a vertex matrix. Transform the object represented through a 90° counterclockwise rotation and use matrices to give the vertices of the object after transformation.

____

10. Graph the following system of linear inequalities on the grid provided.

$$\begin{cases} x \geq 1 \\ y \leq 8 - x \\ x + 2y \leq 13 \end{cases}$$

11. Use the grid provided to determine the feasible region for the system

$$\begin{cases} x + y \leq 8 \\ 2x + y \leq 10 \\ x \geq 0 \\ y \geq 0 \end{cases}$$

12. Find the value that maximizes the objective function $P = 2x + 7y$ for the feasible region in Exercise. 11 ____________

Alternative Assessment

Applications of Matrix Algebra, Chapter 4, Form A

TASK: To solve a real-world problem using matrices

HOW YOU WILL BE SCORED: As you work through the task, your teacher will be looking for the following:

- how well you can show the steps of solving systems of equations using the row reduction method and back substitution or matrix inversion
- whether you can solve systems of equations using matrices

Brett can borrow $800 from a finance company for a used car at $12\frac{1}{2}\%$ interest each year. If Brett borrowed $1200 at 5% interest each year from a family member, how long would it take for the interest and amount on this money to accumulate to the same sum as the accumulated amount from the finance company. What amount would be owed at this time?

1. Write a system of equations for the problem. Use the formula $A = Prt + P$, where A represents the amount Brett will owe, P represents the principal (the amount borrowed), r represents the annual rate of simple interest expressed in decimal form, and t represents the time expressed in years.

____

____

2. Write the system of equations in matrix form.

3. Solve the system of equations.

____

4. Describe how you would check the solution.

____

5. Explain the meaning of the solution in relation to this business application.

____

____

SELF-ASSESSMENT: Describe how to solve this system of equations using graphing. Then explain what the point of intersection represents.

NAME ______________________ CLASS ____________ DATE ____________

Alternative Assessment

Exploring Systems of Linear Inequalities, Chapter 4, Form B

TASK: To solve a real-world problem using matrices

HOW YOU WILL BE SCORED: As you work through the task, your teacher will be looking for the following:

- how well you can identify solutions to a system of linear inequalities
- whether you can identify the feasible region corresponding to a set of constraints
- how well you find minimum and maximum values for an objective function

The graph represents a system of linear inequalities.

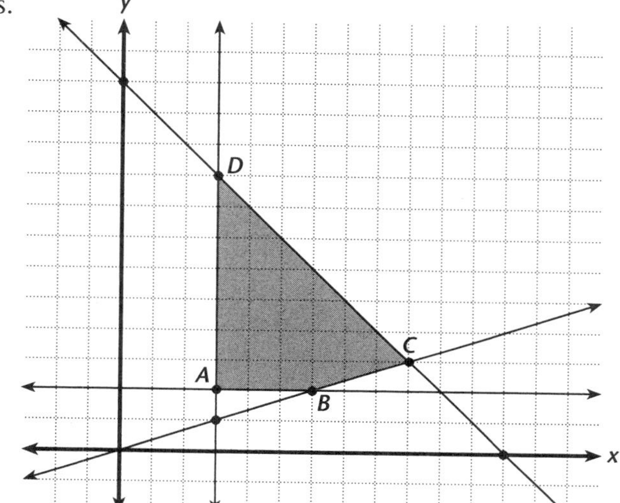

1. Describe the shape of the shaded region.

____

2. Does this system of inequalities have solutions in a bounded or unbounded region? Explain.

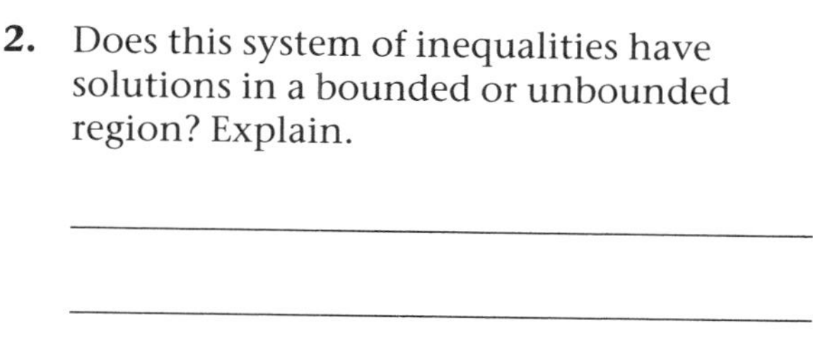

____

____

____

3. Find any 4 points that are solutions to this system of inequalities. What are the ordered pairs for these points? Describe how you know that your 4 points are solution points.

____

____

Suppose this system of inequalities corresponded to a set of constraints with the objective function $P = 6x + 3y$.

4. Describe the feasible region. ______________________

____

5. What are the coordinates of points A, B, C, and D. Explain what the coordinates represent.

____

6. Find the maximum value of the objective function that satisfies the constraints. ____________

SELF-ASSESSMENT: What does it mean when an objective function equation intersects the feasible region?

____

NAME ______________________ CLASS ____________ DATE ____________

Practice & Apply

5.1 Exploring Products of Two Linear Functions

1. Show that $y = (3x + 1)(x - 5) - x^2$ is a quadratic function and identify a, b, and c.

____

Find the x-intercepts and the coordinates of the vertex of the graph of each function.

2. $y = 2(x + 3)(x - 3)$ ______________

3. $y = -2(x + 3)(x - 3)$ ______________

The Junior Class is selling hamburgers at a football game. The profit function $P(x)$ is represented by $P(x) = \frac{-1}{200}(x^2 - 500x + 100)$, where x is the number of hamburgers.

Complete the table.

4.

Number of hamburgers	50	100	150	200	250	300	350	400
Profit ($)								

5. Graph the profit function on the grid provided.

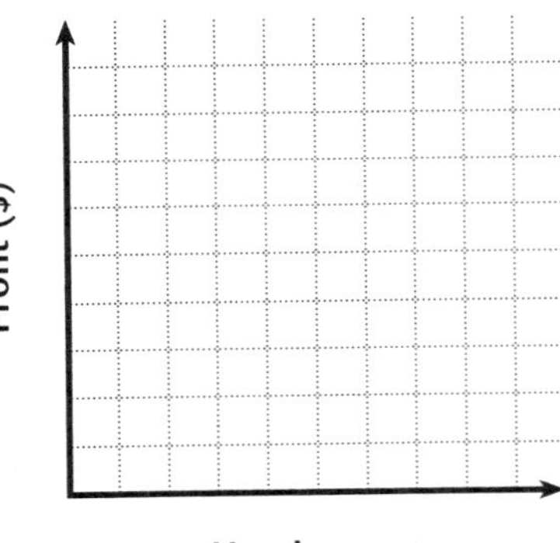

6. What type of function is the profit function?

____

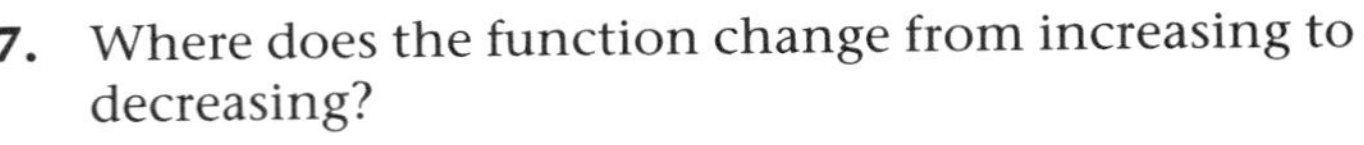

7. Where does the function change from increasing to decreasing?

____

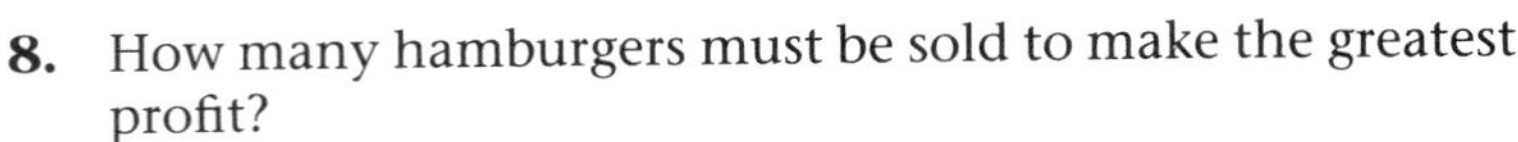

8. How many hamburgers must be sold to make the greatest profit?

____

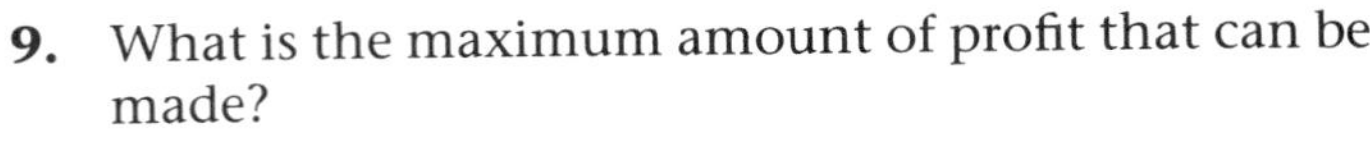

9. What is the maximum amount of profit that can be made?

____

NAME ______________________ CLASS ____________ DATE ____________

Practice & Apply

5.2 Solving Quadratic Equations

1. How many solutions does the equation $\sqrt{x-7} = 10$ have? ____________

2. How many solutions does the equation $x^2 - 7 = 10$ have? ____________

Solve and check.

3. $x^2 = 36$ ____________

4. $(x-2)^2 = 36$ ____________

5. $4(x-2)^2 = 36$ ____________

6. $x^2 = \frac{1}{25}$ ____________

7. $3x^2 = 27$ ____________

8. $\frac{1}{8}x^2 = 8$ ____________

9. $x^2 - 5 = 20$ ____________

10. $\frac{1}{3}x^2 - 1 = 2$ ____________

11. $x^2 - 10 = 39$ ____________

12. $(x-3)^2 = 1$ ____________

13. $\frac{3}{4}(x+4)^2 = \frac{1}{3}$ ____________

14. $\sqrt{x-3} = 5$ ____________

15. $\sqrt{4(x-3)} = 25$ ____________

16. $4\sqrt{x-3} = 16$ ____________

Use the given figure to find each length.

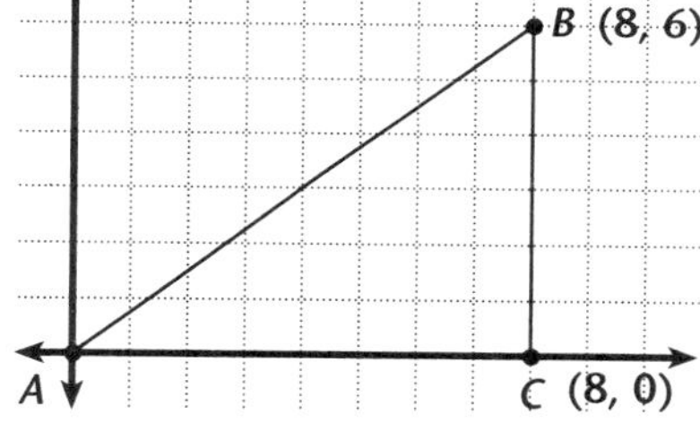

17. AC ____________

18. BC ____________

19. AB ____________

20. Find the length of the hypotenuse for a right triangle with legs of 7 and 9. ____________

Find the distance between the points with the given coordinates.

21. $(-2, 4)$ and $(1, 7)$ ____________

22. $(4, -6)$ and $(12, 3)$ ____________

NAME ______________________ CLASS ____________ DATE ____________

Practice & Apply

5.3 Graphs of Quadratic Functions

Graph each function on the grid provided. Label the vertex and the axis of symmetry.

1. $y = (x - 2)^2$

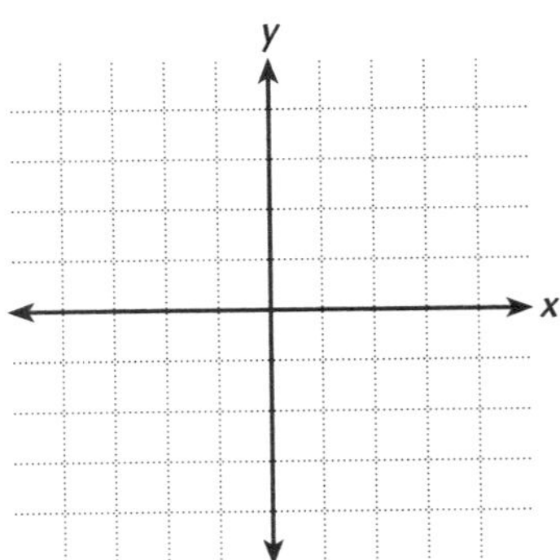

2. $y = -x^2 - 2$

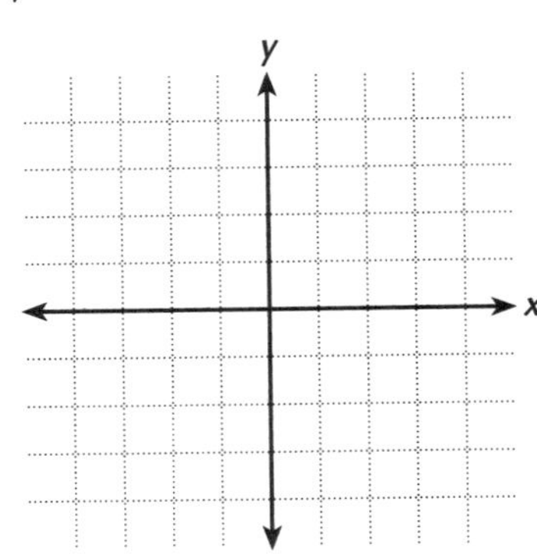

Give the coordinates of the vertex and the equation of the axis of symmetry.

3. $f(x) = \frac{3}{4}(x + 3)^2 - 2$ ____________

4. $f(x) = (x + 4)^2$ ____________

Give the values of x for which the function is increasing and for which the function is decreasing.

5. $f(x) = (x - 1)^2 + 3$ ____________

6. $f(x) = -2(x + 1)^2$ ____________

Find the zeros for each function.

7. $y = 2(x + 1)^2 - 2$

____

8. $y = 2(x - 1)^2$

____

9. If $y = (x + 4)^2 + 6$, what is the smallest possible value of y? ____________

10. Write the quadratic equation to describe the graph shown.

____

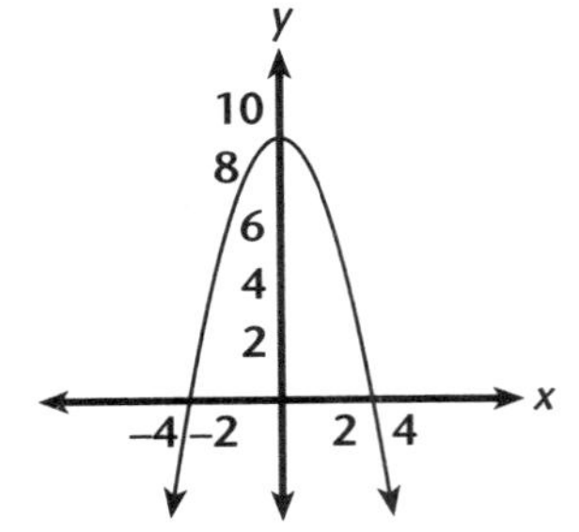

11. A coin is dropped from a balcony 36 feet above the surface of a water fountain. How long does it take for the coin to reach the water?

____

NAME ______________________ CLASS ____________ DATE ____________

Practice & Apply

5.4 Completing the Square

Find the minimum number of additional unit tiles needed to form a square. Then write the perfect-square trinomial modeled by the square.

1.

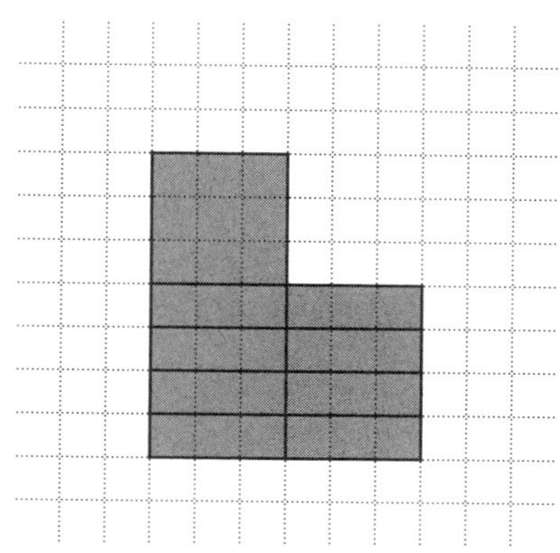

____

2.

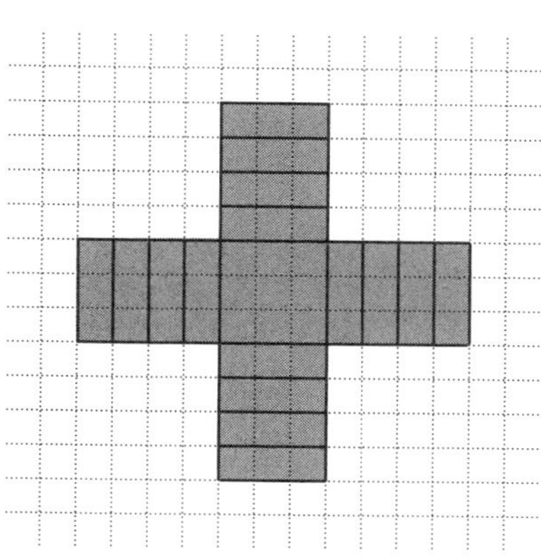

____

Find the term needed to complete each perfect-square trinomial.

3. $x^2 +$ ________ $+ 121$ **4.** $x^2 -$ ________ 225 **5.** $x^2 +$ ________ $+ \frac{1}{4}$

Complete the square for each quadratic expression.

6. $x^2 - 12x$ ________ **7.** $x^2 + 9x$ ________ **8.** $x^2 + \frac{2}{3}x$ ________

Solve by completing the square. Give your answers to the nearest tenth.

9. $x^2 - 24x = -63$ ________ **10.** $x^2 + 3x = 28$ ________

11. $x^2 - 9x = 0$ ________ **12.** $x^2 + x = 0$ ________

13. $x^2 - 11x + 18 = 0$ ________ **14.** $x^2 - x - 8 = -x + 1$ ________

15. $0 = 6x^2 - x - 2$ ________ **16.** $0 = 2x^2 + 13x + 15$ ________

17. $5x = 3x^2 - 12$ ________ **18.** $4x^2 = 6x + 4$ ________

A ball is thrown vertically upward at a rate of 25 m/s. The appropriate height h of the ball (in meters) t seconds later is given by the equation $h = 25t - 5t^2$.

19. In how many seconds will the ball reach 20 meters? ________

20. How high does the ball go? ________

21. At what height is the ball after 4.5 seconds? ________

NAME ______________________ CLASS __________ DATE __________

Practice & Apply

5.5 The Quadratic Formula

Identify *a*, *b*, and *c* for each quadratic equation.

1. $x(x - 3) = 4$

____

2. $-2x^2 + x = -1$

____

3. Write a quadratic equation having the values $a = 2$, $b = -1$, and $c = 1$.

____

4. Use the axis of symmetry formula to find the vertex of $f(x) = 3x^2 - 4x + 1$.

____

Use the quadratic formula to solve the following equations. Give your answers to the nearest tenth.

5. $x^2 + 6x + 6 = 0$

____

6. $3x^2 - 9 = 0$

____

7. $5x^2 = 4x$

____

8. $7x = 6 - 3x^2$

____

9. $12x^2 - 2 = -5x$

____

10. $25x^2 = 90$

____

11. How many x-intercepts could the graph of a quadratic function have? ______________________

Find the *x*-intercepts for each parabola.

12. $y = 3x^2 + x - 2$

____

13. $y = 4x^2 + 4x + 1$

____

Find the *x*-coordinate of the vertex of each parabola, given that the *x*-intercepts are

14. -3 and 5. ______________________

15. 4 and 8. ______________________

NAME ______________________ CLASS __________ DATE __________

Practice & Apply

5.6 The Discriminant and Imaginary Numbers

Write whether the discriminant will be greater than, less than, or equal to zero.

1.

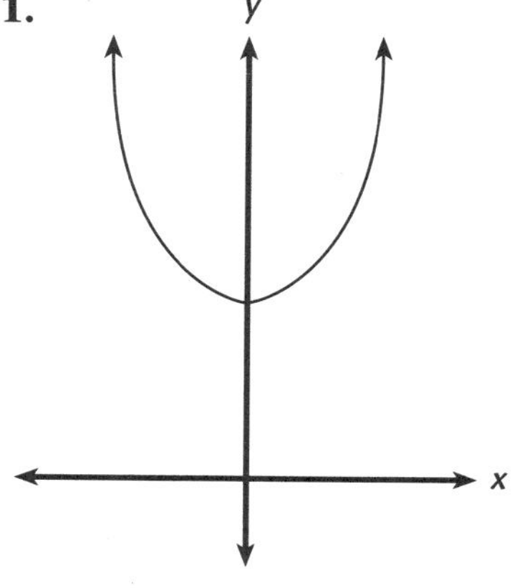

____

2.

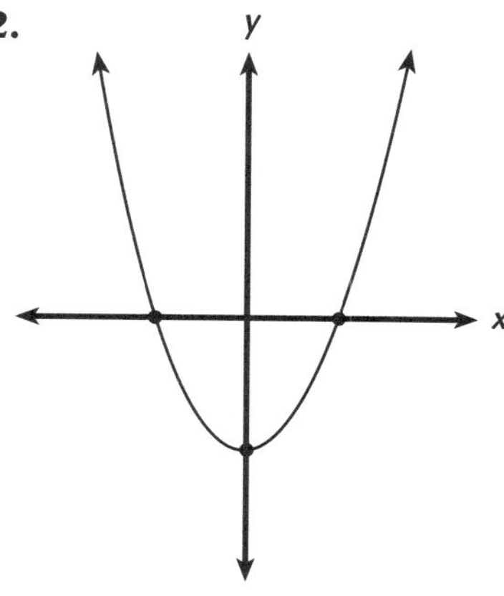

____

3.

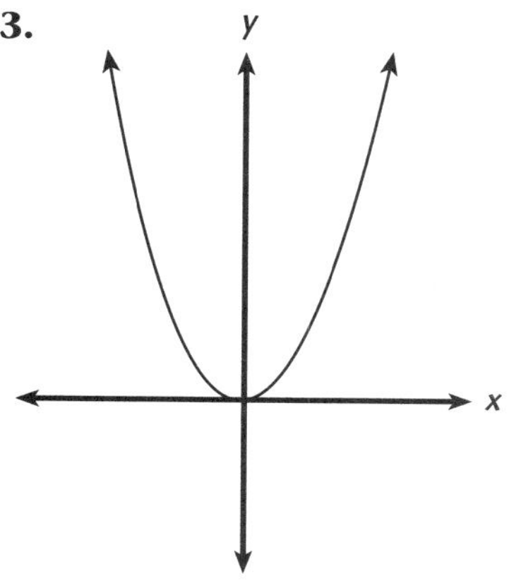

____

Use the discriminant to determine how many real-number solutions each equation has.

4. $2x^2 - 20x + 50 = 0$ ______________ **5.** $x^2 + 2.5x = 0$ ______________

6. $4x^2 - 36 = 0$ ______________ **7.** $5x^2 + 125 = 0$ ______________

Simplify.

8. $\sqrt{-169}$ ________ **9.** $\sqrt{-98}$ ________ **10.** $\sqrt{-27}$ ________

11. $\sqrt{-3} \cdot \sqrt{-27}$ ________ **12.** $-5i \cdot -5i$ ________ **13.** $i(\sqrt{-6})$ ________

14. $(2i)^5$ ________ **15.** $-\sqrt{-100}$ ________ **16.** $i\sqrt{-2} \cdot \sqrt{-8}$ ________

17. Which equation has imaginary solutions? ______________

a. $-x^2 = -16$ **b.** $x^2 - 16 = 0$ **c.** $(2 - x)^2 = 16$ **d.** $x^2 + 16 = 0$

18. Write a quadratic equation that has $7i$ as one of its roots.

____

Solve.

19. $x^2 = -54$ ______________ **20.** $(x - 1)^2 = -54$ ______________

21. $4x^2 = -64$ ______________ **22.** $(4x)^2 = -64$ ______________

NAME ______________________ CLASS ____________ DATE ____________

Practice & Apply

5.7 Complex Numbers

Identify the real part and the imaginary part for each complex number.

1. $4 - 2i$ ____________ **2.** $3 + 2i$ ____________

3. $-2i$ ____________ **4.** 5 ____________

Simplify.

5. $(3 + 4i) + (2 - i)$ ____________ **6.** $(5 - 3i) - (3 + 4i)$ ____________

7. $-4(2 + 3i)$ ____________ **8.** $3i(1 - 7i)$ ____________

9. $(6 + 2i)(2 + i)$ ____________ **10.** $(9 + 3i)(2 - 3i)$ ____________

11. $-5i(-3 - 6i)$ ____________ **12.** $(1 - i)(2 - 8i)$ ____________

13. Write a quadratic equation that has $5 - i\sqrt{3}$ and $5 + i\sqrt{3}$ as solutions. ____________

Write the complex conjugate for each complex number.

14. $2 + 3i$ ____________ **15.** $5i$ ____________ **16.** $-7 - 2i$ ____________

Write the solutions of each equation in *a* + *bi* form.

17. $x^2 + 6x + 15 = 0$ ____________ **18.** $10x^2 + 4x = -1$ ____________

19. $x^2 - x + 1 = 0$ ____________ **20.** $60 = -x^2 - 14x$ ____________

21. $2x^2 = 5x - 6$ ____________ **22.** $x^2 + x = -1$ ____________

Write the letter of the graph that best matches each description.

23. the complex number $2 + 3i$ ____________ **24.** the conjugate of $-2 + 3i$ ____________

25. the complex number with real part 2 and imaginary part -3 ____________

26. the reflection of $2 + 3i$ over the imaginary axis ____________

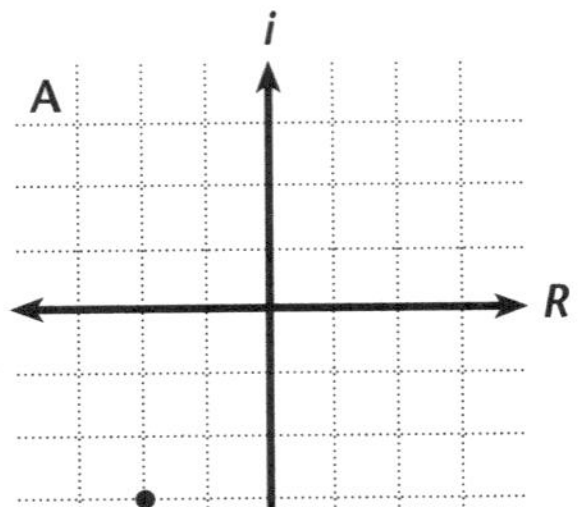

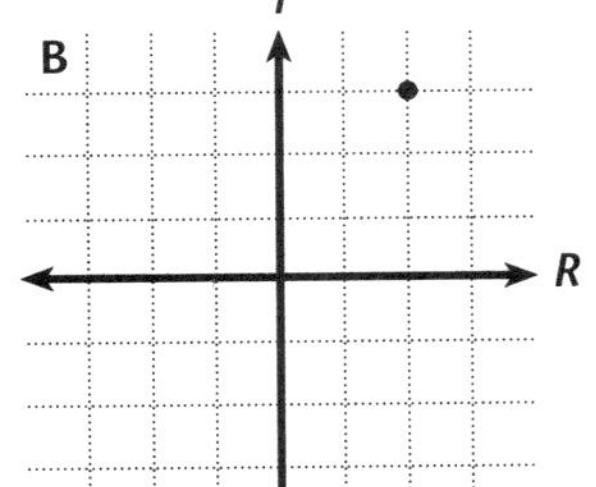

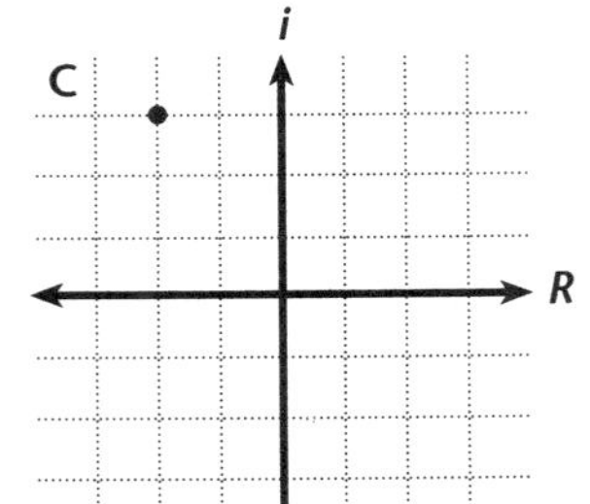

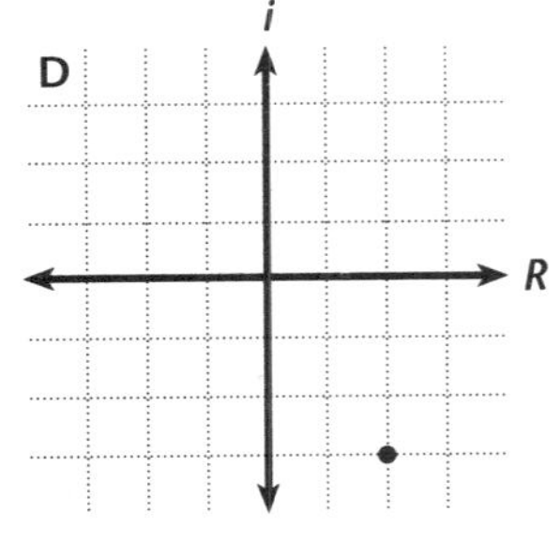

NAME ______________________ CLASS ____________ DATE ____________

Practice & Apply

5.8 Curve Fitting With Quadratic Functions

Solve each system of equations using matrix algebra.

1. $a + b + c = 5$
$a + 2b + 2c = 6$
$a + 2b + 3c = 10$

____

2. $a + 2b = -1$
$2a - b + c = 5$
$a + c = 0$

____

3. Write a quadratic function to describe the graph shown.

4. Show that the point $(-2, 19)$ is on the graph of $y = x^2 - 6x + 3$.

____

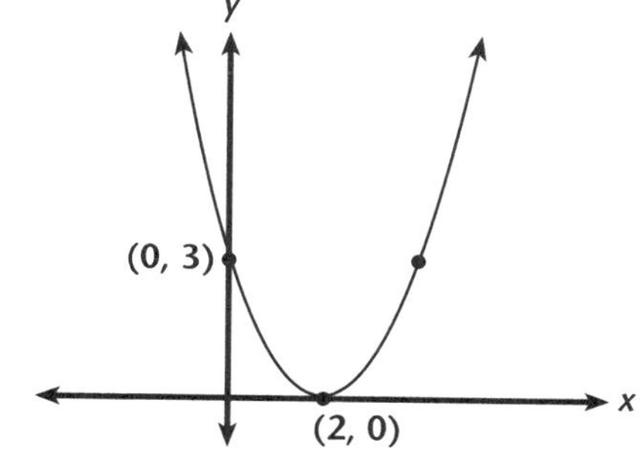

Find a quadratic function that fits each set of data points.

5. (1, 2), (3, 34), (−2, 14)

____

6. (1, 9), (2, 9), (3, 5)

____

7. (0, 25), (3, 1), (6, 49)

____

8. (0, 1), (3, 10), (5, 26)

____

Jill participated in a study to determine the total time required to bring her automobile to a stop after she becomes aware of danger. Jill's reaction time includes the time between recognizing danger and applying the brakes plus the time for stopping after applying the brakes. The data table gives the stopping distance d (in feet) of Jill's automobile traveling at speed s (in miles per hour) from the time that she notices danger.

s (mi/h)	14	50	64
d (ft)	40	202	314

9. Find a quadratic function to model the data. ______________________

10. Use your model to predict the stopping distance when the speed is 30 mi/h. ______________________

11. Use your model to find the speed when the stopping distance is 138.8 ft. ______________________

NAME ______________________ CLASS ____________ DATE ____________

Practice & Apply

5.9 Exploring Quadratic Inequalities

1. Write the quadratic inequality represented in the graph.

____

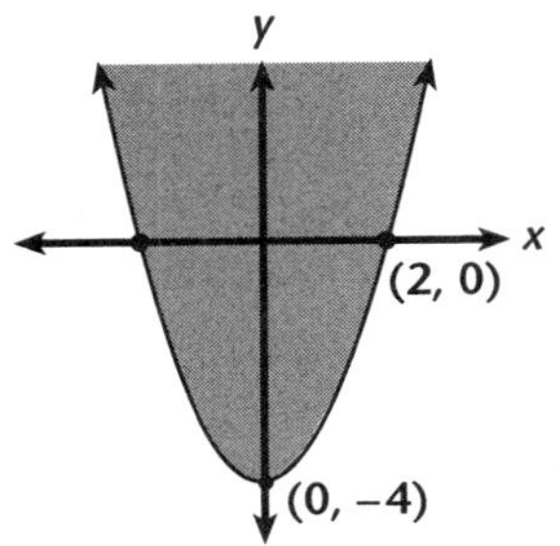

Solve the following inequalities by graphing.

2. $x^2 - 4x - 5 \leq 0$ ______________

3. $(x - 3)(x + 2) > 0$ ______________

4. $(x - 1)^2 < 0$ ______________

5. $x^2 - 3x + 2 \geq 0$ ______________

Fran is applying fertilizer to her vegetable garden. She knows that the yield of a crop may increase with fertilizer up to a point, after which there is a decrease. The amount of fertilizer to apply per square feet is given by $f(p) = -2p^2 + 18p + 64$, where p is the number of pounds of fertilizer, and $f(p)$ is the yield in terms of pounds per square foot.

6. Complete the table.

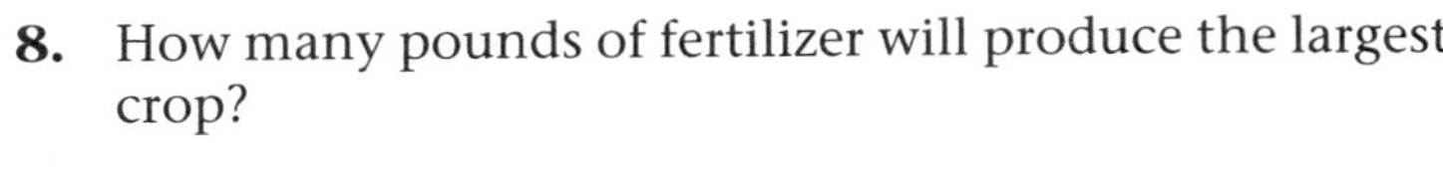

Fertilizer (lb)	1	2	3	4	5	6	7	8	9
Crop Yield (lb/ft^2)									

7. Graph $f(p)$ on the grid provided.

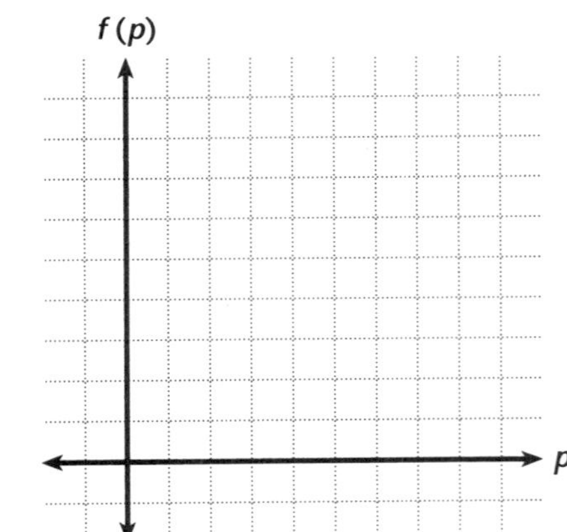

8. How many pounds of fertilizer will produce the largest crop?

____

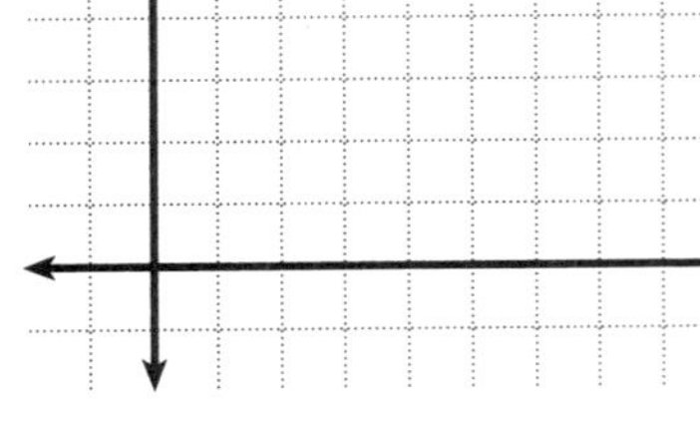

9. What is the possible crop yield?

____

10. At what point will fertilizer use decrease crop yield?

____

NAME ______________________ CLASS __________ DATE __________

Enrichment

5.1 The Max!

Write the letter of the function that represents each situation. Then solve for the values indicated.

1. An orchard grower currently has 200 apple trees, with an average yield of 450 apples per tree. She estimates that for each additional set of 10 trees the average yield per tree would drop by 15 apples. How many trees will produce the maximum yield of her apple crop? What is the maximum number of apples that would be produced?
a. $f(x) = (x + 200)(10)(450 - 15x)$ **b.** $f(x) = (200 + 10x)(450 - 15x)$
c. $f(x) = (200 + 10x)(450 + 15x)$

____

2. The committee for the spring concert is trying to decide on a selling price for the tickets. They know from past experience that they can expect 800 people if the tickets sell for \$7 each. They anticipate losing 10 people for each \$0.25 increase in price. By what amount should the committee increase the price in order to maximize the income from the spring concert? What would be the maximum income at this price?
a. $f(x) = (800 + 10x)(7 + 0.25x)$ **b.** $f(x) = (800 - 10x)(7 - 0.25x)$
c. $f(x) = (800 - 10x)(7 + 0.25x)$

____

3. A library charges an annual fee of \$30 for those who live outside the district and want to check out books. Currently, 200 patrons live outside the district. The library anticipates losing 4 patrons for each \$1 increase in the annual fee. What annual fee will yield the greatest income from these patrons? What is the maximum annual income the increase would yield?
a. $f(x) = (200 - 4x)(30 + x)$ **b.** $f(x) = (200 - 4x)(30 - x)$ **c.** $f(x) = (200 - x)(30 + 4x)$

____

4. A new CD is selling for \$12.95 and the store estimates that they will sell 500 of these CDs. Past experience shows that for each \$1.00 increase in the price of a CD, the store will lose 10 sales. What price should the store charge in order to maximize the income from the CDs? What would be the maximum income at this price?
a. $f(x) = (500 + 10x)(12.95 + x)$ **b.** $f(x) = (500 - x)(12.95 + 10x)$
c. $f(x) + (500 - 10x)(12.95 + x)$

____

NAME ______________________________ CLASS ______________ DATE ______________

Enrichment

5.2 Squares or Roots?

To find the mystery phrase, first solve each equation. Then write the letter variable above its corresponding positive solution. The first one has been done for you.

$\pm 11 = V$ **1.** $V^2 = 121$

_____ **2.** $(Q - 2)^2 = 9$

_____ **3.** $\sqrt{R^2} = 16$

_____ **4.** $\frac{1}{7}A^2 = 7$

_____ **5.** $\sqrt{2H + 5} = 3$

_____ **6.** $3(T^2 + 1) = 1086$

_____ **7.** $\sqrt{N + 6} = 6$

_____ **8.** $-3 + \sqrt{2L + 10} = 5$

_____ **9.** $2(H^2 - 3) + 4 = 390$

_____ **10.** $\frac{1}{3}(H^2 - 4) = 160$

_____ **11.** $15 = 3\sqrt{I}$

_____ **12.** $\frac{1}{4}S^2 - 1 = 35$

_____ **13.** $O^2 - 5 = 284$

_____ **14.** $\sqrt{2W + 1} - 2 = 5$

_____ **15.** $4\sqrt{3E + 7} = 16$

_____ **16.** $5(T^2 + 1) - 20 = 830$

_____ **17.** $O^2 + 4 = 328$

_____ **18.** $\sqrt{3W - 3} + 4 = 13$

_____ **19.** $5W^2 - 2000 = 205$

_____ **20.** $3(U^2 - 6) + 1 = 91$

_____ **21.** $\sqrt{3S + 6} - 4 = 2$

_____ **22.** $(S - 1)^2 = 9$

_____ **23.** $\sqrt{6I + 226} = 20$

_____ **24.** $\sqrt{8R + 17} - 7 = 2$

_____ **25.** $E^2 + 5 = 230$

_____ **26.** $\sqrt{4S + 1} + 7 = 16$

_____ **27.** $\frac{1}{2}L^2 - 128 = 210$

_____ **28.** $\sqrt{O + 98} + 16 = 27$

_____ **29.** $3E^2 + 14 = 257$

_____ **30.** $\sqrt{7T + 9} + 7 = 11$

___	___	___	___	___	___	___	___	___	___
1	2	3	4	5	6	7	8	9	10

V	___	___	___	___	___	___	___	___	___
11	12	13	14	15	16	17	18	19	20

___	___	___	___	___	___	___	___	___	___
21	22	23	24	25	26	27	28	29	30

NAME ______________________ CLASS ____________ DATE ____________

Enrichment

5.3 The Vertex (x, y)

Use a graphics calculator to graph each parabola. Then shade in the box if the vertex of the parabola lies on the line $x + y = 8$.

$y = 2x^2 - 16x + 30$	$y = x^2 - 6x + 10$	$y = 2x^2 - 3x + 4$	$y = -x^2 + 6x - 9$
$y = 3x^2 + 2x - 8$	$y = \frac{1}{2}x^2 - 3x + \frac{19}{2}$	$y = 4x^2 - 8x + 15$	$y = -2x^2 - 16x - 20$
$y = \frac{1}{4}x^2 + \frac{11}{2}x - \frac{7}{4}$	$y = x^2 + 6$	$y = x^2 - 5x + 11.75$	$y = (x - 4)^2$
$y = 7x^2 - 8x + 17$	$y = 3x^2 - 54x + 242$	$y = \frac{1}{2}x^2 - 3x + \frac{1}{4}$	$y = \frac{1}{4}x^2 - 6x + 32$
$y = x^2 + 18$	$y = -x^2 - 8x + 5$	$y = -\frac{1}{3}x^2 + 4x^2 - \frac{5}{3}$	$y = -x^2 - \frac{1}{3}x + 5$
$y = \frac{1}{2}x^2 - 8x + 1$	$y = -x^2 + 10x + 38$	$y = -\frac{2}{3}x^2 + 6x - 7$	$y = -4x^2 + 40x - 97$
$y = 3x^2 - 14x + 11$	$y = -x^2 - 14x - 34$	$y = 3x^2 - 5x + 12$	$y = 4x^2 + 40x + 113$
$y = x^2 + 5x - 18$	$y = x^2 - 8x + 13$	$y = \frac{1}{3}x^2 - \frac{8}{3}x + \frac{28}{3}$	$y = 4x^2 - 12x + 9$
$y = x^2 - 8x + 19$	$y = -x^2 - 11x + 20$	$y = \frac{1}{4}x^2 - 4x + 16$	$y = x^2 + 18x - 20$
$y = x^2 + 16x + 16$	$y = -\frac{1}{2}x^2 + 6x - 6$	$y = x^2 + 8$	$y = x^2 + 4x + 4$

NAME ______________________ CLASS ______________ DATE ______________

Enrichment

5.4 A Complete-the-Square Move

To get through the maze, set each expression equal to zero and solve it by completing the square. Then move either horizontally or vertically only to an equation whose roots are only positive.

START ↓

$x^2 + 5x + 6$	$x^2 - 2$	$x^2 + 2x - 24$	$x^2 + x - 3$	$6x^2 + 5x + 1$
$12x^2 + 9x + 1$	$12x^2 - 17x + 6$	$24x^2 - 38x + 15$	$x^2 + 8x + 14$	$x^2 + 2x - 4$
$x^2 - 6x - 17$	$x^2 + 12x + 27$	$x^2 - 15x + 56$	$40x^2 - 47x + 12$	$21x^2 - 23x + 6$
$x^2 + 14x + 46$	$8x^2 + 22x + 15$	$x^2 + 6x + 7$	$9x^2 - 42x - 31$	$4x^2 - 20x + 21$
$48x^2 - 58x + 15$	$x^2 - 10x + 9$	$x^2 - 19x + 60$	$24x^2 - 10x + 1$	$4x^2 - 12x + 7$
$9x^2 - 48x + 62$	$x^2 + 4x - 12$	$x^2 + 16x + 63$	$12x^2 + 5x - 2$	$9x^2 - 12x - 2$
$x^2 - 21x + 108$	$2x^2 - 11x + 12$	$3x^2 - x - 2$	$5x^2 - 9x - 2$	$2x^2 + 9x + 4$
$16x^2 + 24x + 4$	$25x^2 - 30x + 7$	$x^2 - 21x + 68$	$x^2 + 22x + 57$	$x^2 - 11x - 80$
$x^2 - 4x - 3$	$x^2 - 6x - 3$	$x^2 - 24x + 135$	$2x^2 - 30x + 110$	$30x^2 - 47x + 18$

↓ FINISH

NAME ________________________ CLASS __________ DATE __________

Enrichment

5.5 Finding Maximums and Minimums

Find the value of x for which each function reaches a maximum or minimum, then find the maximum or the minimum value of the function.

1. A business manager estimates that the cost of manufacturing x items each day is represented by $C(x) = 0.5x^2 - 8x + 1500$. How many items should the business manufacture each day in order to minimize its cost?

____

2. When a ball is thrown, the height (in feet) of the ball can be modeled by $h(x) = -16x^2 + 80x + 6$, where x represents the number of seconds that the ball is in the air. What is the maximum height that the ball reaches?

____

3. A company's daily profits are modeled by the equation $P(x) = -x^2 + 50x + 200$, where x is the number of items sold. What is the maximum profit the company can achieve per day, according to this model?

____

4. A football is kicked into the air. Its height (in feet) after x seconds is modeled by the equation $h(x) = -16x^2 + 64x$. What is the maximum height reached by the football?

____

5. The amount, in milligrams, of a certain drug in the bloodstream after x hours is modeled by $d(x) = -0.1x^2 + 2x + 300$. What is the maximum level that the drug reaches?

____

6. The cost of operating a shop is modeled by the function $s(x) = 5x^2 - 150x + 2000$, where x represents the number of items sold each day. What is the minimum daily cost of operating the shop?

____

7. The Art Depot approximates its monthly income from sales of paintings by using the equation $I(x) = -x^2 + 500x + 200$, where x represents the number of paintings sold. What is the maximum income for the Art Depot?

____

NAME ______________________ CLASS ____________ DATE ____________

Enrichment

5.6 How Many?

For each quadratic equation, determine the number of real solutions. If two, choose the letter in Column A; if one, choose the letter in Column B; if none, choose the letter in Column C. Write the letter you chose on the blank provided.

		Column A Two	Column B One	Column C None
______	**1.** $x^2 - 3x + 2 = 0$	I	T	R
______	**2.** $x^2 + 4x - 6 = 0$	M	E	O
______	**3.** $x^2 - 6x + 9 = 0$	M	A	P
______	**4.** $9x^2 - 12x + 4 = 0$	S	G	K
______	**5.** $x^2 - 4x + 13 = 0$	T	W	I
______	**6.** $8x^2 + 2x - 3 = 0$	N	R	O
______	**7.** $16x^2 - 72x + 81 = 0$	V	A	W
______	**8.** $x^2 - 4x + 1 = 0$	R	P	N
______	**9.** $x^2 - 2x + 5 = 0$	B	D	Y
______	**10.** $64x^2 - 112x + 49 = 0$	E	N	L
______	**11.** $25x^2 - 40x + 16 = 0$	T	U	R
______	**12.** $6x^2 + 11x + 4 = 0$	M	B	N
______	**13.** $x^2 - 2x - 2 = 0$	B	V	K
______	**14.** $9x^2 + 30x + 25 = 0$	L	E	P
______	**15.** $x^2 - 8x + 41 = 0$	S	T	R
______	**16.** $16x^2 - 16x + 13 = 0$	E	N	S
______	**17.** $25x^2 + 30x + 9 = 0$	S	E	C
______	**18.** $12x^2 - x - 35 = 0$	X	V	R
______	**19.** $9x^2 - 42x + 25 = 0$	I	M	T
______	**20.** $9x^2 - 48x + 64 = 0$	D	S	G
______	**21.** $16x^2 - 24x + 58 = 0$	N	O	T

NAME ______________________ CLASS ____________ DATE ____________

Enrichment

5.7 The *i*'s Have It

Start where indicated. Move down one space if the answer shown is correct; move right one space if the answer shown is incorrect. Continue in this matter until you reach the finish.

START ↓ (first column)

$(4 + 3i) + (6 - 2i) = 10 + i$	$(2 - 3i) - (4 + i) = 2 - 4i$	$2i(3 + 5i) = 6i - 10$	$7(2i + 1) = 14 - 7i$	$(2 + i)(3 - i) = 7 + i$
$4i(3 - 5i) = 20 + 12i$	$2(3 - 4i) + 6i = 6 - 2i$	$3i(2 - i) = 6i - 3$	$(5 + 3i) - 2(1 - i) = 3 + 5i$	$(4 + 2i) - i(3) = 4 + i$
$(5 - 3i) - (2 + 4i) = -3 + i$	$4(-3 + i) = -12 + 4i$	$2i(i + 6) = -2 + 12i$	$(2 + i)(2 - i) = 5$	$(5 + 6i)(3 - i) = 21 + 13i$
$(5i + 3)i = 5 - 3i$	$(8 - 5i) - (3 - 4i) = 5 - i$	$6i(3 - i) = -6 + 18i$	$-5i(3 + 7i) = -35 - 15i$	$(6 + 4i) - (2 + 7i) = 4 - 3i$
$(8 - 3i) - (9 + i) = 1 - 4i$	$3(2 + 4i) - (1 + i) = 5 + 11i$	$5(7 + 3i) = 35 + 15i$	$(8i + 3)(2i - 4) = -28 - 26i$	$3i(51) = 153i$
$-7i(8 - 15i) = 105 - 56i$	$(3 - 5i) - (-4 + 6i) = 7 - 11i$	$5i(3i - 6) = -15 - 30i$	$(9 - 11i)(2 + i) = 29 - 13i$	$2i(3i) = 6$
$15(3 - 4i) = 60 - 45i$	$(8 + 3i)(8 - 3i) = 55$	$(4 - 3i) - (7 + 6i) = 3 - 9i$	$(2 + 3i) + 3(6 - i) = 20$	$78i(4 - 9i) = 63 + 28i$
$-4i(9 - 3i) = 12 - 36i$	$14(3 - 6i) = 42 - 84i$	$8(9 - 2i) = 72 - 16i$	$(9 + 3i)(4 - 2i) = 42 - 6i$	$-4i(18 + i) = -4 - 72i$
$3(2 + 4i) + 2(3 - i) = 12 + 10i$	$3(4 - 2i) - (2 + i) = 12 - 7i$	$(3 - 2i)(3 + 2i) = 13$	$(4 - 3i)(5 + i) = 23 - 11i$	$(4 + 7i) + (-1 + 2i) = 3 + 9i$
$(5 + 3i) - (8 - 4i) = -3 - i$	$3(6 - i) + 4i = 6 + i$	$(5 - 3i)(-4i) = 12 - 20i$	$(8 + 3i) - (-4 - i) = 12 + 4i$	$(7 - 5i) - (-2 - 3i) = 9 - 8i$
$(12 - 3i)(-5i) = 15 - 60i$	$(5 - 9i) + (4 - 6i) = 9 - 15i$	$-7i(4 - 5i) = 35 - 28i$	$(6i - 3) - 4(-2 + i) = 5 + 2i$	$2(3 + 4i) + (-5 + i) = 1 + 9i$
$(2 + 4i)(3 - 5i) = 26 - 2i$	$3(9i - 1) + 12i = -3 + 24i$	$-12i(4 - 5i) = 60 - 48i$	$(9 - 5i)(-8 + 2i) = 62 + 58i$	$17i(3 - i) = 17 + 51i$

FINISH ↑ (last column)

NAME ______________________ CLASS ____________ DATE ____________

Enrichment

5.8 Spotting Quadratic Functions

Not all sets of points will fit a quadratic model. When the *x*-coordinates are consecutive integers, the simple test shown in the example can be used to determine if the ordered pairs will lie on a parabola.

Example: Do the points (2, 5), (3, 10), (4, 17), (5, 26), and (6, 37) lie on the same parabola?

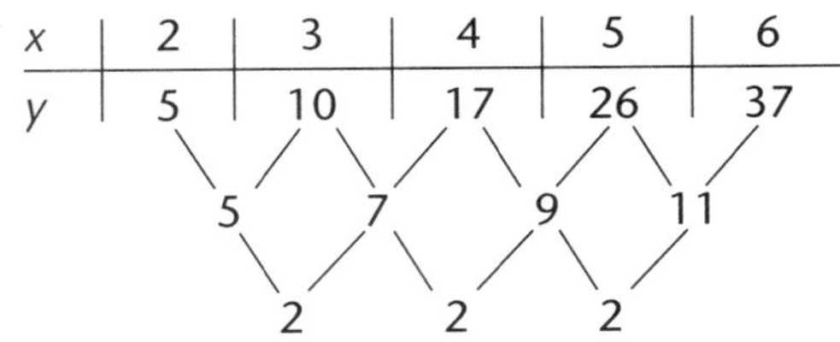

First differences: Find the differences of the *y*-values.
Second differences: Find the differences of the first differences.

If the second differences are equal, the points all lie on the same parabola.

Determine whether each set of points lie on the same parabola. If they do, find the equation of that parabola.

1. (1, 0), (2, 3), (3, 10), (4, 21), (5, 36) ______________________

2. (−2, 0), (−1, 3), (0, $\frac{19}{9}$), (1, 24), (3, 56) ______________________

3. (−2, 0), (−1, −3), (0, −2), (1, 0), (2, 6) ______________________

4. (−3, −14), (−2, −2), (−1, 8), (0, 16), (1, 22) ______________________

5. (−1, 12), (0, 8), (1, 10), (2, 18), (3, 32) ______________________

6. (−4, −78), (−3, −55), (−2, −36), (−1, −21), (0, −10) ______________________

7. (−3, 32), (−2, −13), (−1, 8), (0, −1), (1, −9) ______________________

8. (2, 9), (3, 20), (4, 37), (5, 60), (6, 89) ______________________

9. (−1, $-15\frac{1}{2}$), (0, 9), (1, $-\frac{1}{2}$), (2, 7), (3, $16\frac{1}{2}$) ______________________

10. (4, −48), (5, −59), (6, −68), (7, −75), (8, −80) ______________________

NAME ______________________ CLASS __________ DATE __________

Enrichment

5.9 Writing Quadratic Inequalities

Suppose the roots of a quadratic equation are a and b. Then the quadratic equation can be written as $(x - a)(x - b) = 0$, or $x^2 - (a + b)x + ab = 0$, and the associated function is $y = x^2 - (a + b)x + ab$. Use these facts plus the given graph to write the quadratic inequality that fits each situation.

1.

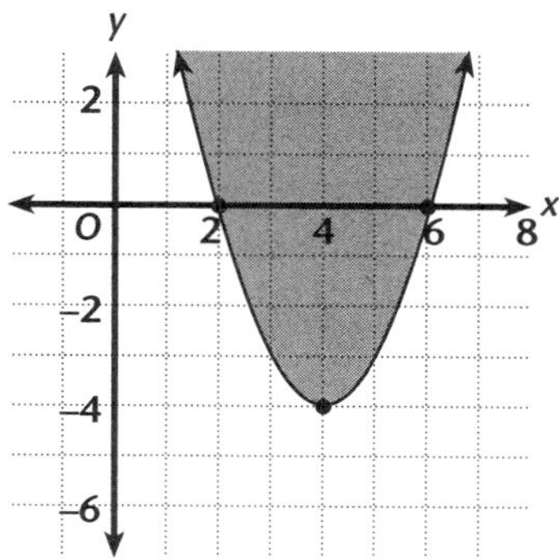

____

2.

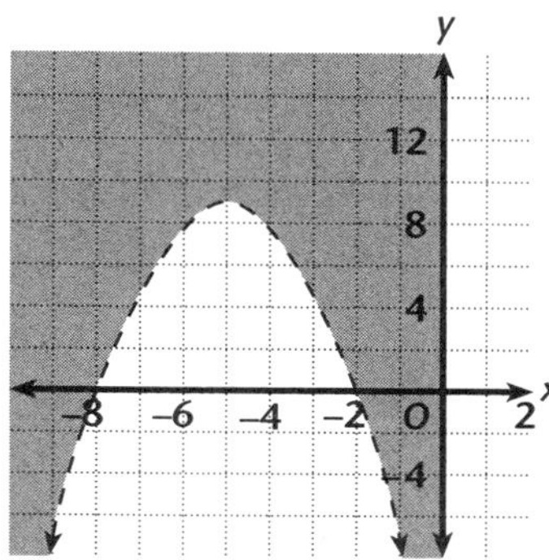

____

3.

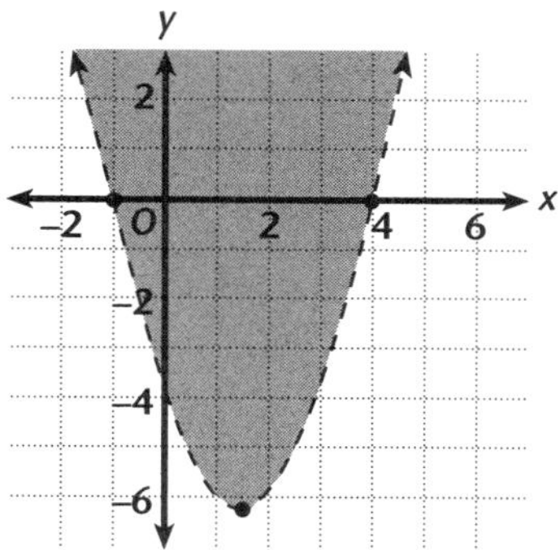

____

4.

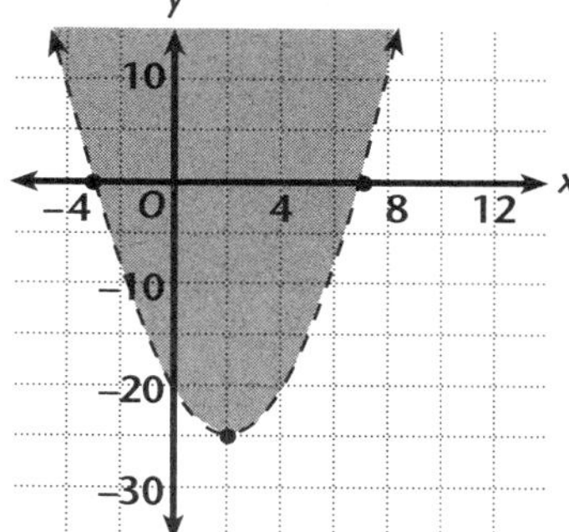

____

5.

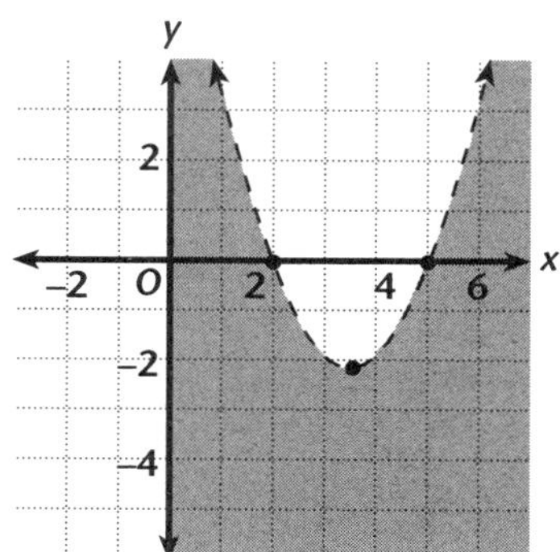

____

6.

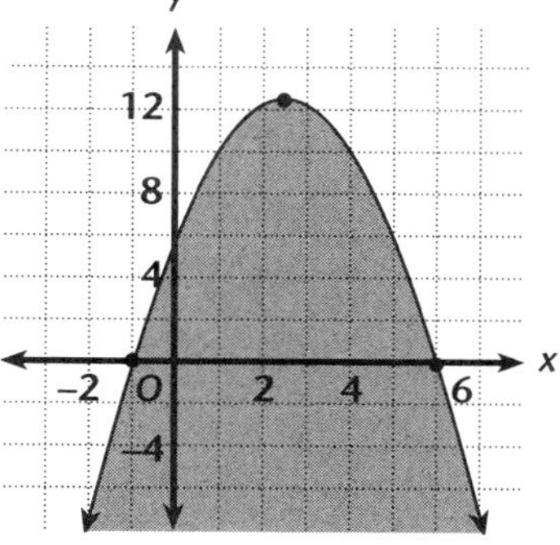

____

NAME ______________________ CLASS ____________ DATE ____________

Technology

5.1 The Families $h(x) = -x^2 + bx$ and $j(x) = x^2 + bx$

With a graphics calculator, you can study the behavior of the products of functions such as $f(x) = -x$ and $g(x) = x - b$ for various real numbers b.

Let $h(x) = f(x)g(x) = -x(x - b) = -x^2 + bx$. Using a graphics calculator and $b = 1, 2, 3,$ and 4, graph the product functions in the same coordinate plane. With all the graphs displayed, you can begin to see a pattern.

Use a graphics calculator for each exercise.

1. Graph $h(x) = -x^2 + bx$ for $b = 1, 2, 3,$ and 4.

2. Graph $h(x) = -x^2 + bx$ for $b = -1, -2, -3,$ and -4.

3. Write a brief conclusion that you can draw from Exercise 1.

____

4. Write a brief conclusion that you can draw from Exercise 2.

____

5. Graph $j(x) = x^2 + bx$ for $b = 1, 2, 3,$ and 4.

6. Graph $j(x) = x^2 + bx$ for $b = -1, -2, -3,$ and -4.

7. Write a brief conclusion that you can draw from Exercise 5.

____

8. Write a brief conclusion that you can draw from Exercise 6.

____

9. How is the distance between the x-intercepts of h related to the x-value of the highest point on the graph of h?

____

10. How is the distance between the x-intercepts of j related to the x-value of the lowest point on the graph of j?

____

NAME ______________________ CLASS ____________ DATE ____________

Technology

5.2 Lengths of Paths

The diagram shows a path from point A to point Z with points B and C along the way. Suppose you wish to find the length of path $ABCZ$. You know, first of all, that you can read the coordinates of each point. You also know that there is a formula that gives you the length of a line segment. Lastly, you also know that a spreadsheet will help you find each distance and the sum of those distances.

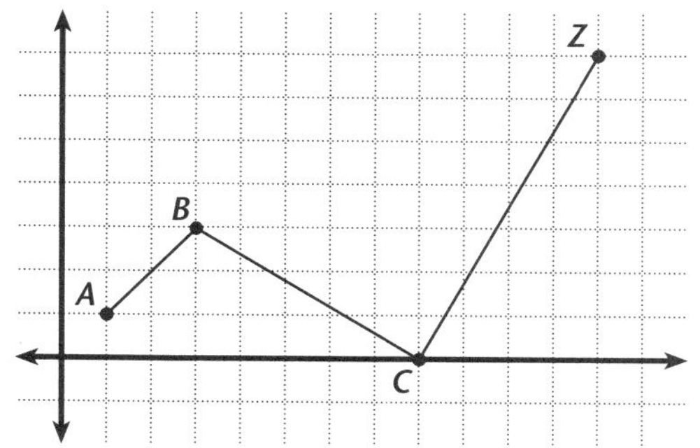

The spreadsheet gives the coordinates of each point, a space for the distance between successive points, and a space for the sum of the distances.

	A	B	C	D
1	POINT	X	Y	DISTANCE
2	A	1	1	
3	B	3	3	
4	C	8	0	
5	Z	12	7	
6			SUM:	

Cell D3 contains SQRT((B3−B2)^2+(C3−C2)^2).
Cell D6 contains D3+D4+D5.

Find the length of each path given its starting point, ending point, and intermediate points.

1. $A(0, 0)$, $B(3, 1)$, $C(4, 7)$, $D(11, 0)$; Path $ABCD$ ____________

2. $L(0, 0)$, $M(4, 2)$, $N(0, 7)$, $P(-5, 3)$; Path $LMNP$ ____________

3. $A(4, 4)$, $X(4, -4)$, $Y(-4, -4)$, $Z(-4, 4)$; Path $AXYZ$ ____________

4. $G(0, 0)$, $H(3,5)$, $I(7, 2)$, $J(-5, 2)$; Path $HJIG$ ____________

5. $A(7, 1)$, $B(-7, 1)$, $V(0, 0)$, $W(6, 5)$; Path $ABVWA$ ____________

6. $T(1, 2)$, $S(2, 1)$, $R(0, 0)$, $N(-5, 5)$; Path $TRSNT$ ____________

7. $A(-7, 0)$, $B(0, 7)$, $C(7, 0)$; Path $ABCA$ ____________

8. $N(0, 4)$, $M(4, 0)$, $T(0, -4)$, $R(-4, 0)$; Path $NMTRN$ ____________

9. $W(0, 0)$; $Y(2, 4)$, $Z(6, 4)$, $X(4, 0)$; Path $WYZXW$ ____________

10. Given $A(-5, 0)$, $B(-5, 5)$, $C(0, y)$, $D(5, 5)$, and $E(5, 0)$, find y such that the length of path $ABCDEA$ is $20 + \sqrt{29} \approx 25.3852$ units long.

____

NAME ______________________________ CLASS ______________ DATE ______________

Technology

5.3 The Parent Function $f(x) = x^2$

The function $f(x) = x^2$ can be used to generate all functions that can be written in the form $f(x) = a(x - h)^2 + k$ for some real numbers a, h, and k. You can consider $f(x) = x^2$ as a parent function and each other quadratic function as an offspring of it. The table indicates how a, h, and k give an offspring from the parent.

	Effect
a	deep/shallow
h	← →
k	↑ ↓

You can graph $g(x) = 2(x - 3)^2 + 1$ by making the parent deeper or shallower, shifting left or right, and shifting up or down.

Roughly sketch each offspring. Check with a graphics calculator.

1. Shift the parent left two units and up three units.

2. Shift the parent right two units and down three units.

3. Shift the parent left one unit.

4. Shift the parent down two units.

5. Make the parent twice as deep.

6. Make the parent one-fourth as deep.

7. Make the parent one-half as deep and then shift the result two units left.

8. Make the parent twice as deep and then shift the result two units down.

9. Describe the graph of $f(x) = a(x - h)^2 + k$ if a is greater than 1, h is negative, and k is positive.

____

10. Describe the graph of $f(x) = a(x - h)^2 + k$ if a is greater than 1, h is positive, and k is negative.

____

NAME ______________________________ CLASS ____________ DATE ____________

Technology

5.4 Completing Many Squares

A square is a planar geometric shape with four right angles and sides that have equal length. The spreadsheet shown illustrates four shapes in columns A, B, C, and D that you can use in some combination to complete a square, part of which is given to you.

If, for example, you are given $x^2 + x$, you could complete a square by making the picture shown in columns G through K and rows 1 through 4. The complete square would represent $x^2 + 2x + 1$, which can also be written as $(x + 1)^2$. See columns G through K and rows 6 through 10.

In the following exercises, you can start with $x^2 + x$ and complete larger squares. As you do so, you will begin to see patterns.

	A	B	C	D	E	F	G	H	I	J	K	L
1												
2												
3												
4												
5												
6												
7												
8												
9												
10												
11												

Use a spreadsheet and the diagram shown.

1. Make the next complete square larger than the one shown.

2. Make the next complete square larger than Exercise 1.

3. Make the next complete square larger than Exercise 2.

4. Make the next complete square larger than Exercise 3.

5. Based on Exercises 1–4, how many x tiles and unit tiles must you add to $x^2 + x$ to represent $(x + n)^2$, where n is a positive integer.

____

6. Repeat Exercise 5 if you are originally given $x^2 + 2x$.

____

7. Repeat Exercise 5 if you are originally given $x^2 + 3x$.

____

8. Repeat Exercise 5 if you start with $x^2 + 1$.

____

9. Repeat Exercise 5 if you start with $x^2 + 2$.

____

NAME ______________________ CLASS ____________ DATE ____________

Technology

5.5 Reading Information from a Graph

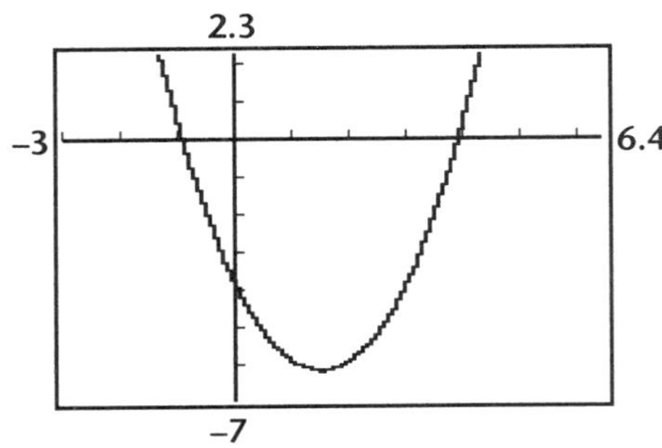

It is often said that a picture is worth a thousand words. The graph, for example, provides a wealth of information about the function $f(x) = x^2 - 3x - 4$. You can read, for example, that the function has two roots. It has an axis of symmetry and a vertex that is a minimum point. In addition, the y-intercept is -4.

The graph is a U shape that opens upward. As you move along the graph to the right, the x-values increase and so do the y-values. As you move along the graph to the left, the x-values decrease but the y-values increase.

Tell as much as you can about the graph of each function. Use a graphics calculator and the quadratic formula to justify your conclusions.

1. $f(x) = x^2 + 3x + 4$

____

2. $f(x) = x^2 + 6x + 9$

____

3. $f(x) = x^2 - 4x + 4$

____

4. $f(x) = x^2 + 5x - 6$

____

5. $f(x) = 3x^2 - 9$

____

6. $f(x) = -3x^2 - 1$

____

7. $f(x) = 2x^2 + x$

____

8. $f(x) = -x^2 + 5x + 4$

____

9. Tell all you can about the graph of $f(x) = ax^2 + bx$.

____

____

10. Tell all you can about the graph of $f(x) = ax^2 + c$.

____

____

NAME ______________________ CLASS __________ DATE __________

Technology

5.6 Organizing Your Work

You can use a spreadsheet to sort out the various solution possibilities for $ax^2 + bx + c = 0$. Enter the coefficients into columns A, B, and C. In the spreadsheet shown, the coefficients of $x^2 - 7x + 12 = 0$ are entered into cells A2, B2, and C2. Cells C5, D5, C6, D6, C7, D7, C8, and D8 will contain conditional statements that give the real and imaginary parts of the solutions for the two cases involving the discriminant calculated in D2. For example, cells C5 and C7 contain, respectively:
IF(D2>=0,(−B2+SQRT(D2))/(2*A2)," ") and IF(D2>=0," ",−B2/(2*A2))

Worksheet 1

	A	B	C	D
1	A	B	C	Discriminant
2	1	−7	12	1
3				
4			Real Part	Imaginary Part
5	Nonnegative	Solution 1	4	0
6		Solution 2	3	0
7	Negative	Solution 1		
8		Solution 2		
9				

Use the discussion above and a spreadsheet to answer the following exercises.

1. Write a conditional for cell C6. ______________________

2. Write a conditional for cell D5. ______________________

3. Write a conditional for cell C8. ______________________

4. Write a conditional for cell D7. ______________________

5. Solve $x^2 + 6x + 11 = 0$.

____

6. Solve $x^2 + 5x + 1 = 0$.

____

7. Solve $3x^2 - 5x + 13 = 0$.

____

8. Solve $x^2 + 4x + 4 = 0$.

____

9. Solve $(x^2 - 12x + 11)(x^2 + 8x + 12) = 0$.

____

NAME ______________________ CLASS ____________ DATE ____________

Technology

5.7 Powers of Complex Numbers

The product of two complex numbers $z = r + si$ and $w = t + ui$, is $zw = (rt - su) + (st + ru)i$. If $z = w$, $zw = z^2 = (r^2 - s^2) + (2rs)i$. With a spreadsheet, you can explore how the powers of a given complex number behave. In cells A2 and B2, enter the real and the imaginary parts of a given number. In cell C2, enter A2^2−B2^2 and in cell D2, enter 2*A2*B2. Set cells A3 and B3 equal to C2 and D2. By using the FILL DOWN command, you can obtain the first several powers of the given number.

	A	B	C	D
1	REAL	IMAGINARY	REAL	IMAGINARY
2				
3				
4				

Use a spreadsheet to compute the first several positive integral powers of z. Keep a record of your spreadsheets.

1. $z = 1 + 0i$

____

2. $z = 2 + 0i$

____

3. $z = 0 + 0.5i$

____

4. $z = 0 - 2.5i$

____

5. $z = 0.45 + 0.20i$

____

6. $z = 1.01 + 1.15i$

____

7. $z = -0.50 - 0.24i$

____

8. $z = -1.2 + 1.3i$

____

9. Draw some conclusions about the behavior of z^n from your work in Exercises 1–4.

____

10. Draw some conclusions about the behavior of z^n from your work in Exercises 5–8.

____

NAME ______________________ CLASS ____________ DATE ____________

Technology

5.8 Too Little, Just Enough, and Too Much Information

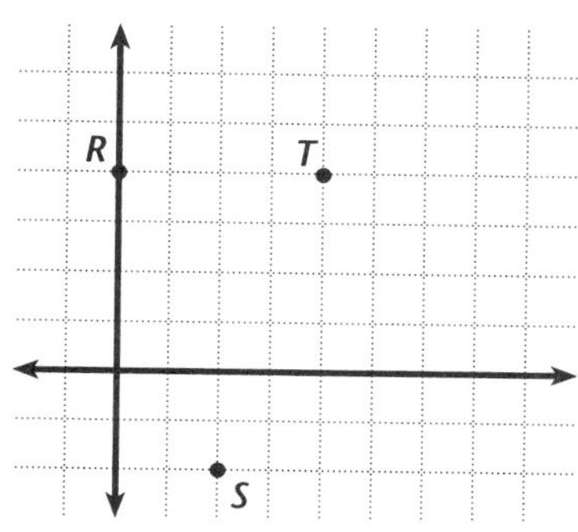

The diagram shows the graph of the set Z of points R, S, and T. Using the techniques explained in Lesson 5.8, you can find a quadratic function, $f(x) = ax^2 + bx + c \ (a \neq 0)$, whose graph contains Z. What happens if you remove one point from the set, for example, S? The system you need to solve now is:

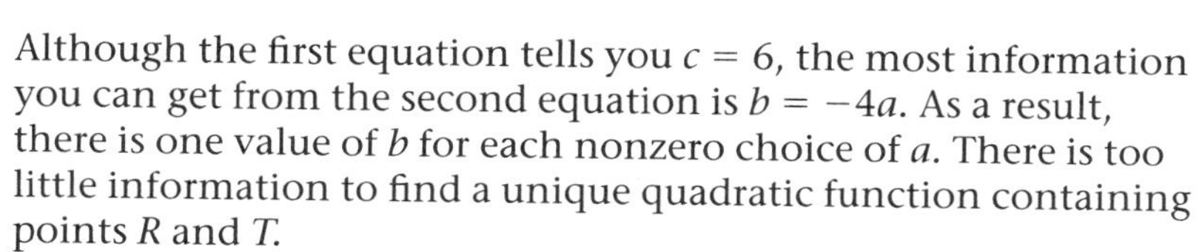

$$0a + 0b + c = 6$$
$$16a + 4b + c = 6$$

Although the first equation tells you $c = 6$, the most information you can get from the second equation is $b = -4a$. As a result, there is one value of b for each nonzero choice of a. There is too little information to find a unique quadratic function containing points R and T.

Use a graphics calculator along with row reduction or matrix inverses to find an equation in the form $f(x) = ax^2 + bx + c \ (a \neq 0)$ whose graph contains the given points. If no such function exists, say so. If many functions exist, give one example.

1. $A(1, 0)$ and $B(3, 10)$

   ____

2. $L(1, -6)$, $M(-2, -4)$, $N(3, 6)$

   ____

3. $M(0, -5)$, $B(-1, 0)$, $C(2, -9)$, $T(1, -8)$

   ____

4. $A(1, 3)$, $F(-2, 15)$, $G(3, 35)$

   ____

5. $N(-1, 4)$, $A(0, 1)$, $T(1, 16)$

   ____

6. $G(2, 9)$ and $H(3, 9)$

   ____

7. $X(4, 18)$ and $Y(-3, 39)$

   ____

8. $A(1, -4)$, $C(-1, 10)$, $T(3, 30)$, $P(2, 6)$

   ____

9. Show that $A(0, 0)$, $B(3, 2)$, and $C(6, 4)$ are collinear.

   ____

10. What happens when you try to find an equation of the form $f(x) = ax^2 + bx + c \ (a \neq 0)$ that contains A, B, and C from Exercise 9?

   ____

NAME ______________________ CLASS ____________ DATE ____________

Technology

5.9 Variations on Quadratic Inequalities

In many situations, you may have to deal with compound inequalities that involve one or more quadratic expressions. The inequality shown is an example of a compound quadratic inequality.

$$3 < x^2 + 3x + 4 < 5$$

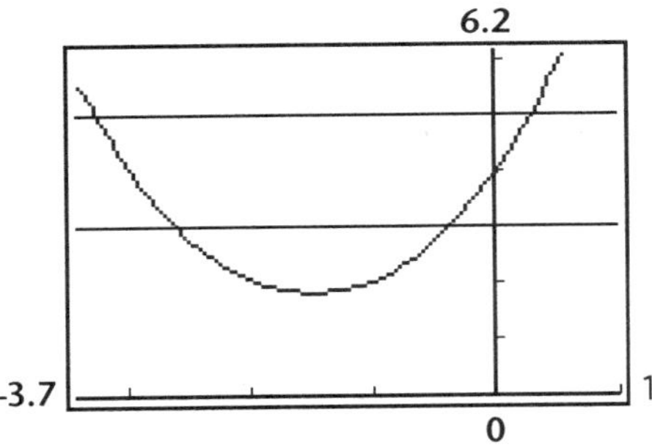

The inequality is illustrated and the diagram is obtained by using a graphics calculator with $Y_1 = 3$, $Y_2 = x^2 + 3x + 4$, and $Y_3 = 5$.

With the trace feature you will find that when $-3.3 < x < -2.6$ or $-0.4 < x < 0.3$, $3 < x^2 + 3x + 4 < 5$.

Use a graphics calculator to solve each inequality.

1. $-3 \leq x^2 \leq 5$

____

2. $-2 < x^2 - 3 < 2$

____

3. $-x + 0.5 \leq x^2 < 0.5x + 1$

____

4. $x + 3 < x^2 \leq 6$

____

5. $-(x - 1)^2 + 3 < x^2 < x^2 + 3$

____

6. $(x - 1)^2 \leq -x + 2 \leq x^2 + 3$

____

7. $-0.4x + 2 \leq (x - 2)^2 - 1 < 0.5x + 2$

____

8. $0.5x + 2 < x^2 + 3x + 6 \leq 0.5x - 3$

____

9. Write an inequality involving $x^2 + x - 2$, $0.5x + 1$, and 4 that has two intervals for x as its solution.

____

10. Write an inequality involving $2x - 3$, $-2.5x$, and $x^2 - 4x + 1$ that has no solution.

____

NAME ______________________ CLASS __________ DATE __________

Lesson Activity

5.1 Small Business Venture

Christen decided to enter the lawn-mowing business. In order to be successful, Christen has outlined her daily operating costs as follows:

Overhead		Costs (per standard size lawn)	
Advertising	$25	Labor	$ 90
Rent (storage)	$40	Energy	$ 5
Administration	$15	Freight	$ 5
Total	$80	Total	$100

Use this data set and these two basic economic principles:

Revenue = Price charged (per average size lawn) × number of lawns mowed

Profit = Revenue − Costs

Christen arrived at the following rule for daily profit $P(x)$ as a function of the price of mowing a lawn, x.

$$P(x) = -x^2 + 60x - 780$$

1. Use this rule to generate a function table for lawn prices ranging from $0 to $45 in increments of $5.

x	0	5	10	15	20	25	30	35	40	45
$P(x)$										

2. Describe the trend in profits as prices increase.

____

3. Use a graphics calculator to graph $P(x)$. What properties of the relation between price and profit can be learned from the graph?

____

4. Approximate $P(x) = 0$. ______________________________

5. What information do the roots of the equation reveal to Christen?

____

6. What price for mowing a standard size lawn should Christen charge? Support your recommendations.

____

____

NAME ______________________ CLASS ____________ DATE ____________

Lesson Activity

5.2 Real-Number Solutions

To find out how many real-number solutions a quadratic equation $ax^2 + bx + c = 0$ has, you can graph $y = ax^2$ and $y = -bx - c$ on the same coordinate grid. Then visually determine the number of points of intersection.

1. Solve $y = x^2 - 6x + 5 = 0$. ______________________

2. How many real-number solutions exist for Exercise 1? ______________________

3. Graph $y = x^2$ and $y = 6x - 5$ on the grid provided. At how many points do the graphs intersect and what are the coordinates of these points?

____

4. Solve $y = x^2 + 4x + 4 = 0$. ____________

5. How many real-number solutions exist for Exercise 4?

____

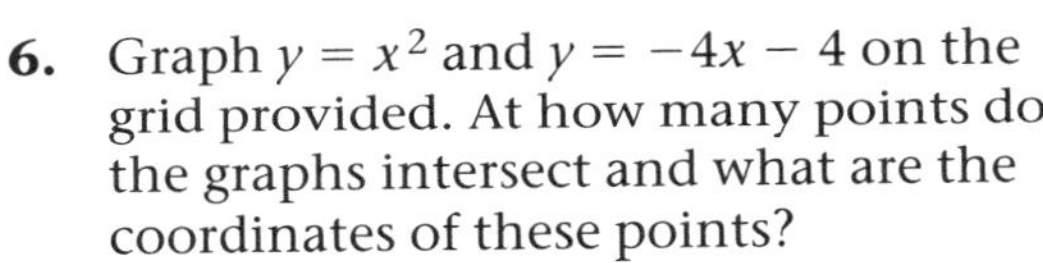

6. Graph $y = x^2$ and $y = -4x - 4$ on the grid provided. At how many points do the graphs intersect and what are the coordinates of these points?

____

7. Solve $y = x^2 + 4 = 0$. ____________

8. How many real-number solutions exist for Exercise 7?

____

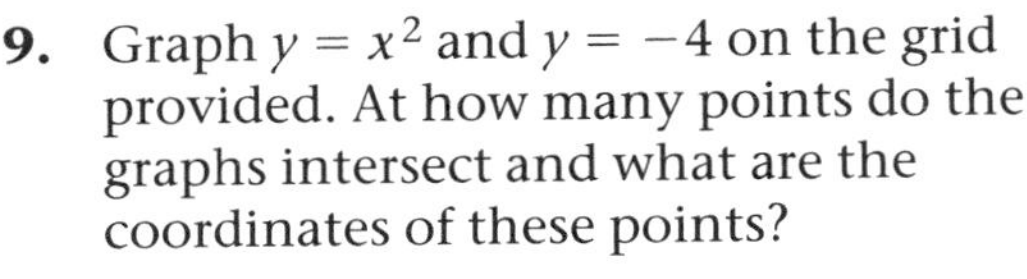

9. Graph $y = x^2$ and $y = -4$ on the grid provided. At how many points do the graphs intersect and what are the coordinates of these points?

____

10. Make a generalization about the number of points of intersections and the number of real-number solutions.

____

Lesson Activity

5.3 Transformations and Functions

Use your knowledge of transformations to write a function that matches a graph. In the case of quadratic functions, the vertex of the parabola is the place to look to determine the vertical and the horizontal shifts.

To determine the amount of vertical stretch or shrink, place your pencil at the vertex of the parabola and move one unit to the right. Then move upward until you are on the parabola. This gives you an estimate of the vertical stretch or shrink.

1. Determine the vertical shift and horizontal shift for the parabola shown.

____

____

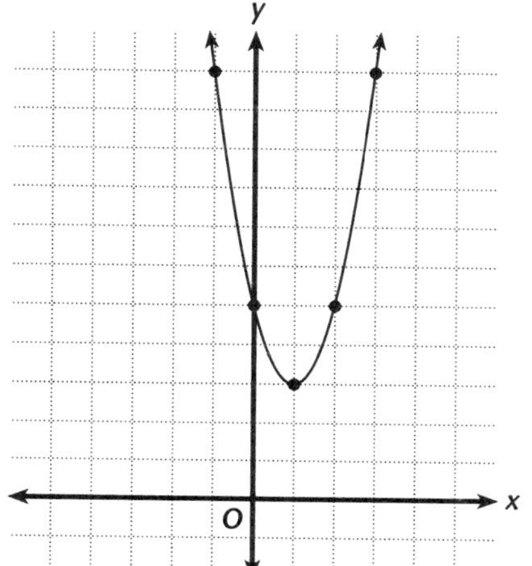

2. What is the amount of vertical stretch or shrink?

____

3. What is the equation of the parabola?

____

Write a function for each graph.

4.

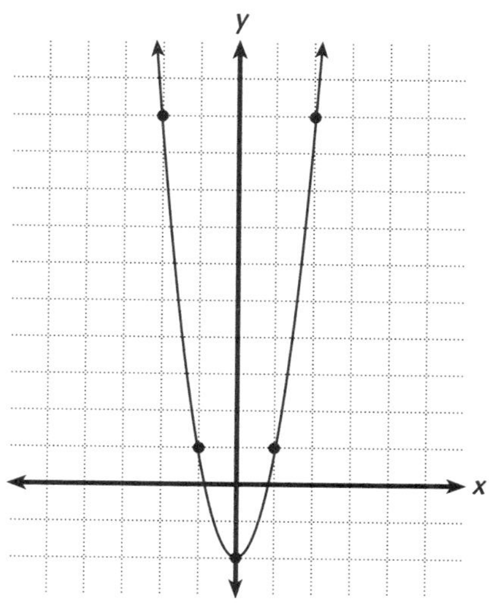

____

5.

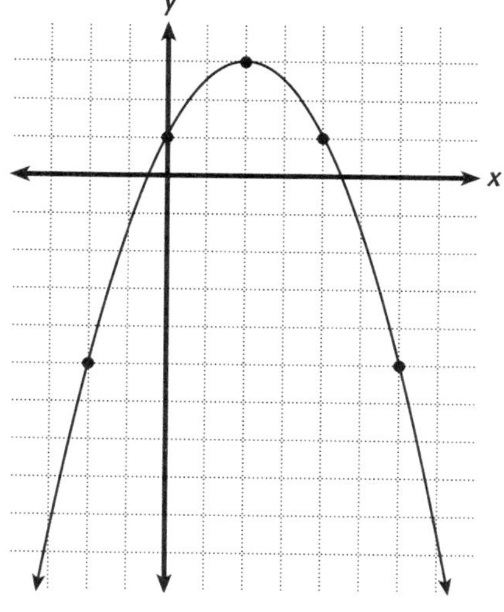

____

NAME ________________ CLASS ________ DATE ________

Lesson Activity

5.4 Modeling Expressions

Square grids can be used to model quadratic expressions.

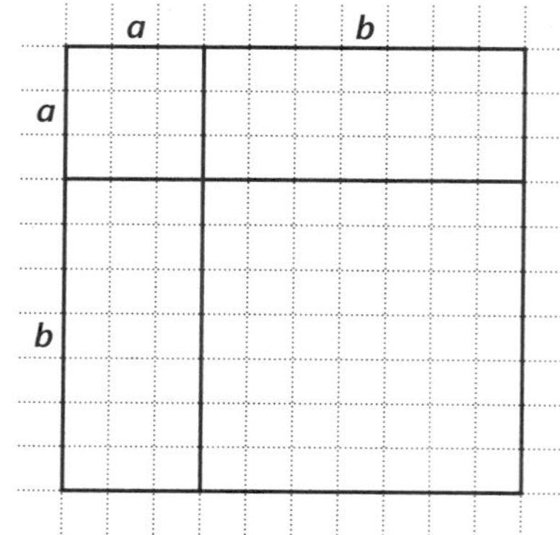

1. Write the quadratic expression modeled by the diagram.

____

2. Write the perfect square trinomial modeled by the diagram.

____

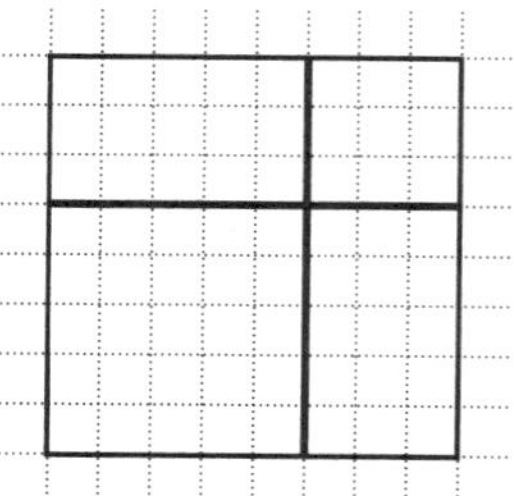

This whole square grid represents 8^2. The bottom left square is $(8 - 3)^2$. Model $(8 - 3)^2 = 8^2 - (8)(3) - (8)(3) + 3^2$ by doing the following: represent (8)(3) being subtracted twice by shading one 8 by 3 rectangle red, then by shading the other 8 by 3 rectangle green. Notice that an area is shaded twice.

3. Why is 3^2 added to the quadratic expression?

____

4. Develop a generalized rule for the quadratic expression equivalent to the perfect square trinomial $(a - b)^2$.

____

Fill in the missing parts of each model. Then write the perfect square trinomial modeled by the heavily outlined square grid.

5.

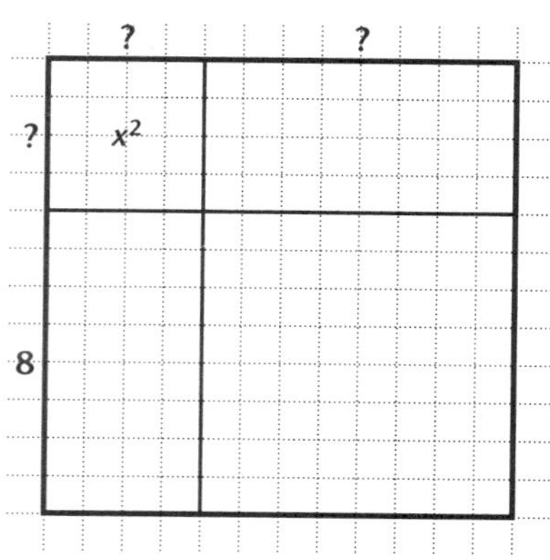

____

6.

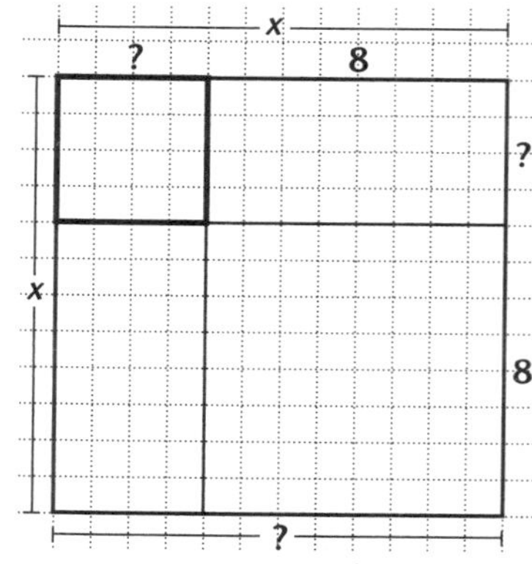

____

NAME ______________________ CLASS ____________ DATE ____________

Lesson Activity

5.5 Quadratic Equations

Numerical analysts use an alternate form of the quadratic formula to solve quadratic equations using calculators or computers. This form is obtained by rationalizing the numerator.

Alternate Form of the Quadratic Formula

Let $ax^2 + bx + c = 0$, where $a \neq b$.

If $b > 0$, then $x_1 = \dfrac{2c}{-b + \sqrt{b^2 - 4ac}}$ and $x_2 = \dfrac{-b - \sqrt{b^2 - 4ac}}{2a}$

If $b < 0$, then $x_1 = \dfrac{-b + \sqrt{b^2 - 4ac}}{2a}$ and $x_2 = \dfrac{2c}{-b - \sqrt{b^2 - 4ac}}$

1. Use a calculator to solve $2x^2 + 5x + 1 = 0$, using the alternate form of the quadratic equation. Do not round off.

____

2. Use a calculator to solve $2x^2 - 5x + 1 = 0$, using the alternate form of the quadratic equation. Do not round off.

____

Solve $x^2 - x - 10^{-15} = 0$ by using each of the following methods and a calculator. Do not round off.

3. Using the quadratic formula. ______________________

4. Using the alternate form of the quadratic formula. ______________________

5. Compare the two solutions. ______________________

6. Why do you think the solutions are not always the same?

____

Lesson Activity

5.6 Cityscape

Suppose your are walking through a city built with square blocks. At every intersection you have four choices. You can turn left (L), turn right (R), go forward (F), or make a U-turn and go backwards (B).

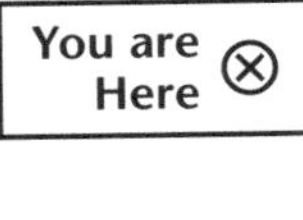

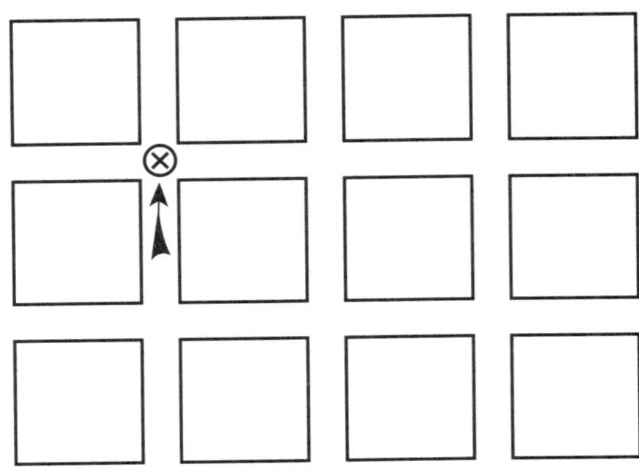

This table shows pairs of choices and the resultant direction you go. For example, LB means turn left, then go backwards (make a U-turn) which takes you in the original right direction.

	L	B	R	F
L	B	R	F	L
B	R	F		
R				
F				

1. Complete the table.

2. What direction acts like the identity element?

____

3. What direction acts like an additive inverse?

____

Let F represent 1, and let B represent -1. Since going left twice in succession is the same as going backwards, let $L = i$. Finally, let $R = -i$.

4. Replace L, B, R, and F with i, -1, $-i$, and 1. Complete the table.

	i	-1	$-i$	1
i				
-1				
$-i$				
1				

5. What does the table represent?

____

NAME ______________________________ CLASS ______________ DATE ______________

Lesson Activity

5.7 Complex Numbers

You can add to complex numbers geometrically in a complex plane by writing each complex number as an ordered pair, then plotting the ordered pair.

For example, to find the sum of $(-2 + 3i)$ and $(7 - 5i)$, plot the points $P(-2, 3)$, $Q(7, -5)$, and $O(0, 0)$ on the complex plane. The points are three vertices of a parallelogram. The sum of the two complex numbers is the fourth vertex, $S(5, -2)$, of the parallelogram.

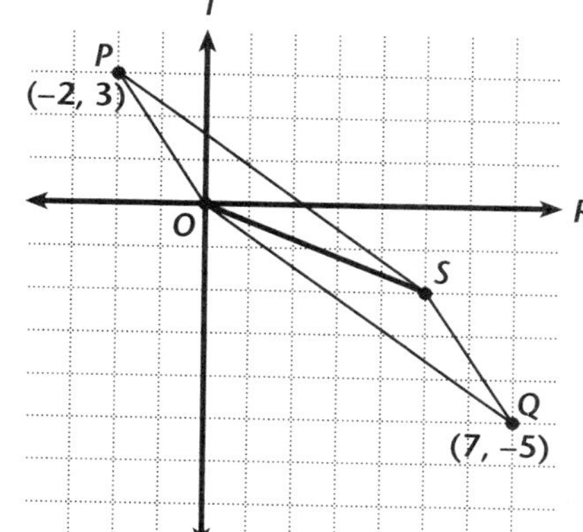

Verify that *POQS* is a parallelogram by finding the slope of each of the following.

1. $\overline{PS}$ ______________ 2. $\overline{PO}$ ______________

3. $\overline{OQ}$ ______________ 4. $\overline{SQ}$ ______________

5. How do you know *POQS* is a parallelogram? ______________

6. Find $(3 + 2i) + (-1 + i)$ geometrically using the grid provided.

7. Describe how you could find $(4 + 3i) - (2 - i)$ geometrically and find the result.

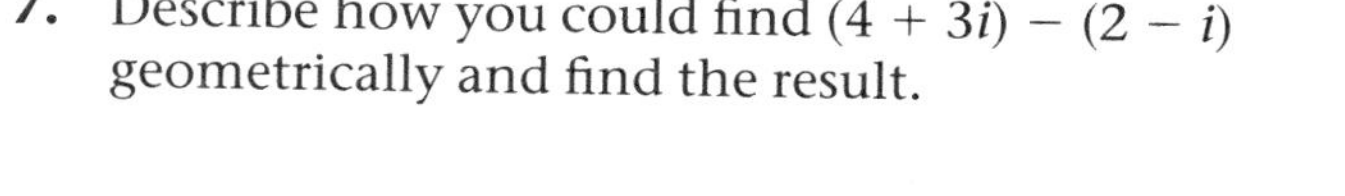

____

____

8. How is the geometric sum of two numbers similar to adding forces in physics?

____

NAME ______________________ CLASS __________ DATE __________

Lesson Activity
5.8 The Lorenz Curve

In economics, the Lorenz Curve is used to estimate the inequality in distribution of income. The curve shows the relationship between the percent of families earning a certain income and the percent cumulative income earned by all families, on the data given in this table.

Percent Families	Percent Income Earned	Percent Cumulative Income Earned
20 (lowest fifth)	5	5
40 (second fifth)	15	20
60 (third fifth)	20	40
80 (fourth fifth)	25	65
100 (highest fifth)	35	100

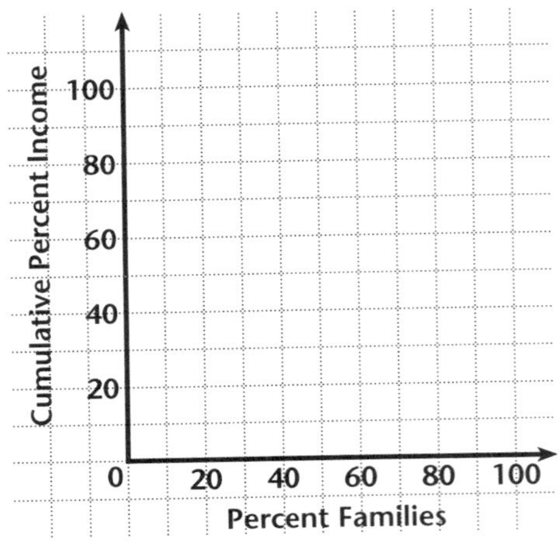

1. Plot the Lorenz Curve on the grid provided.

2. According to the Lorenz Curve, 40% of the families earned about what percent of the cumulative income?

   ____

3. What percent of the families earned 65% of the cumulative income?

   ____

4. Use a graphics calculator to find the quadratic regression equation for the Lorenz Curve. ______________________

5. The line of "perfect equality" is the line $y = x$. Graph this line on the same set of axes as the Lorenz Curve. Then shade the area between the line $y = x$ and the Lorenz Curve.

6. The shaded area indicates the degree of income inequality. Describe what your graph shows about income and inequality.

   ____

7. What happens to income and equality as the Lorenz Curve moves away from the line of perfect equality?

   ____

NAME ______________________ CLASS ____________ DATE ____________

5.9 Bus Fare

A tour bus with a 42-passenger capacity is to be chartered for a day trip. The bus cannot be chartered unless there are at least 21 passengers. The cost of the trip is $28 per person. The bus company agrees to reduce the price of all tickets by $0.75 for each ticket sold in excess of 30.

1. Complete the following table.

Number of tickets in excess of 30	Number of tickets sold	Price of ticket	Bus company income
1			
2			
3			
4			
5			
6			
7			
8			
9			
10			
11			
12			

2. Which value of x results in the bus company's maximum income?

____

3. Plot the points that model the bus company's income in excess of 30 tickets on a separate sheet of paper. What type of function do the points seem to fit?

____

4. Write a quadratic function that models the bus company income for the tickets sold in the data table.

____

NAME ______________________ CLASS __________ DATE __________

Assessing Prior Knowledge

5.1 Exploring Products of Two Linear Functions

Find each product.

1. $(3x - 1)(x - 4)$ ______________________

2. $-(x + 5)(4x - 3)$ ______________________

NAME ______________________ CLASS __________ DATE __________

Quiz

5.1 Exploring Products of Two Linear Functions

1. Write $f(x) = 2(x + 5)(x - 8)$ in the form $f(x) = ax^2 + bx + c$ and identify a, b, and c.

____

2. Find the maximum area of a rectangular garden that you can enclose with 120 feet of fencing material.

____

For Exercises 3–5, let $f(x) = (x + 5)(3x - 4)$.

3. What are the zeros of the function? ______________________

4. Find the vertex of the graph and identify it as a minimum or a

maximum point. ______________________

5. What are the values of x for which f is increasing and for which f is decreasing?

____

6. Find the x-intercepts and coordinates of the vertex of the graph of $f(x) = (2x + 3)(3x - 4)$.

____

NAME ______________________ CLASS __________ DATE __________

Assessing Prior Knowledge

5.2 Solving Quadratic Equations

Simplify.

1. $\sqrt{81}$ __________
2. $-\sqrt{144}$ __________
3. $\sqrt{99}$ __________
4. $-\sqrt{68}$ __________

NAME ______________________ CLASS __________ DATE __________

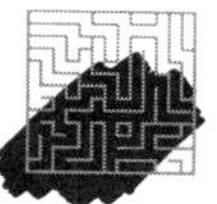

Quiz

5.2 Solving Quadratic Equations

Solve and check.

1. $x^2 = 169$ __________
2. $-8 + \sqrt{2x + 4} = 6$ __________
3. $4x^2 - 7 = 8$ __________
4. $3\sqrt{2x - 5} + 4 = 0$ __________

Find the length of the hypotenuse of a right triangle with legs of indicated lengths.

5. 3 and 8 __________
6. 6 and 10 __________

Find the distance between the points with the given coordinates.

7. $(4, -3)$ and $(-4, 6)$ __________
8. $(-3, 5)$ and $(7, -2)$ __________
9. Find the perimeter of the triangle with vertices at $(-3, 2)$, $(-3, -7)$, and $(4, -3)$. __________

NAME ______________________ CLASS ____________ DATE ____________

Assessing Prior Knowledge

5.3 Graphs of Quadratic Functions

Graph each equation on the grid provided.

1. $y = x + 1$

2. $y = 2(x + 1)$

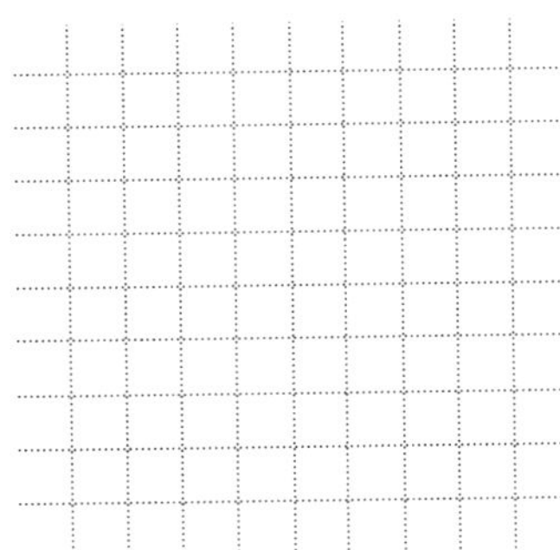

NAME ______________________ CLASS ____________ DATE ____________

Quiz

5.3 Graphs of Quadratic Functions

Use $f(x) = 4(x - 3)^2 + 5$ to answer Excercises 1–2.

1. Give the coordinates of the vertex of $f(x)$.

____

2. Write the equation of the axis of symmetry for $f(x)$.

____

3. Create a quadratic function whose vertex $(-3, 2)$ is a maximum point.

____

4. Using $f(x) = -(x - 4)^2 + 8$, find the values of x for which f is increasing and for which f is decreasing.

____

5. Find the x-intercepts of $f(x) = 2(x - 4)^2 + 6$.

____

NAME ______________________ CLASS __________ DATE __________

Assessing Prior Knowledge

5.4 Completing the Square

Solve and check.

1. $x^2 = 100$ __________

2. $2x^2 - 7 = 281$ __________

3. $(x - 1)^2 - 2 = 7$ __________

NAME ______________________ CLASS __________ DATE __________

Quiz

5.4 Completing the Square

1. Write the quadratic expression modeled by the tiles shown.

____

2. Draw a model to represent $x^2 + 6x + 5$.

3. Find the minimum number of additional unit tiles needed to form a square. Then write the perfect-square trinomial modeled by the square.

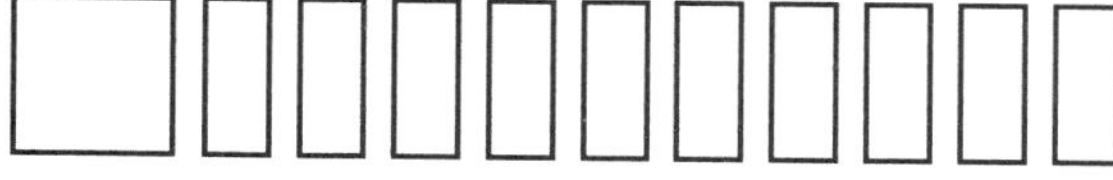

____

Complete the square for each quadratic expression.

4. $x^2 - 4x$ __________

5. $x^2 + 5x$ __________

Solve by completing the square. Give your answers to the nearest tenth.

6. $x^2 + 5x = 12$ __________

7. $3x^2 - 6 = -12x$ __________

NAME ______________________ CLASS __________ DATE __________

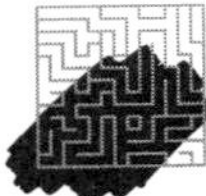

Assessing Prior Knowledge

5.5 The Quadratic Formula

Solve for *x*.

1. $x^2 = 49$ ______________

2. $x^2 - 4 = 0$ ______________

3. $9x^2 - 25 = 0$ ______________

4. $x^2 + 8x + 16 = 0$ ______________

NAME ______________________ CLASS __________ DATE __________

Quiz

5.5 The Quadratic Formula

Use the quadratic formula to solve the following equations. Give your answers to the nearest tenth.

1. $x^2 - 8x = 0$

____

2. $x^2 + 6x + 12 = 0$

____

3. $(x - 3)(x + 2) = 8$

____

4. $4x^2 - 8 = 2x + 3$

____

Find the *x*-intercepts and the vertex of each parabola.

5. $y = 3x^2 - 6x + 7$

____

6. $y = 5x^2 + 3x + 2$

____

7. $y = -x^2 - 5x + 7$

____

8. $y = -3x^2 + 18x - 26$

____

NAME ______________________ CLASS ____________ DATE ____________

Mid-Chapter Assessment

Chapter 5 (Lessons 5.1 – 5.5)

Write the letter that best answers the question or completes the statement.

______ **1.** The distance between the points with coordinates $(-5, -2)$ and $(-7, 4)$.

a. $2\sqrt{37}$ **b.** $2\sqrt{2}$ **c.** $6\sqrt{5}$ **d.** $2\sqrt{10}$

______ **2.** The length of the hypotenuse of a right triangle whose legs are 7 and 12.

a. 19 **b.** $\sqrt{95}$ **c.** $\sqrt{193}$ **d.** 381

______ **3.** The equation of the axis of symmetry of $y = \frac{1}{2}(x - 3)^2 + 4$.

a. $x = 3$ **b.** $y = -3$ **c.** $y = 3$ **d.** $x = -3$

______ **4.** What is the quadratic expression modeled by the following tiles?

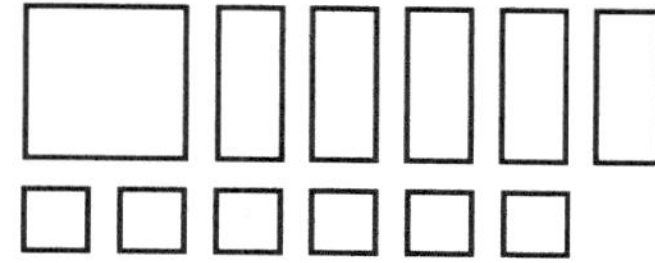

a. $x^2 - 7x + 10$ **b.** $x^2 + 5x + 12$
c. $x^2 + 7x + 10$ **d.** $x^2 + 2x + 17$

5. Solve and check: $\sqrt{6x - 3} = 9$ ______________________

6. Find the coordinates of the vertex of $f(x)$, given $f(x) = x^2 + 9x + 8$.

____

7. Solve by completing the square. Give your answer to the nearest tenth.
$0 = 2x^2 - 8x + 4$.

____

8. Find the x-intercepts for $y = 3x^2 + 12x - 81$. ______________________

NAME ____________ CLASS ________ DATE ________

Assessing Prior Knowledge

5.6 The Discriminant and Imaginary Numbers

Find the value of each expression.

1. $b^2 - 4ac$, if $a = -3$, $b = 4$, and $c = 2$ ____________

2. $b^2 - 4ac$, if $a = 2$, $b = 5$, and $c = 3$ ____________

3. $b^2 - 4ac$, if $a = -2$, $b = -6$, and $c = 1$ ____________

4. $b^2 - 4ac$, if $a = -3$, $b = -2$, and $c = -2$ ____________

NAME ____________ CLASS ________ DATE ________

Quiz

5.6 The Discriminant and Imaginary Numbers

Without solving, determine how many real-number solutions each equation has.

1. $x^2 - 2x + 5 = 0$

____

2. $3x^2 - 5 = -7x$

____

Simplify.

3. $(5i)(5i)$ ____________

4. $i^3 \cdot i^4$ ____________

5. $i\sqrt{3} \cdot (-5i)$ ____________

6. $-i\sqrt{-9} \cdot i^2$ ____________

Solve.

7. $t^2 = -8$

____

8. $y^2 = -25$

____

NAME ______________________ CLASS __________ DATE __________

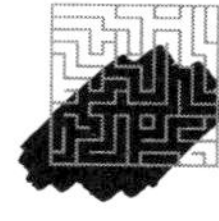

Assessing Prior Knowledge

5.7 Complex Numbers

Simplify.

1. $(3 + 4a) + (2 + 5a)$ ______________

2. $(4 + 2x) - (7 + 3x)$ ______________

3. $-3y(4 + 3y)$ ______________

4. $(5 - 2m)(-3 + 4m)$ ______________

NAME ______________________ CLASS __________ DATE __________

Quiz

5.7 Complex Numbers

Write the solutions of each equation in *a* + *bi* form.

1. $3x^2 - 3x + 5 = 0$

____

2. $x^2 - 5x = -8$

____

3. $-x^2 + 7x - 13 = 0$

____

4. $5x^2 - 3x = -7$

____

Identify the real part and the imaginary part for each complex number.

5. $-3 + 7i$ ______________

6. $2i$ ______________

Find the complex conjugate for each complex number.

7. $5 - 2i$ ______________

8. $-3 + i$ ______________

Use the grid provided to plot each complex number and its conjugate together on a complex plane.

9. $-7 + 6i$

10. $-2 - 3i$

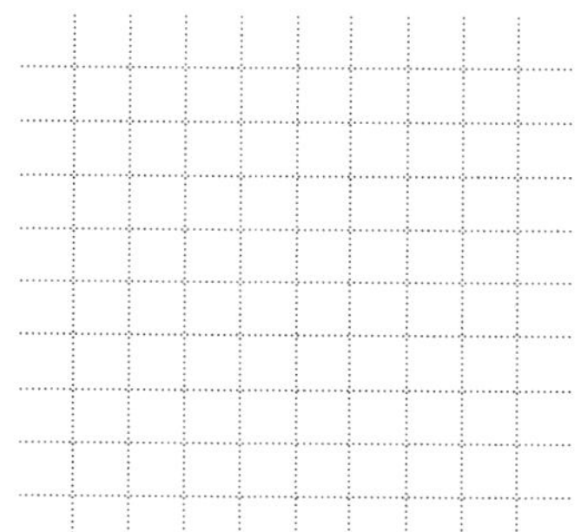

NAME ______ CLASS ______ DATE ______

Assessing Prior Knowledge

5.8 Curve Fitting With Quadratic Functions

Use a graphics calculator to find the inverse of each matrix.

1. $\begin{bmatrix} 3 & 5 \\ 1 & 2 \end{bmatrix}$ ______

2. $\begin{bmatrix} -1 & 0 \\ 0 & 1 \end{bmatrix}$ ______

3. $\begin{bmatrix} 4 & -1 & 3 \\ 2 & -2 & 4 \\ 3 & -1 & 1 \end{bmatrix}$ ______

NAME ______ CLASS ______ DATE ______

Quiz

5.8 Curve Fitting With Quadratic Functions

Use your graphics calculator to find a quadratic model that fits each set of data points.

1. $(-1, -7)$, $(2, 14)$, $(3, 29)$ ______

2. $(-2, 22)$, $(1, -2)$, $(3, -8)$ ______

An object thrown into the air was 14 feet high at 1 second, 23 feet high at 2 seconds, and 9 feet high at 3 seconds.

3. Find a quadratic model for the data.

______

4. What was the maximum height reached by the object? ______

5. When did the object reach its maximum height? ______

6. Use your model to predict the height of the object at 2.5 seconds.

______

7. Use your model to predict the time(s) when the object was at a height of 19 feet. ______

NAME ______________________ CLASS __________ DATE __________

Assessing Prior Knowledge

5.9 Exploring Quadratic Inequalities

Find the zeros of each function.

1. $f(x) = (x + 1)(2x - 5)$ ______________

2. $f(x) = x^2 - 9$ ______________

3. $f(x) = x^2 - 2x + 1$ ______________

4. $f(x) = x^2 - 2x - 3$ ______________

NAME ______________________ CLASS __________ DATE __________

Quiz

5.9 Exploring Quadratic Inequalities

Let $f(x) = x^2 - 5x + 6$.

1. Find the zeros of $f(x)$. ______________

2. For which values of x is $f(x) < 0$? ______________

3. For which values of x is $f(x) > 0$? ______________

4. Draw the graph of $f(x) \leq 0$ on the number line provided.

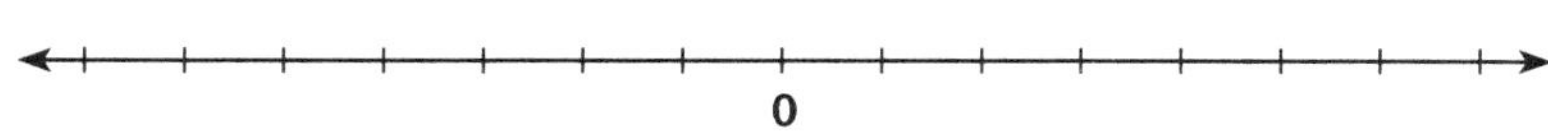

5. Solve $-4x^2 + 5x - 2 \geq 0$.

____

6. Solve $x^2 + 8x - 9 \leq 0$. Graph the solution on the grid provided.

____

NAME ______________________ CLASS __________ DATE __________

Chapter Assessment

Chapter 5, Form A, page 1

Write the letter that best answers the question or completes the statement.

______ **1.** Let $f(x) = (x + 3)(5 - x)$. What are the zeros of 8?

a. 3 and 5 **b.** -3 and 5 **c.** -3 and -5 **d.** 3 and -5

______ **2.** The coordinates of the vertex of the graph of $f(x) = (x - 6)(x - 2)$ are

a. (6, 4) **b.** (6, 2) **c.** $(4, -4)$ **d.** $(4, -2)$

______ **3.** The solution(s) of $5x^2 = 36$ is (are)

a. $\pm\frac{6\sqrt{5}}{5}$ **b.** $\pm\frac{36}{5}$ **c.** $\frac{6}{5}$ **d.** ± 9

______ **4.** The length of the hypotenuse of a right triangle with legs of 3 and 7 is

a. -58 **b.** 10 **c.** $\sqrt{58}$ **d.** 58

______ **5.** The distance between the points having coordinates $(-4, -5)$ and $(-2, 7)$ is

a. 12 **b.** $2\sqrt{2}$ **c.** $6\sqrt{5}$ **d.** $2\sqrt{37}$

______ **6.** What is the quadratic expression modeled by the following tiles?

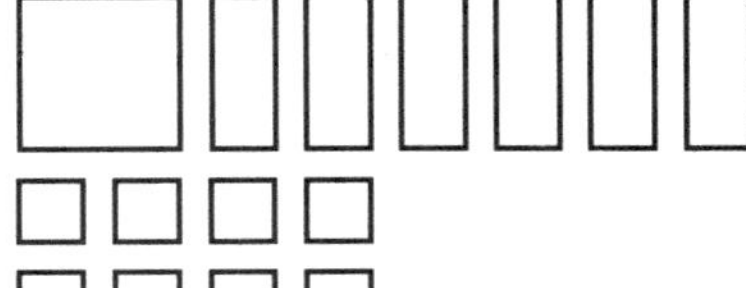

a. $x^2 + 4x + 8$ **b.** $x^2 + 2x + 8$
c. $x^2 + 6x + 8$ **d.** $x^2 + 8x + 6$

______ **7.** How many real-number solutions does $5x^2 + 3x - 1 = 0$ have?

a. 0 **b.** 2 **c.** -1 **d.** 3

______ **8.** Simplify $(-4i^2) \cdot i$

a. $4i$ **b.** $-4i^3$ **c.** -4 **d.** 4

Chapter Assessment
Chapter 5, Form A, page 2

_____ **9.** The solution(s) of $x^2 = -25$ is (are)

a. $\pm 5i$ **b.** -5 **c.** 5 **d.** ± 25

_____ **10.** What is the complex conjugate of $-5 - 8i$?

a. $5 + 8i$ **b.** $5 - 8i$ **c.** 89 **d.** $-5 + 8i$

_____ **11.** The equation $y = 3x^2 - 2x + 7$ is a quadratic model for which set of points?

a. $(-1, 12), (0, 7), (2, 23)$ **b.** $(0, 7), (2, 9), (3, 28)$
c. $(-1, 12), (1, 8), (3, 28)$ **d.** $(-1, 12), (0, 5), (1, 12)$

_____ **12.** Which of the following is a solution for $-x^2 + 8x - 7 > 0$?

a. $-7 < x < -1$ **b.** $1 < x < 7$ **c.** $-1 < x < 7$ **d.** $-7 < x < 1$

_____ **13.** Which represents the graph of $-2 + 3i$ in the complex plane?

a.

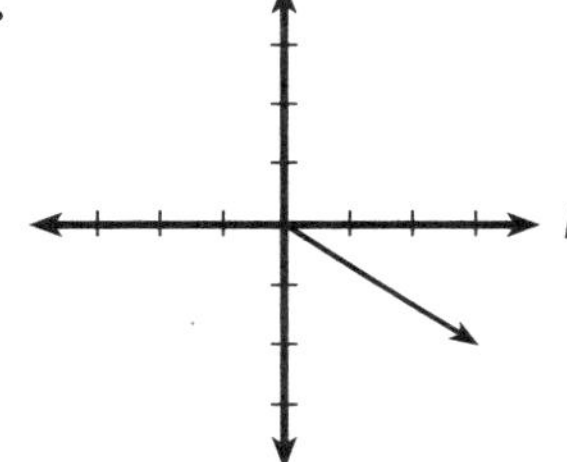

b.

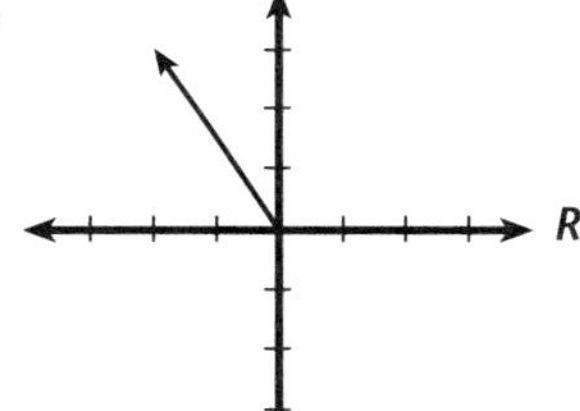

c.

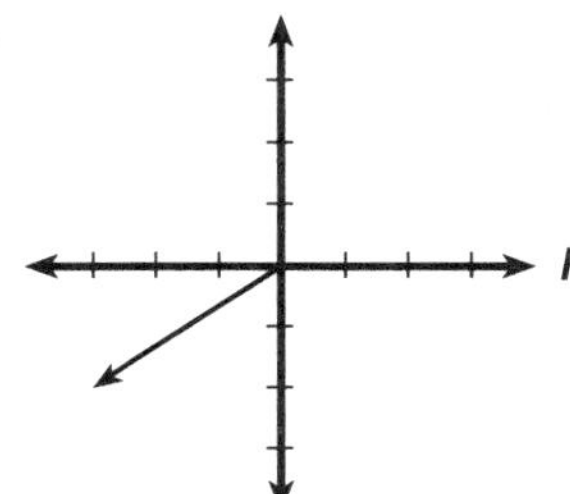

d.

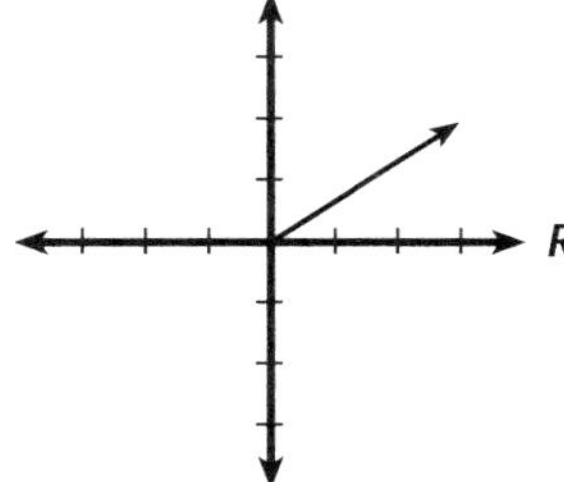

_____ **14.** Given: $f(x) = x^2 - 7x + 6$. For what values of x is the function increasing?

a. $x \leq -\frac{7}{2}$ **b.** $x > \frac{7}{2}$ **c.** $x \geq -\frac{7}{2}$ **d.** $x < \frac{7}{2}$

_____ **15.** The solution(s) of $m^2 = -75$ is (are)

a. $\sqrt{-75}$ **b.** $\pm i\sqrt{75}$ **c.** $5i\sqrt{3}$ **d.** $\pm 5i\sqrt{3}$

_____ **16.** Simplify: $i\sqrt{-16} \cdot i\sqrt{-8}$

a. $8i\sqrt{2}$ **b.** $-8i\sqrt{2}$ **c.** $8\sqrt{2}$ **d.** $-8\sqrt{2}$

NAME ______________________ CLASS __________ DATE __________

Chapter Assessment

Chapter 5, Form B, page 1

1. What are the zeros of $f(x) = 4(x + 3)(2x - 5)$?

____

2. Find the vertex of the graph of $f(x) = -(x + 3)(x - 7)$.

____

3. Solve and check. $3\sqrt{2x - 6} = 12$

____

4. Find the length of the hypotenuse of a right triangle with legs of length 5 and 10.

____

5. Find the distance between the points with coordinates $(-2, 9)$ and $(4, 3)$.

____

6. Find the equation for the axis of symmetry for $f(x) = -\frac{1}{2}(x - 6)^2 - 4$.

____

7. Draw a model to represent $x^2 + 5x + 6$.

8. Solve by completing the square. Give your answer to the nearest tenth.
$x^2 - 4x - 2 = 5 + 2x$

____

9. Use the quadratic formula to solve. Give your answer to the nearest tenth. $x^2 - 4x - 11 = 7$

____

NAME ______________________ CLASS __________ DATE __________

Chapter Assessment

Chapter 5, Form B, page 2

10. How many real-number solutions does $-2x^2 + 5x - 3 = -6x^2 - 11$ have?

____

11. Find a quadratic model that fits the set of data points $(-1, 14)$, $(1, 4)$, $(3, 10)$.

____

12. Solve: $x^2 - 12x + 20 \leq 0$.

____

13. Find the complex conjugate of $-5-i$.

____

14. Plot $-3 - 4i$ in the complex plane.

15. Given: $f(x) = -x^2 + 5x - 6$. For what values of x is the function decreasing?

____

16. Simplify: $-i\sqrt{-9} \cdot i^3$.

____

17. Solve: $m^2 = -54$. ______________________

18. Given: $g(x) = 3x^2 - 5x - 4$. For what values of x is the function increasing?

____

19. Simplify: $i\sqrt{-4} \cdot i\sqrt{-12}$. ______________________

NAME ______________________ CLASS __________ DATE __________

Alternative Assessment

Graphs of Quadratic Functions, Chapter 5, Form A

TASK: To identify the transformations that result from changing the terms of the quadratic function

HOW YOU WILL BE SCORED: As you work through the task, your teacher will be looking for the following:

- how well you can find the vertex and axis of symmetry for a quadratic function
- how well you can describe the transformation that results from changing the terms of a quadratic function
- whether you can determine the values of x for which a quadratic function is increasing or decreasing

1. Find the coordinates of the vertex and the equation of the axis of symmetry of the graph of $f(x) = -(x - 2)^2 + 1$. Then find and explain how to find the x-intercepts.

____

____

2. Describe how you can obtain the graph of $g(x) = 3(x - 2)^2 + 1$ from the graph of $f(x) = 3x^2$.

____

____

3. How does the graph of $f(x) = 4(x - 3)^2 + 1$ differ from the graph of $g(x) = 2(x - 3)^2 + 1$. Explain.

____

____

4. Compare the functions $f(x) = 2(x + 1)^2$ and $g(x) = -2(x + 1)^2$. For what values of x is f increasing and for what values of x is g increasing? For what values of x is f decreasing and for what values of x is g decreasing? Explain.

____

____

5. Create a quadratic function of the form $f(x) = ax^2$ that passes through the point (1, 9).

____

SELF-ASSESSMENT: Explain why the quadratic function is written in the form $f(x) = ax^2 + bx + c$, where $a \neq 0$.

NAME ______________________ CLASS ____________ DATE ____________

Alternative Assessment

Writing a Quadratic Model, Chapter 5, Form B

TASK: To write a quadratic model that fits three data points from real-world data

HOW YOU WILL BE SCORED: As you work through the task, your teacher will be looking for the following:

- whether you can find the quadratic model that fits three data points
- how well you can solve real-world problems

Different sizes of staples are used to hold together different amounts of paper. A $\frac{1}{4}$-inch staple can fasten about 30 sheets of paper, a $\frac{1}{2}$-inch staple can fasten about 100 sheets, a $\frac{5}{8}$-inch staple can fasten about 120 sheets of paper.

1. Explain why a quadratic function models this data.

____

____

2. Find the quadratic model that fits the points.

____

____

3. Find an approximation for the number of sheets a $\frac{3}{8}$-inch staple can fasten. ______________________

4. What is the maximum size for a staple? How many sheets of paper will the staple fasten? Explain.

____

____

5. Describe what restricted domain would be reasonable for this situation.

____

____

SELF-ASSESSMENT: Describe how you would check your solution to this real-world problem.

NAME ______________________ CLASS ____________ DATE ____________

Practice & Apply

6.1 Exploring Products of Linear Functions

Identify whether the function is a polynomial function, a linear function, or neither.

1. $f(x) = x^2 + x - 1$ ____________

2. $f(x) = \frac{(x-3)(x^2-1)}{x}$ ____________

3. $f(x) = 3x - 2$ ____________

4. $f(x) = (2x-1)(x+2)$ ____________

Use the function $f(x) = -(x+2)(x-3)(x-1)$ to answer the following exercises.

5. What is the degree of the function? ____________

6. What are the zeros of the function? ____________

7. Write the function in expanded form.

____

8. What is the leading coefficient of the function? ____________

9. What is the constant term of the function? ____________

Determine the degree of each polynomial function.

10. $f(x) = x(x-5)^2(x+2)$ ____________

11. $f(x) = 4x^2 - 5x^3 + 1$ ____________

12. $f(x) = x(1 - x^4)$ ____________

13. $f(x) = x^2(x-4)^4$ ____________

Determine the degree of each function and create an equation of a polynomial function having the same zeros as the functions graphed.

14.

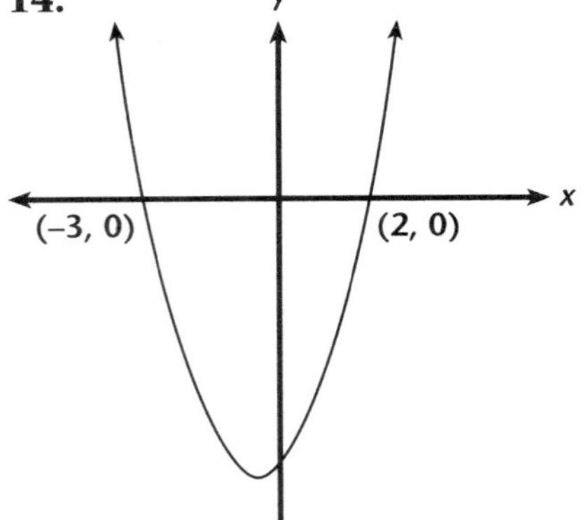

____

15.

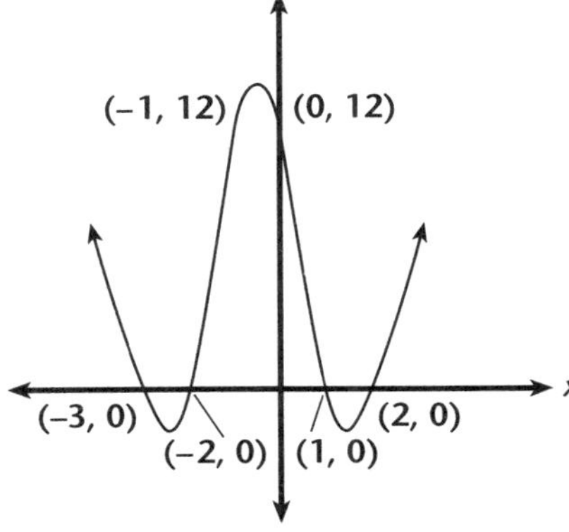

____

16.

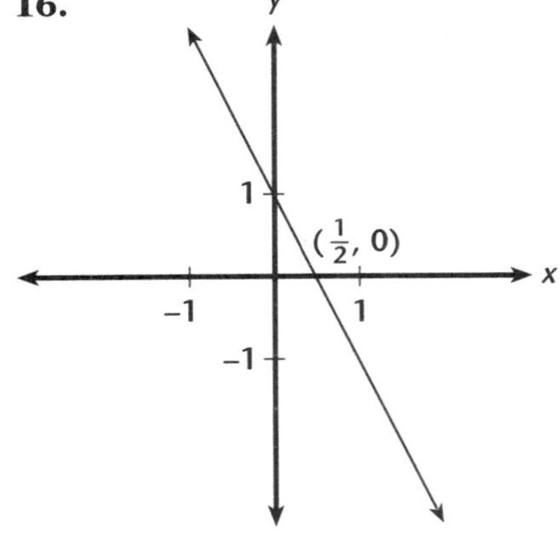

____

Practice & Apply

6.2 The Factored Form of a Polynomial

Find each product.

1. $(3x - 5)(3x + 5)$ ______

2. $x(x - 7)(x + 7)$ ______

3. $(0.4x + 1)(0.4x - 1)$ ______

4. $\left(\frac{1}{3}x + \frac{1}{8}\right)\left(\frac{1}{3}x - \frac{1}{8}\right)$ ______

5. $(4x + 9)(4x + 9)$ ______

6. $(5x + 2)(2x - 5)$ ______

Write each polynomial expression in factored form.

7. $(2x^2 - 11x + 15)$ ______

8. $6x^2 - 5x - 6$ ______

9. $2x^2 - 5x + 3$ ______

10. $3x^2 + 11x + 6$ ______

11. $9x^2 - 6x + 1$ ______

12. $144x^2 - 25$ ______

13. Which expressions cannot be written as a product of two linear factors? ______

a. $x^2 + 16$ **b.** $x^2 - 16$ **c.** $x^2 - 8x + 16$ **d.** $x^2 + 8x + 16$

14. Write a polynomial having zeros at 0, 1, and 4, and a leading coefficient of 2.

______

15. Write $g(x) = x^3 - 5x^2 + 4x$ as a product of linear factors.

______

16. What are the zeros of the polynomial in Exercise 15?

______

NAME ______________________ CLASS ______________ DATE ______________

Practice & Apply

6.3 Dividing Polynomials

1. If $(x - 4)$ is a factor of a polynomial, which of the following points must be on the graph? ______________

a. $(0, -4)$ **b.** $(0, 4)$ **c.** $(4, 0)$ **d.** $(-4, 0)$

2. Fill in the blanks from the information given.

$$
\begin{array}{r}
2x^2 + \underline{\quad\quad} x + \underline{\quad\quad} \\
3x - 2 \overline{)\, 6 \underline{\quad\quad} + \underline{\quad\quad} - \underline{\quad\quad} - \underline{\quad\quad}} \\
6x^3 - 4x^2 \\
\hline
15x^2 - 4x \\
\underline{\quad\quad} - \underline{\quad\quad} \\
\hline
\underline{\quad\quad} - \underline{\quad\quad} \\
\underline{\quad\quad} - \underline{\quad\quad} \\
\hline
0
\end{array}
$$

Divide using synthetic division. Write the quotient and the remainder.

3. $\dfrac{2x^3 + 5x^2 - x + 6}{x - 3}$

____

4. $\dfrac{x^3 - 1}{x - 1}$

____

5. $\dfrac{x^4 - 6x^3 + x^2 - 3x + 7}{x + 1}$

____

6. $\dfrac{x^2 - 6x + 3}{x + 3}$

____

7. Determine whether $x + 2$ is a factor of $x^2 + 4$. ______________

8. Find the quotient when $2x^3 - 5x^2 + 18$ is divided by $2x + 3$. ______________

9. Given $f(x) = x^3 - 5x^2 + 2x + 8$ find the zeros of $f(x)$. ______________

10. Find the linear factors of $f(x)$. ______________

11. Find the common factors of the polynomials $f(x) = x^3 - 5x^2 + 2x + 8$

and $g(x) = x^4 - 4x^3 + 2x^2 + 4x - 3$. ______________

NAME ________________ CLASS ________ DATE ________

Practice & Apply

6.4 Exploring Polynomial Function Behavior

Match each graph with the function it represents.

A $f(x) = (x - 2)^2$ **B** $f(x) = (x - 2)^3$ **C** $f(x) = 3(x - 2)^3$

1.

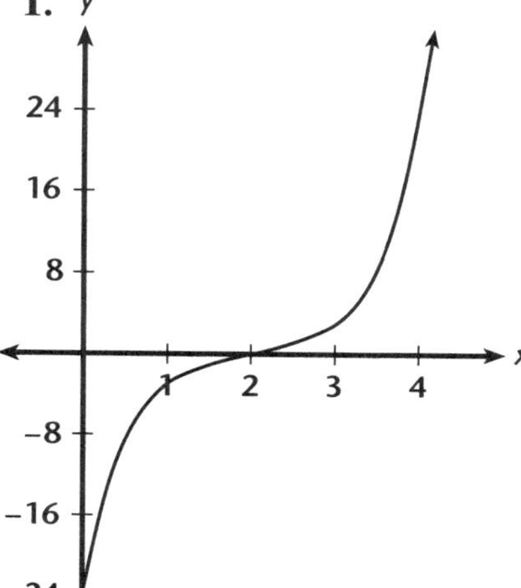

2.

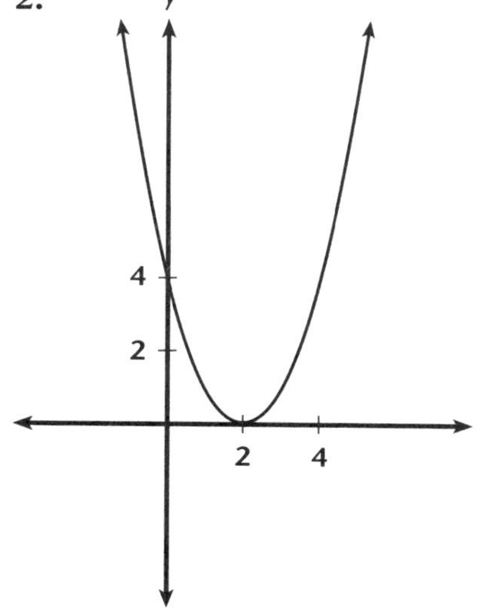

3.

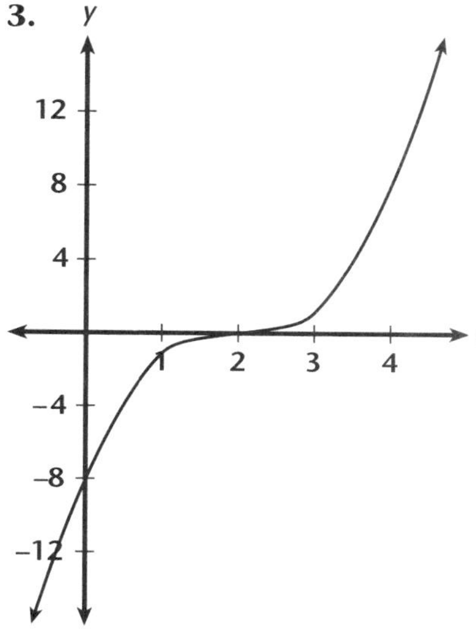

________ ________ ________

4. Write a polynomial in factored form that has degree 4, one zero of multiplicity 2 at -1, one zero at $\frac{1}{2}$, and one zero at 0.

____

Given $f(x) = x^4 - 10x^2 + 9$.

5. Graph $f(x)$ on the grid provided.

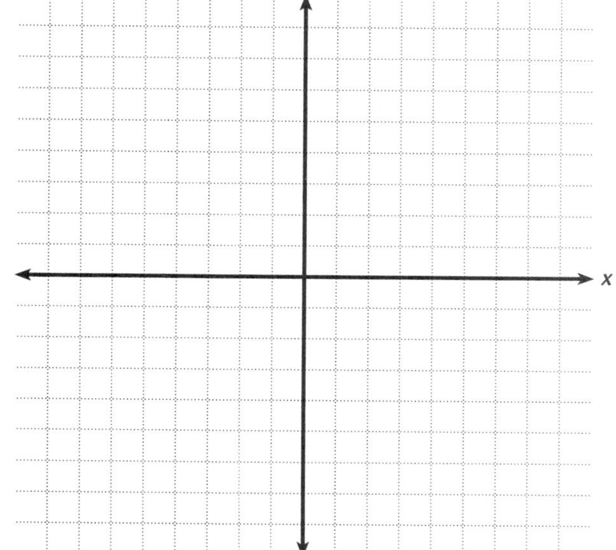

6. How many turning points does $f(x)$ have?

____

7. Where does the function change direction?

____

8. What are the zeros of $f(x)$? ________

9. Where is the function positive?

____

10. Where is the function increasing?

____

NAME ______________________ CLASS __________ DATE __________

Practice & Apply

6.5 Applications of Polynomial Functions

A box with an open top is constructed from a rectangular piece of cardboard that measures 12 in. by 16 in. by cutting out from each corner a square of side x, and then folding up the sides.

1. Find a formula for the volume of the box V as a function x.

____

2. Find the realistic domain for the volume function.

____

3. Find the maximum volume of the box.

____

Jorge is able to deposit $600, $700, $800, and $900 at the end of his freshman, sophomore, junior, and senior years of high school, respectively.

4. How much money will he have at the end of his senior year if the money is invested at 7% interest compounded annually?

____

5. What is the smallest interest rate Jorge can receive to have $3206 at the end of his senior year?

____

In a trade discount, each discount is applied successively to the declining balance in order to arrive at the final invoice price. When purchasing jackets for the senior class, Jane and Andy receive a trade discount of 10% for volume purchase, a trade discount of 10% for transportation, and a trade discount of 10% for storage.

6. If the list price of each jacket is $30, what is the final price Jane and Andy will pay for each jacket? __________

7. Write a polynomial equation that models the trade discounts. __________

8. Graph the polynomial function on a graphics calculator and describe the behavior of the function.

____

NAME ______________________ CLASS ____________ DATE ____________

Enrichment

6.1 Writing Polynomial Functions

If you are given the zeros of a polynomial function and the multiplicity of the zeros (the number of times a zero appears), then you can write the polynomial function in factored form.

For example, to write a polynomial function in factored form whose zeros are −2, 3, and 7, where 7 has a multiplicity of 3, use the zeros to write a product of linear factors.

$$f(x) = (x - (-2))(x - 3)(x - 7)(x - 7)(x - 7)$$
$$f(x) = (x + 2)(x - 3)(x - 7)^3$$

Write a polynomial function in factored form for each of the following.

1. zeros: 1; 3, multiplicity 2; 5

____

2. zeros: −5, multiplicity 2; −1; −12, multiplicity 3

____

3. zeros: −2, multiplicity 6; $-\frac{1}{2}$; 4, multiplicity 4

____

4. zeros: −7, multiplicity 11; −3, multiplicity 5; 0, multiplicity 6; 5

____

5. zeros: $-\frac{2}{3}$, multiplicity 2; $-\frac{1}{2}$; 0, multiplicity 3; 2, multiplicity 5

____

6. zeros: −10, multiplicity 2; −8; −4, multiplicity 7; −2, multiplicity 5

____

7. zeros: −16; −13; −12, multiplicity 2; 0; 5; multiplicity 3; 8

____

8. zeros: −43; −27; −19, multiplicity 3; 0, multiplicity 6; 7, multiplicity 2; 15

____

9. zeros: 4, multiplicity 8; 9, multiplicity 2; 15, multiplicity 8

____

NAME ______________________ CLASS ____________ DATE ____________

Enrichment

6.2 Factor, Factor, Which is the Factor?

If a polynomial is not factored completely, shade in that box. Write the complete, correct factorization under the incorrect one.

1. $x^2 - 9 =$ $(x + 3)(x - 3)$	**2.** $x^2 + x - 6 =$ $(x + 5)(x - 1)$	**3.** $x^2 + x - 12 =$ $(x - 4)(x + 3)$	**4.** $3x^2 - 9x + 6 =$ $3(x - 1)(x - 2)$
5. $4x^2 - 4x - 3 =$ $(4x - 3)(x - 1)$	**6.** $6x^2 - 5x - 4 =$ $(2x + 1)(3x - 4)$	**7.** $4x^3 - x =$ $x(2x + 1)(2x - 1)$	**8.** $5x^2 - 15x + 10 =$ $(5x - 5)(x - 2)$
9. $14x^2 + 49x - 28 =$ $(2x - 1)(7x + 28)$	**10.** $3x^3 + 6x^3 - 105x =$ $3x(x - 5)(x + 7)$	**11.** $6x^2 - 7x - 20 =$ $(3x + 4)(2x - 5)$	**12.** $12x^2 + x - 63 =$ $(3x - 7)(4x + 9)$
13. $4x^2 - 1 =$ $(2x - 1)(2x + 1)$	**14.** $4x^2 + x =$ $(2x + 1)(2x + 1)$	**15.** $6x^2 - 15x =$ $3(2x^2 - 5x)$	**16.** $3x^2 - 26x - 9 =$ $(3x + 1)(x - 9)$
17. $4x^2 + 5x - 6 =$ $(2x - 1)(2x + 1)$	**18.** $6x^3 - 11x^2 - 72x =$ $x(3x + 8)(2x - 9)$	**19.** $6x^2 + 7x - 55 =$ $(3x + 11)(2x - 5)$	**20.** $24x^2 + 29x - 4 =$ $(6x - 2)(4x + 2)$
21. $x^2 + 3x - 108 =$ $(x - 9)(x + 12)$	**22.** $9x^2 - 3x - 2 =$ $(3x - 1)(3x + 2)$	**23.** $5x^2 + 30x - 80 =$ $(5x - 10)(x + 8)$	**24.** $9x^3 - x =$ $x(3x + 1)(3x - 1)$
25. $14x^2 + 29x - 15 =$ $(12x - 3)(7x + 5)$	**26.** $9x^2 - 16 =$ $(3x + 4)(3x - 4)$	**27.** $18x^2 + 33x - 40 =$ $(3x + 8)(6x - 5)$	**28.** $24x^2 + 34x - 45 =$ $(8x - 50)(3x + 9)$
29. $6x^2 - x - 40 =$ $(3x + 8)(2x - 5)$	**30.** $x^3 - 10x^2 + 25x =$ $x(x - 5)^2$	**31.** $9x^2 - 24x + 16 =$ $(3x - 4)^2$	**32.** $25x^2 - 70x + 49 =$ $(5x + 7)^2$
33. $8x^2 + 26x - 99 =$ $(4x - 9)(2x + 11)$	**34.** $42x^2 + 31x - 21 =$ $(3x - 7)(12x + 3)$	**35.** $64x^2 + 48x + 9 =$ $(8x + 3)(8x - 3)$	**36.** $56x^2 + 3x - 20 =$ $(7x - 4)(8x + 5)$
37. $18x^2 + 24x - 40 =$ $(8x - 8)(2x + 5)$	**38.** $24x^2 + 34x - 45 =$ $(6x - 5)(4x + 9)$	**39.** $10x^2 - 3x - 27 =$ $(5x - 9)(2x + 3)$	**40.** $2x^3 - 4x^2 - 38x =$ $2(x^2 + 4x)(x - 6)$
41. $12x^2 + 12x - 24 =$ $12(x - 1)(x + 2)$	**42.** $16x^2 - 64x + 48 =$ $4(x - 1)(4x - 12)$	**43.** $50x^2 - 300x + 450 =$ $50(x^2 - 6x + 9)$	**44.** $6x^2 + 35x - 209 =$ $(2x + 19)(3x - 11)$

NAME ______________________ CLASS __________ DATE __________

Enrichment

6.3 Using Synthetic Division to Evaluate Polynomial Functions

You can use synthetic division to evaluate a polynomial. For example, to find $f(2)$, for $f(x) = 5x^4 - 3x^3 + 6x^2 - 7x + 8$, using substitution you would simplify $5(2)^4 - 3(2)^3 + 6(2)^2 - 7(2) + 8 = 5(16) - 3(8) + 6(4) - 14 + 8 = 80 - 24 + 24 - 14 + 8 = 74$.

When you use synthetic division, notice that the remainder is the same as the value obtained using substitution.

2	5	−3	6	−7	8
		10	14	40	66
	5	7	20	33	74

Evaluate each polynomial using synthetic division.

1. $f(3)$; $f(x) = 5x^6 - 2x^5 - 4x^4 + 7x^3 - 2x^2 + x - 8$ ______

2. $f(-1)$; $f(x) = 6x^5 - 8x^4 + 10x^3 - 15x^2 + 7x - 10$ ______

3. $f(6)$; $f(x) = 9x^6 - 10x^5 + 12x^4 + 7x^3 - 9x^2 + 14x + 18$ ______

4. $f(-5)$; $f(x) = -8x^5 + 17x^4 + 12x^3 - 8x^2 + 9x + 11$ ______

5. $f(12)$; $f(x) = x^7 - x^6 + x^5 - 2x^4 + 3x^3 - x^2 + x + 8$ ______

6. $f(-4)$; $f(x) = 12x^6 - 4x^5 + 7x^4 + 9x^3 - 12x^2 - 17x - 14$ ______

7. $f(7)$; $f(x) = 4x^5 + 8x^4 + 12x^3 - 9x^2 - 17x + 2$ ______

8. $f(-4)$; $f(x) = 21x^4 - 18x^3 + 24x^2 - 19x + 17$ ______

9. $f(11)$; $f(x) = 3x^4 - 8x^3 + 10x^2 + 20$ ______

10. $f(-8)$; $f(x) = 5x^4 - 9x^3 + 12x^2 - 8x + 25$ ______

11. $f(14)$; $f(x) = 5x^5 + 8x^4 - 10x^3 + 14x^2 - 10x + 28$ ______

12. $f(-9)$; $f(x) = -4x^5 + 12x^4 + 18x^3 - 24x^2 + 12x - 18$ ______

13. $f(-8)$; $f(x) = x^7 - 15x^6 + 12x^5 - 40x^4 + 35x^3 - 23x^2 - 18x + 6$

______

14. $f(7)$; $f(x) = 4x^7 - 18x^6 + 4x^5 - 28x^4 - 16x^3 + 104x^2 - 280x + 14$

______

NAME ______________________ CLASS ____________ DATE ____________

Enrichment

6.4 Zeros Maze

Move only horizontally or vertically one space at a time to a box in which the equation has at least two real zeros. If necessary, use a graphics calculator to help you determine the zeros. Watch out for dead ends!

START ↘

$P(x) = 2x^2 - 5x - 3$	$P(x) = x^3 - x^2 + 4x - 4$	$P(x) = 6x^2 + 7x - 3$	$P(x) = 2x^3 - 5x^2 - 4x + 3$
$P(x) = 2x^3 + 5x^2 + x + 2$	$P(x) = x^3 - 7x^2 + 17x - 15$	$P(x) = x^3 - 6x^2 - 27x + 140$	$P(x) = x^4 + x^3 - 7x^2 - x + 6$
$P(x) = x^4 + 2x^3 + x^2 + 8x - 12$	$P(x) = x^3 - 2x^2 - 11x + 52$	$P(x) = x^3 + 4x^2 - 7x + 30$	$P(x) = 6x^2 + 7x - 20$
$P(x) = 18x^3 - 27x^2 + x + 4$	$P(x) = x^4 - 5x^3 - 7x^2 + 41x - 30$	$P(x) = x^4 - x^3 + 14x^2 - 16x - 32$	$P(x) = 6x^4 + 25x^3 - 29x - 20$
$P(x) = x^4 - 7x^2 + 6x$	$P(x) = x^3 - 10x^2 + 37x - 52$	$P(x) = 2x^3 + 3x^2 + 50x + 75$	$P(x) = 6x^2 - 11x - 35$
$P(x) = 3x^3 - 19x^2 + 67x - 91$	$P(x) = 18x^4 + 69x^3 - 286x^2 - 207x + 630$	$P(x) = 6x^3 + 11x^2 - 72x$	$P(x) = 6x^2 + 13x - 110$
$P(x) = 3x^2 + 23x - 36$	$P(x) = x^4 + 5x^3 - 17x^2 - 21x$	$P(x) = 16x^3 + 8x^2 - 38x + 20$	$P(x) = x^3 + 25x$
$P(x) = x^3 + 2x^2 + 4x + 8$	$P(x) = 2x^3 - x^2 - 3x$	$P(x) = x^3 - 8x^2 + 38x - 218$	$P(x) = x^3 - 2x^2 - 5x + 6$
$P(x) = x^3 - 4x^2 - 3x + 12$	$P(x) = x^2 - 2x - 1$	$P(x) = x^3 + 3x^2 + 2x + 6$	$P(x) = 2x^4 - 7x^3 + 9x$
$P(x) = 8x^3 + 20x^2 - 20x + 28$	$P(x) = 16x^3 + 28x^2 - 18x - 4$	$P(x) = 9x^2 - 12x - 3$	$P(x) = 36x^4 - 36x^3 - 29x^2 + 20x + 5$

↘ FINISH

NAME ______________________ CLASS ____________ DATE ____________

Enrichment
6.5 Polynomial Problems

Solve each of the following.

1. What is the maximum volume of the box that can be formed using the drawing shown?

____

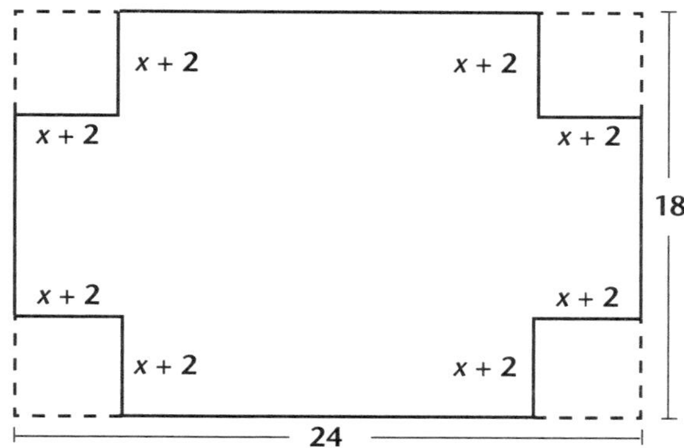

Margaret Smith deposited \$500 in a savings account on her son's first birthday. She deposited \$500 in the account each year after that, again on his birthday. After depositing the money on his eighth birthday, Margaret asked for the account balance.

2. If the interest rate was always $5\frac{1}{4}$%, compounded annually, what balance was Margaret given?

____

3. Suppose Margaret begins depositing \$600 on her son's ninth birthday and deposits \$600 on each birthday after that. If the interest rate does not change, how much more money would there be after the deposit on his twelfth birthday than there would have been if she would have continued to deposit \$500 on each birthday?

____

4. Suppose that Margaret deposits \$500 each year at $6\frac{1}{4}$% interest, beginning on her son's first birthday and continuing until her son turns 18. If the interest is compounded annually, how much will be in the account?

____

5. If her son needs \$19,700 on his eighteenth birthday in order to go to college, how much should Margaret deposit each year starting with his first birthday and continuing through his eighteenth birthday? Assume that the interest rate is 5%, compounded annually.

____

NAME ______________________ CLASS ____________ DATE ____________

Technology

6.1 Symmetric Zeros

Suppose that $f_a(x) = (x + a)(x)(x - a)$ and that a is a nonnegative real number. If, for example, $a = 1$, $f_a(x) = (x + 1)(x)(x - 1)$. The graph has three zeros, two of which are an equal distance from the origin. Notice that the graph is symmetric about the origin and that if you expand the polynomial $(x + 1)(x)(x - 1)$ you will get $x^3 - x$, a polynomial with only odd powers of x.

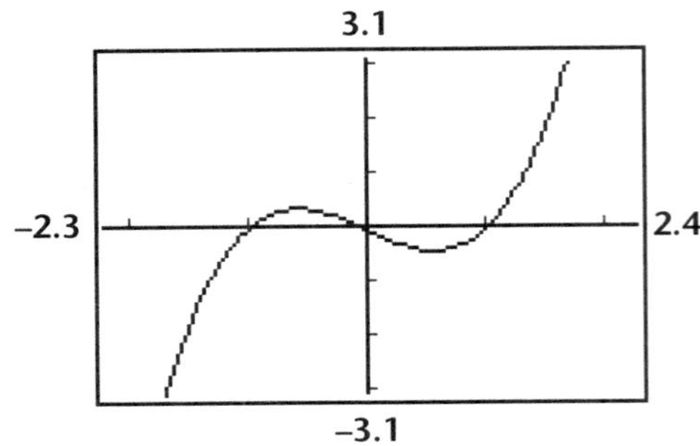

With a graphics calculator, you can classify all the different graphs that arise from different values of a.

Use a graphics calculator to graph $f_a(x) = (x + a)(x)(x - a)$ for each value of a.

1. $a = 0$

2. $a = 1.5$

3. $a = 2$

4. Classify and describe the graphs of $f_a(x) = (x + a)(x)(x - a)$.

____

____

Let $g_{a,b}(x) = (x + b)(x + a)(x)(x - a)(x - b)$. Use a graphics calculator to graph g for each value of a and b.

5. $a = 0; b = 0$

6. $a = 1.5; b = 0$

7. $a = 2; b = 1$

8. Describe the graph of $g_{a,b}(x) = (x + b)(x + a)(x)(x - a)(x - b)$ if a and b are not 0 and $a = b$.

____

____

9. Describe the graph of $g_{a,b}(x) = (x + b)(x + a)(x)(x - a)(x - b)$ if a and b are not 0 and $a \neq b$.

____

____

10. Describe the graph of $h_{a,b}(x) = (x + b)(x + a)(x - a)(x - b)$.

____

____

NAME ______________________ CLASS ____________ DATE ____________

Technology

6.2 Factoring Over the Integers

A polynomial is said to be factorable over the integers if it can be decomposed into factors whose coefficients are integers. The polynomial $x^2 + 7x + 6$, for example, is factorable over the integers since it can be written as $(x + 1)(x + 6)$. However, the polynomial $x^2 + x + 1$ cannot be so factored.

For what integers n will $x^2 + (n + 1)x + n$ be factorable? To find out, you can test different values of n with a graphics calculator. The display shows the graph of $f(x) = x^2 + 3x + 2$. It corresponds to $n = 2$. The graph indicates that the zeros of f are integers, -1 and -2. So $x^2 + 3x + 2$ is factorable over the integers:

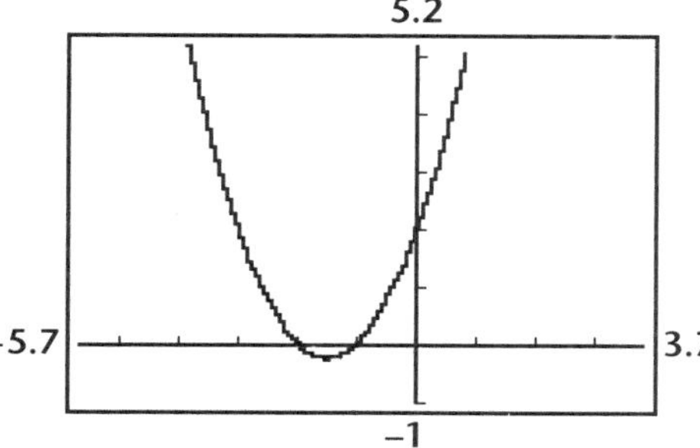

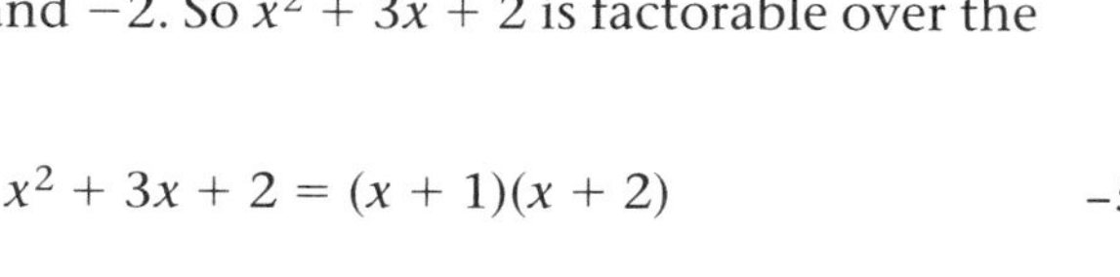

$$x^2 + 3x + 2 = (x + 1)(x + 2)$$

For what nonnegative integral values of *n* will the given polynomial be factorable over the integers?

1. $x^2 + (n + 1)x + n$

____

2. $x^2 + (n - 1)x + n$ (n is also prime.)

____

3. $x^2 + 2nx + n$

____

4. $x^2 + (2n - 1)x + n$

____

5. $x^2 + n$ $(n > 0)$

____

6. $x^2 - n$ $(n > 0)$

____

7. $x^2 - nx + n$

____

8. $x^2 + nx + n$

____

Let $f(x) = x^2 + 2nx + n$ for Exercises 9–10.

9. Use the quadratic formula to show that $x = -n \pm \sqrt{n^2 - n}$.

____

10. Use the graphs of $x = -n \pm \sqrt{n^2 - n}$ to find values of n that make x an integer.

____

NAME ______________________ CLASS ____________ DATE ____________

Technology

6.3 Synthetic Spreadsheet Division

Suppose you have the polynomial $P(x) = rx^3 + sx^2 + tx + u$ and the polynomial $D(x) = x - z$. In the spreadsheet, cell A1 contains the z from $x - z$. Cells B1, C1, D1, and E1 contain the coefficients of $rx^3 + sx^2 + tx + u$.

	A	B	C	D	E
1	Z	R	S	T	U
2		R			

The following steps define a procedure involving the given polynomials.

- Copy the leading coefficient of $P(x)$ into cell B2.
- In cell C2, enter A1*B2+C1.
- In cell D2, enter A1*C2+D1.
- In cell E2, enter A1*D2+E1.

In the exercises, you can discover a relationship between the procedure shown, polynomial division, remainders, and factors.

In Exercises 1–4, carry out the procedure outlined above.

1. $P(x) = x^3 - 5x^2 + 10x - 3$
$D(x) = x - 2$

____

2. $P(x) = x^3 - 3x^2 - 9x + 2$
$D(x) = x + 2$

____

3. $P(x) = x^3 - 3x^2 - 26x + 9$
$D(x) = x + 4$

____

4. $P(x) = 2x^3 - x^2 - 12x - 9$
$D(x) = x - 3$

____

In Exercises 5–8, find the quotient and remainder when $P(x)$ is divided by $D(x)$.

5. Exercise 1

____

6. Exercise 2

____

7. Exercise 3

____

8. Exercise 4

____

9. Briefly describe the relationship between the spreadsheet procedure and the division of a polynomial by a linear polynomial.

____

NAME ______________________ CLASS __________ DATE __________

Technology

6.4 Sign Graphs and Polynomial Functions

The sign graph shown characterizes a polynomial with zeros at $x = -2, 0,$ 1, and 3 and whose sign changes from + to − as x passes through −2. As x passes through 0, the sign changes from − to +, and so on.

If the sign changes as x passes through a zero, the polynomial's factor corresponding to it must have an odd degree. If the sign does not change the factor must have an even degree.

+ + − − + + + +
−3 −2 −1 0 1 2 3

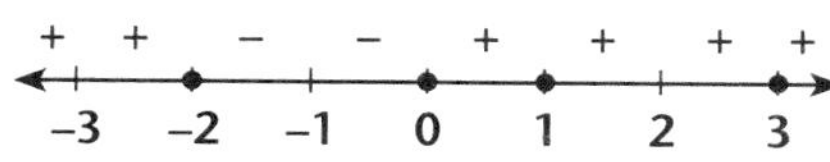

With some experimentation on a graphics calculator, you can find a polynomial function given its sign graph.

Write an equation in factored form for a polynomial function that behaves according to the given sign graph. Your polynomial function should have the lowest positive degree possible.

1.

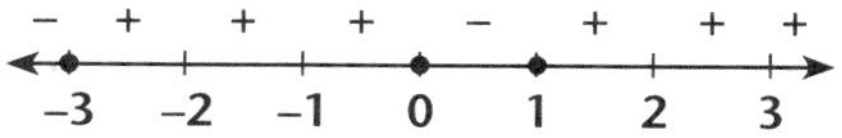

− + + + − + + +
−3 −2 −1 0 1 2 3

____

2.

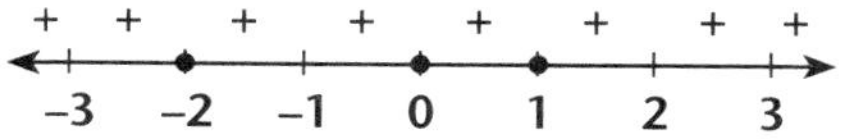

+ + + + + + + +
−3 −2 −1 0 1 2 3

____

3.

− − − − − − − −
−3 −2 −1 0 1 2 3

____

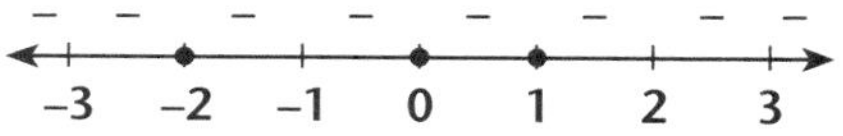

4.

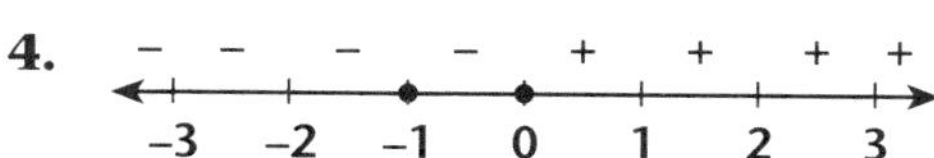

− − − − + + + +
−3 −2 −1 0 1 2 3

____

5.

− − − − − + + +
−3 −2 −1 0 1 2 3

____

6.

+ + + + + + + +
−3 −2 −1 0 1 2 3

____

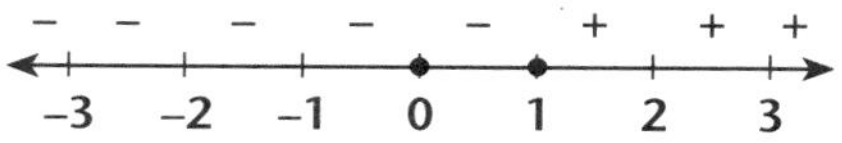

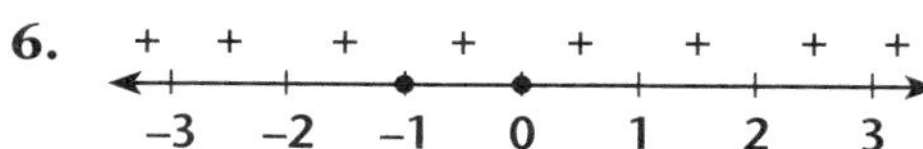

7.

+ + + + + + + +
−3 −2 −1 0 1 2 3

____

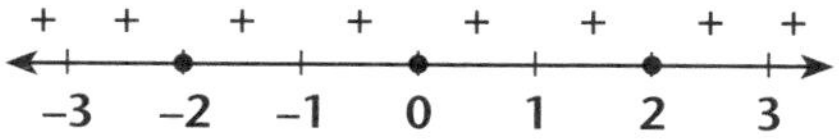

8.

+ + − − + + + +
−3 −2 −1 0 1 2 3

____

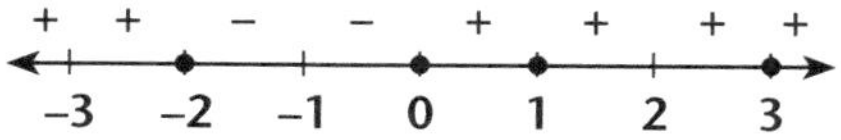

9. Refer to the sign graph in Exercise 8. Find a polynomial function with degree 8 that satisfies the sign graph.

____

NAME ______________________ CLASS __________ DATE __________

Technology
6.5 Annuity Tables

An annuity is the periodic deposit of a fixed amount of money at a given annual interest rate for a certain number of years.

Suppose $100 is deposited at the end of each year for 30 years into an account that pays 5% annually. How much money will be in the account after 30 years and how much interest will be gained? Use a spreadsheet like the one shown.

	A	B	C	D
1	PERIOD	DEPOSIT	AMOUNT	INTEREST
2	1	100		
3	2	100		
...	...	100		
30	29			
31	30			
32		TOTALS>		

Cell C2 contains B2*(1+5/100)^(30−A2). Cell D2 contains C2−B2. Cells C32 and D32 contain the totals from columns C and D.

Use the spreadsheet.

1. Write a formula for the amount the *j*th payment will become at maturity.

____

2. Write a formula for the interest earned by the *j*th payment. ______________

Find the total amount and the interest gained.

3. $100 deposit at 5% over 30 years

____

4. $1000 deposit at 4.40% over 24 years

____

5. $150 deposit at 6.25% over 25 years

____

6. $1000 deposit at 4.40% over 25 years

____

Write a polynomial to represent the total amount in the account.

7. for Exercise 3

____

8. for Exercise 4

____

9. To the nearest dollar, what yearly amount would have to be deposited at 6.25% so that the annuity after 25 years will be $12,000? ______________

Lesson Activity

6.1 Generating a Polynomial Function

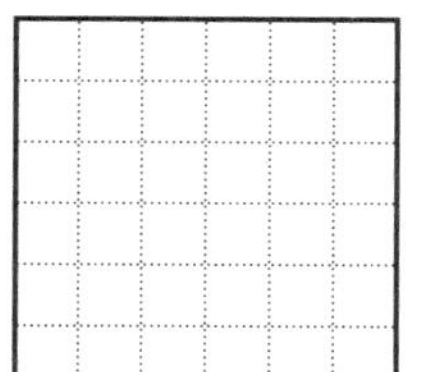

1. How can you determine the number of different squares this 6×6 square contains if the units along each side of the given square are of equal measure?

____

To solve this problem you probably first broke down the given problem into a set of simpler, related problems. Consider the number of different squares in each of the squares shown here.

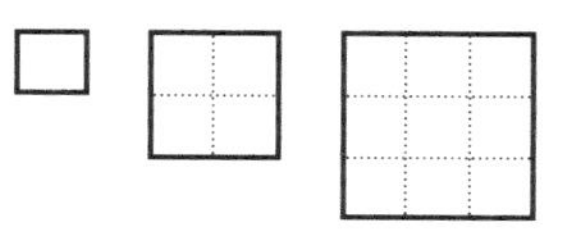

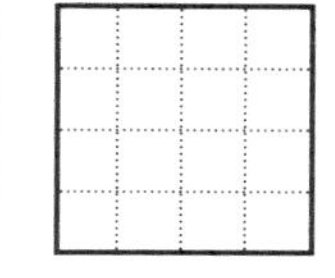

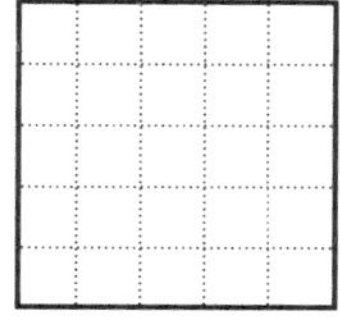

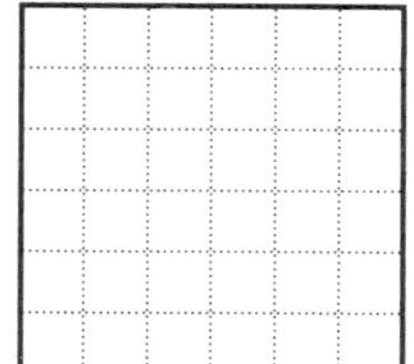

2. Complete the table showing the total number of different squares of each size the 6×6 square contains.

Size of inner squares	Number of units along one side					
	1	2	3	4	5	6
1×1	1	4	9	16	25	36
2×2	0	1	4	9	____	____
3×3	0	0	1	____	____	____
4×4	0	0	____	____	____	____
5×5	0	____	____	____	____	____
6×6	____	____	____	____	____	____
Total	1	5	14	____	____	____

3. How many different squares are contained in a 6×6 square? ______________________

Let n represent the number of equal units on each side of a $n \times n$ square. Let S = the total number of different squares each size square contains.

4. Use your results from Exercise 2 to make a table of values of S. Find the first, second, and third differences.

5. Describe in words why the polynomial $S(n) = an^3 + bn^2 + cn + d$ defines the relationship between the size of the squares and the sum.

____

NAME ______________________ CLASS __________ DATE __________

Lesson Activity

6.2 Modeling Unit Cubes

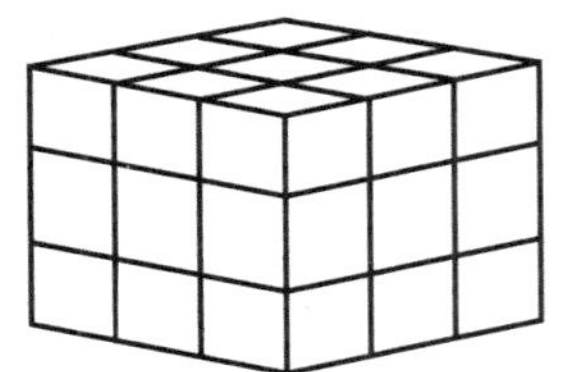

Different-sized cubes are constructed from unit cubes. The surface area is painted, and the large cube is then taken apart. How many of the unit cubes are painted on three faces, two faces, one face, and no faces?

Each of the eight corner unit cubes have three faces painted. The unit cubes with two faces painted occur on the edges between two corners. Each of the central squares on the six faces of the original cube have one face painted. The zero-faced unit cubes are the internal cubes.

Examine the table showing the dimensions of the cube and the number of cubes with paint on them.

Dimensions	Number of $1 \times 1 \times 1$ Cubes Needed	Number of cubes with paint on them			
		3 faces	2 faces	1 face	0 faces
$2 \times 2 \times 2$	8	8	0	0	0
$3 \times 3 \times 3$	27	8	12	6	1
$4 \times 4 \times 4$	64	8	24	24	8
$5 \times 5 \times 5$	125	8	36	54	27
⋮	⋮	⋮	⋮	⋮	⋮
$n \times n \times n$	n^3	8	$12(n-2)$	$6(n-2)^2$	$(n-2)^3$

Use your graphics calculator to examine the graphs of the functions.

1. Graph the zero face function $Z(n) = (n-2)^3$.

2. At which zero does the function touch the x-axis? ______________________

3. What restriction on n is necessary?

____

4. How does a cubed linear factor affect the graph of the function?

____

5. Graph the one face function $E(n) = 6(n-2)^2$. Find the zero(s) of E. ______________

6. How do the zero(s) of the function relate to the linear factor(s)?

____

NAME ______________________ CLASS __________ DATE __________

Lesson Activity

6.3 The Nested Form of Polynomial Functions

The nested form for a polynomial is another way of writing a polynomial. It is used by programmers because a computer can graph a function faster in nested form.

Follow the example to see how to write the polynomial $2x^4 - 4x^2 - 7x + 30$ in nested form. First write the polynomial in descending order of powers. Then factor x out of all terms in which it appears. Continue factoring out x until no power of x greater than one appears.

$$\begin{aligned} 2x^4 - 4x^2 - 7x + 30 &= 2x^4 + 0x^3 - 4x^2 - 7x + 30 \\ &= (2x^3 + 0x^2 - 4x - 7)x + 30 \\ &= ((2x^2 + 0x - 4)x - 7)x + 30 \\ &= (((2x + 0)x - 4)x - 7)x + 30 \end{aligned}$$

1. Write the function $x^6 + x^5 - 2x^4 - 5x^3 + 2x^2 - 8x + 3$ in nested form.

____

2. Rewrite the function $((4x + 3)x - 1)x + 1$ in expanded form.

____

You can use the nested form to divide polynomials. For example, to divide $P(x) = 2x^3 - 8x^2 + 4x - 1$ by $x - 3$, write $P(x)$ in nested form, then evaluate $P(3)$. Notice the relationship between the coefficients of the quotient and the leading values obtained in each step using the nested form.

$$\begin{aligned} P(x) = 2x^3 - 8x^2 + 4x - 1 &= ((2x - 8)x + 4)x - 1 \\ P(3) &= (((2)(3) - 8)(3) + 4)(3) - 1 \\ &= ((-2)(3) + 4)(3) - 1 \\ &= -2(3) - 1 \\ &= -7 \end{aligned}$$

The quotient is $2x^2 - 2x - 2$ and the remainder is -7.

3. Divide $P(x) = x^4 - 5x^3 + x^2 - 2x + 1$ by $x + 1$ using the nested form. Write the quotient and remainder.

____

4. Write $f(x) = x^4 + x^3 - 11x^2 - 31x - 20$ in nested form.

____

5. Show that 4 and -1 are the zeros of this polynomial.

____

NAME ______________________ CLASS __________ DATE __________

Lesson Activity

6.4 Exploring Patterns in Polynomial Coefficients

Consider the coefficients in the expansion of each of the following polynomial products.

$$(x + 1)(x + 2) = x^2 + 3x + 2$$
$$(x + 1)(x + 2)(x + 3) = x^3 + 6x^2 + 11x + 6$$
$$(x + 1)(x + 2)(x + 3)(x + 4) = x^4 + 10x^3 + 35x^2 + 50x + 24$$
$$(x + 1)(x + 2)(x + 3)(x + 4)(x + 5) = x^5 + 15x^4 + 85x^3 + 225x^2 + 274x + 120$$
$$(x + 1)(x + 2)(x + 3)(x + 4)(x + 5)(x + 6) = x^6 + 21x^5 + 175x^4 + 735x^3 + 1624x^2 + 1764x + 720$$

Starting with the polynomial of degree 4, the coefficients of the fourth terms (the term of the fourth-highest degree) in each polynomial can be found using the sums of triple products. For example,

$$50 = (1)(2)(3) + (1)(2)(4) + (1)(3)(4) + (2)(3)(4)$$

and $225 = (1)(2)(3) + (1)(2)(4) + (1)(3)(4) + (2)(3)(4) + (1)(2)(5) + (1)(3)(5) + (1)(4)(5) + (2)(3)(5) + (2)(4)(5) + (3)(4)(5)$

1. How many different triple products are needed to form the sum of 50? ______________

2. How many different triple products are needed to form the sum of 225? ______________

3. What additional triple products are needed to form the sum of 735?

____

4. Find the coefficient of the fourth term in the expansion of

$(x + 1)(x + 2)(x + 3)(x + 4)(x + 5)(x + 6)(x + 7)$ ______________

Write the different pair products needed to form the provided coefficients of the third term in each polynomial of degree 3 or higher.

5. 11 = ____

6. 35 = ____

7. 85 = ____

8. 175 = ____

9. Find coefficient of the third term in the expansion of $(x - 1)(x + 2)(x + 4)$.

____

NAME ______________________ CLASS ____________ DATE ____________

Lesson Activity

6.5 Polynomial Models and Finance

Brittany borrowed \$1000 at a 12% annual interest rate. By making equal monthly payments of \$88.85, the loan will be completely paid back after 12 equal installments.

On most loans, an interest rate of 12% per year means 1% of the unpaid balanced is charged each month. The balance after the first month, then, is the amount borrowed plus the interest minus the amount paid. So Brittany's balance at the end of month 1 is: $1000 + 1000(0.01) - 88.85 = 1000(1 + 0.01) - 88.85 = 1000(1.01) - 88.85$.

At the end of month 2 she owes $(1000(1.01) - 88.85)(1.01) - 88.85 =$ \$761.08.

1. What is Brittany's balance at the end of month 5? ____________

2. Substitute r for 1.01, then write a polynomial in nested form for the balance at the end of month 8. ____________

3. What is Brittany's balance at the end of month 8? ____________

4. Rewrite the polynomial for the balance at the end of month 8 in expanded form.

____

5. Write a polynomial for the balance at the end of month 12 in expanded form.

____

Mike borrows \$2000 at an annual interest rate of 12% to be repaid in 12 equal monthly installments of \$177.70 each.

6. Write a polynomial in nested form for the balance at the end of month 6.

____

7. What is Mike's balance at the end of month 6? ____________

8. Write a polynomial for the balance at the end of month 12 in expanded form.

____

9. What is the relationship between the degree of the polynomial and the balance?

____

NAME ______________________ CLASS ______________ DATE ______________

Assessing Prior Knowledge

6.1 Exploring Products of Linear Functions

The width and length of the base of a carton are 5 feet and 8 feet, respectively. The height of the carton is 3 feet.

1. What is the perimeter of the base of the carton? ______________

2. What is the area of the base of the carton? ______________

3. What is the volume of the carton? ______________

NAME ______________________ CLASS ______________ DATE ______________

Quiz

6.1 Exploring Products of Linear Functions

Determine the degree of each polynomial function.

1. $f(x) = 3(x + 1)(2x - 3)(x + 5)$ ______________

2. $f(x) = 2x^2 + 3x^4 - x^5 + 8$ ______________

3. For the polynomial $f(x) = 7x^6 + 3x^5 - 2x^4 + 5x^3 - x + 5$, identify the coefficients a_1, a_3, and a_4. ______________

Determine the zeros of each polynomial function.

4. $f(x) = (x - 3)(2x + 5)(x + 2)(3x - 1)$

____

5. $f(x) = (5x - 2)(x + 3)(2x - 1)$

____

Use a graphics calculator to graph $f(x) = (x + 2)^2(x + 3)$.

6. What are the zeros of this function? ______________

7. At which zero(s) does the graph cross the x-axis? ______________

8. At which zero(s) does the graph touch the x-axis? ______________

9. For what domain is the function greater than zero? ______________

10. Find the maximum value of the function between the zeros. ______________

NAME ______________________ CLASS __________ DATE __________

Assessing Prior Knowledge

6.2 The Factored Form of a Polynomial

Solve each quadratic equation using the quadratic formula.

1. $x^2 - 4x - 21 = 0$ __________

2. $4x^2 - 8x - 32 = 0$ __________

3. $3x^2 + 6x = 0$ __________

4. $6x^2 - 150 = 0$ __________

NAME ______________________ CLASS __________ DATE __________

Quiz

6.2 The Factored Form of a Polynomial

Write each polynomial expression in factored form.

1. $x^2 - 5x + 6$ __________

2. $x^2 + 6x + 8$ __________

3. $x^2 - 4x - 21$ __________

4. $x^2 + 7x - 18$ __________

5. $2x^2 + 5x - 12$ __________

6. $9x^2 - 4$ __________

Use a graphics calculator to find the zeros of each function to the nearest hundredth. Then write each function in factored form with integer factor components.

7. $f(x) = 3x^2 - 5x - 2$

____

8. $f(x) = 2x^2 + x - 1$

____

9. $f(x) = 2x^3 - 5x^2 - 3x$

____

10. $f(x) = 3x^3 + 7x^2 - 6x$

____

11. How do the zeros of a function relate to the linear factors?

____

12. Write a quadratic function with zeros at 3 and -1; then use the quadratic formula to verify that the roots are 3 and -1.

____

NAME ______________________ CLASS ____________ DATE ____________

Assessing Prior Knowledge

6.3 Dividing Polynomials

Arrange each polynomial in descending order of *x*. If a term is missing, write that term with a coefficient of 0.

1. $x^2 + 3 - 4x + x^4$ ____________ **2.** $-100 + x^3$ ____________

3. $6 - 3x + 2x^3$ ____________ **4.** $x^2 + 2 + 3x^4$ ____________

NAME ______________________ CLASS ____________ DATE ____________

Quiz

6.3 Dividing Polynomials

1. What is the largest number of complex factors a fifth degree polynomial can have? ____________

2. If $(x + 3)$ is a factor of a polynomial, name a zero of the polynomial. ____________

3. The numbers -1 and 2 are zeros of the polynomial $f(x) = 6x^4 - 5x^3 - 15x^2 + 4$. Find the other zero(s) using long division. ____________

4. The numbers 1 and 2 are zeros of the polynomial $f(x) = x^4 - 4x^3 - x^2 + 16x - 12$. Find the other zero(s) using synthetic division. ____________

5. Given $(x + 3)$ is a factor of *P*, a polynomial of degree 5, determine whether the quotient of $P \div (x + 3)$ is a polynomial. If so, give its degree. ____________

6. Use a graphics calculator to factor the polynomial $f(x) = x^4 - 2x^3 - 13x^2 + 14x + 24$ by graphing. ____________

7. Find the common factors of $f(x) = x^3 - 3x^2 - x + 3$ and $g(x) = x^4 - 2x^3 - 7x^2 + 8x + 12$.

____

NAME ______________________ CLASS ____________ DATE ____________

Mid-Chapter Assessment

Chapter 6 (Lessons 6.1–6.3)

Write the letter that best answers the question or completes the statement.

______ **1.** Determine the degree of $f(x) = -3x^2 + 2x^3 - x^5 + 9$.

a. -3 **b.** 2 **c.** 5 **d.** 9

______ **2.** Find the factored form of $f(x) = x^3 + 2x^2 - 9x - 18$.

a. $(x + 2)^2(x - 3)$ **b.** $(x - 2)(x + 3)^2$
c. $(x + 2)(x - 3)(x + 3)$ **d.** $(x - 2)(x - 3)(x + 3)$

______ **3.** Which quadratic function has zeros at 3 and -2?

a. $f(x) = x^2 + x + 6$ **b.** $f(x) = x^2 + x + 6$
c. $f(x) = x^2 - x + 6$ **d.** $f(x) = x^2 - x - 6$

______ **4.** The numbers 2 and -1 are zeros of the polynomial $f(x) = x^3 - 4x^2 + x + 6$. Find the other zero.

a. $x = -3$ **b.** $x = 3$ **c.** $x = -2$ **d.** $x = 1$

Tell whether each statement is true or false.

5. The function $f(x) = -5x^3 + 7x^2 - 12x + 6$ has at least one real root. ______

6. The function $f(x) = 9x^5 - 3x^4 + 5$ has exactly 9 zeros (counting multiplicities). ______

Write each polynomial expression as a constant multiple of linear factors.

7. $x^2 + 12x + 35$ ______ **8.** $x^2 + 4x - 21$ ______

9. Find the common factors of $f(x) = x^3 + 6x^2 + 5x - 12$ and $g(x) = x^3 + 9x^2 + 26x + 24$. ______

The height of a crate is 2 meters more than its width. The perimeter of the base of the crate cannot exceed 14 meters.

10. Write a function for the volume of a crate in terms of the width. ______

11. Graph this function using a graphics calculator.

12. Which values of the width give realistic volumes? ______

13. Find the dimensions that result in the maximum volume. ______

14. Give the maximum volume to the nearest tenth. ______

NAME ______________________ CLASS __________ DATE __________

Assessing Prior Knowledge

6.4 Exploring Polynomial Function Behavior

Use your graphics calculator to graph each function. Determine the real zeros of each function.

1. $f(x) = 3x^3$ ______________

2. $g(x) = x^2 + 2x + 1$ ______________

3. $h(x) = 2x^2 - 8$ ______________

4. $P(x) = x^3 - 27$ ______________

NAME ______________________ CLASS __________ DATE __________

Quiz

6.4 Exploring Polynomial Function Behavior

Let $f(x) = (x + 1)(x - 3)(x + 2)(x - 1)^2$.

1. What are the zeros and degree of f? ______________

2. Is f always a decreasing function? Explain. ______________

3. Which direction (upward of downward) will the ends of the graph turn?

____

4. Which zeros, if any, are turning points? ______________

5. How many crossing points does f have? ______________

6. Write down an example of a polynomial in factored form that has degree 4, 2 crossing point zeros, and 1 turning point zero.

____

Let $f(x) = 4x^3 + 14x^2 + 10x - 3$.

7. Using a graphics calculator, estimate all the zeros of f. ______________

8. Use the approximate zeros of f to factor f into its linear factors. ______________

9. Where is f increasing? ______________

10. Where is f decreasing? ______________

11. Where are the turning points? ______________

NAME ______________________ CLASS __________ DATE __________

Assessing Prior Knowledge

6.5 Applications of Polynomial Functions

A missile is fired from a nozzle that is 10 feet above the ground. Its initial velocity is 500 feet per second. Its height, $h(t)$, t seconds after being launched, is given by $h(t) = -16t^2 + 500t + 10$.

1. When will the missile reach its maximum height? ______________________

2. How high will the missile be at is maximum height? ______________________

NAME ______________________ CLASS __________ DATE __________

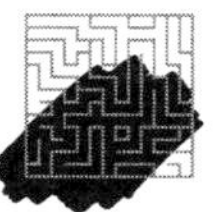

Quiz

6.5 Applications of Polynomial Functions

A piece of cardboard measures 10 in. by 12 in.

1. What would the volume be in terms of the length of the sides of the squares cut from each corner? ______________________

2. Use a graphics calculator to graph this function making sure to include the x- and y-intercepts.

3. What is the maximum volume of the box? ______________________

4. What is the realistic domain of the volume function? ______________________

Tony is able to deposit $400 into his savings account at the end of each of his four years in high school.

5. How much money will Tony have by the end of his senior year if the money is invested at 7% interest? ______________________

6. If his college tuition is $800 for each of the two freshman semesters, how much will he have left over after he pays freshman tuition? ______________________

7. If Tony's total freshman tuition is $2500, will putting off going to college for one year increase his investment return enough to meet the tuition? Why or why not?

____

NAME ______________________________ CLASS ______________ DATE ______________

Chapter Assessment

Chapter 6, Form A, page 1

Write the letter that best answers the question or completes the statement.

______ **1.** What are the zeros of $f(x) = x(x + 3)(x - 7)$?

a. $x = 1, x = 3, x = -7$ **b.** $x = 0, x = 3, x = -7$
c. $x = 0, x = -3, x = 7$ **d.** $x = -3, x = 7$

______ **2.** At which zero(s) does the graph of $f(x) = (x - 2)^2(x + 1)$ touch the x-axis?

a. $x = -1$ **b.** $x = 2$
c. $x = -2, x = 1$ **d.** $x = 2, x = -1$

______ **3.** For what domain is the function $f(x) = (x - 3)(x - 1)$ greater than zero?

a. $x > 3$ or $x < 1$ **b.** $1 < x < 3$
c. $x > -3$ or $x < -1$ **d.** $-1 < x < -3$

______ **4.** When $2x^2 - 11x + 15$ is written as a constant multiple of linear factors, the result is

a. $(x - 5)(2x - 3)$ **b.** $(2x - 5)(x - 3)$
c. $(2x - 5)(2x - 3)$ **d.** $(2x - 5)(x + 3)$

______ **5.** Which represents $36x^2 - 25$ written as a constant multiple of linear factors?

a. $(6x - 5)(6x - 5)$ **b.** $(6x + 5)(6x + 5)$
c. $(5x + 6)(5x - 6)$ **d.** $(6x - 5)(6x + 5)$

______ **6.** Which quadratic function has zeros at 5 and 2?

a. $x^2 - 7x + 10$ **b.** $x^2 + 7x - 10$
c. $x^2 - 7x - 10$ **d.** $x^2 + 7x + 10$

______ **7.** Which polynomial function has two roots at -2 and two roots at 3?

a. $f(x) = (x - 2)^2(x - 3)^2$ **b.** $f(x) = (x - 2)^2(x + 3)^2$
c. $f(x) = (x + 2)^2(x - 3)^2$ **d.** $f(x) = (x + 2)^2(x + 3)^2$

______ **8.** The numbers 5 and -3 are zeros of the polynomial $f(x) = x^3 - x^2 - 17x - 15$. What are the other zero(s)?

a. $x = -5, x = 3$ **b.** $x = 3, x = -1$ **c.** $x = 1$ **d.** $x = -1$

______ **9.** Use any method to factor the polynomial $f(x) = x^3 - 2x^2 - 5x + 6$.

a. $(x + 2)(x - 1)(x - 3)$ **b.** $(x - 2)(x + 1)(x + 3)$
c. $(x + 2)(x + 5)(x - 4)$ **d.** $(x + 1)(x - 1)(x - 6)$

Chapter Assessment

Chapter 6, Form A, page 2

______ **10.** What are the common factor(s) of $f(x) = x^3 - x^2 - 6x$ and $g(x) = x^4 - 5x^2 + 4$?

a. $(x + 2)(x - 1)$ **b.** $(x + 2)$ **c.** $(x + 1)$ **d.** $x(x + 2)$

______ **11.** Which polynomial has degree 5, 1 crossing point zero, and 2 turning point zeros?

a. $f(x) = (x - 5)(x + 1)(x - 2)^2$
b. $f(x) = (x - 1)(x + 2)(x - 3)(x + 4)(x - 5)$
c. $f(x) = (x + 1)(x - 2)^2(x + 3)^2$
d. $f(x) = (x - 1)(x - 2)(x + 3)(x + 1)^2$

______ **12.** Which polynomial has degree 3 and turns down at the right end?

a. $f(x) = x^3 - 7x - 6$
b. $f(x) = -x^3 + 7x + 6$
c. $f(x) = -6x^2 + 14x - 12$
d. $f(x) = 5x^2 + 9x + 6$

______ **13.** Which polynomial could be represented by the given graph?

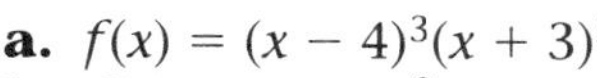

a. $f(x) = (x - 4)^3(x + 3)^2$
b. $f(x) = (x - 3)^2(x + 4)^3$
c. $f(x) = x(x - 3)^2(x + 4)^2$
d. $f(x) = (x - 3)^2(x + 4)$

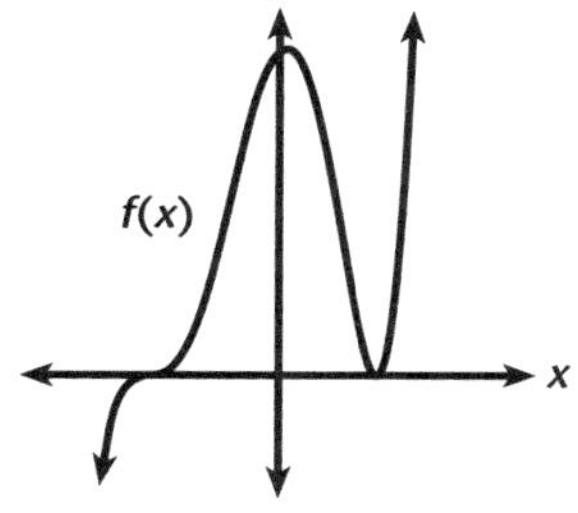

______ **14.** A piece of cardboard measures 6 in. by 14 in. Which function represents the volume in terms of the length of the sides of the squares cut from each corner?

a. $V(x) = 4x(x - 3)(x - 7)$
b. $V(x) = 2x(x - 14)(x - 6)$
c. $V(x) = (x - 14)(x - 6)$
d. $V(x) = 4x(x + 7)(x + 3)$

______ **15.** What is the maximum volume of the box described by $V(x) = 4x(x - 2)(x - 6)$?

a. 67 **b.** 94 **c.** 20 **d.** 24

______ **16.** Soren is able to deposit \$500 at the end of each year in high school. How much money will he have by the end of his senior year if the money is invested at 4% interest?

a. \$2208.16 **b.** \$584.93 **c.** \$2080 **d.** \$2123.23

______ **17.** Cindy is able to deposit \$400 at the end of each year in high school. At what interest rate must she receive to cover her college freshman tuition costs of \$1750?

a. 8% **b.** 7% **c.** 6% **d.** 5%

NAME ______________________ CLASS __________ DATE __________

Chapter Assessment

Chapter 6, Form B, page 1

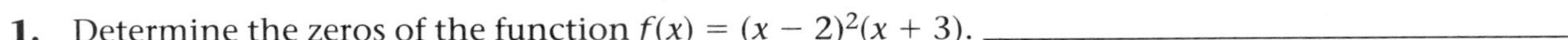

1. Determine the zeros of the function $f(x) = (x - 2)^2(x + 3)$. ______________

2. At which zero(s) does the graph of this function cross the x-axis? ______________

3. At which zero(s) does the graph touch the x-axis? ______________

4. For what domain is this function greater than zero? ______________

5. Find the maximum value of this function between the zeros. ______________

6. The perimeter of the base of a crate cannot exceed 10 ft. The height of the crate is 2 ft less than the width. Write a function for the volume of the crate in terms of the width.

____

Write each polynomial expression as a constant multiple of linear factors.

7. $3x^2 + 5x + 2$ ______________

8. $9x^2 - 16$ ______________

9. Write a quadratic function with zeros at 5 and -1.

____

10. The numbers -2 and -4 are zeros of the polynomial $f(x) = x^3 - x^2 - 34x - 56$. Find the other zeros(s).

____

11. Use any method to factor the polynomial $f(x) = x^3 + x^2 - 80x - 300$.

____

12. Find the common factor(s) of $f(x) = x^3 + 4x^2 - 3x - 18$ and $g(x) = x^4 - 13x^2 + 36$.

____

13. Write a quadratic polynomial with real zeros at 2 and -4 whose y-intercept is 24.

____

Chapter Assessment

Chapter 6, Form B, page 2

Let $f(x) = -(x - 3)(x - 1)(x + 2)^2$.

14. What is the degree of f? ______________________________

15. Which zeros, if any, are turning points? ______________________________

16. What crossing points if any, does f have? ______________________________

Let $f(x) = x^4 + 5x^3 - 24x^2 - 63x + 137$.

17. Use a graphics calculator to estimate all the zeros of f to the nearest tenth.

____

18. Where are the turning points? ______________________________

19. Where is f increasing? ______________________________

20. Where is f decreasing? ______________________________

A piece of cardboard measures 12 in. by 8 in.

21. Express the volume in terms of the length of the sides of the squares cut from each corner. ______________________________

22. What is the realistic domain of the volume function? ______________________________

23. What is the maximum volume of the box? ______________________________

Sue is able to deposit \$300, \$400, \$500, and \$600 at the end of her freshman, sophomore, junior, and senior years, respectively.

24. How much money will she have by the end of her senior year if the money is invested at 6% interest? ______________________________

25. If her tuition is \$1200 for the first two freshman semesters, how much will she have left over after she pays freshman college tuition? ______________________________

26. Joe plans to deposit the same amount of money at the end of each of his four years in high school. He will invest the money at 7% interest. He wants to have at least \$1500 for his first year in college. How much should he deposit each year?______________________________

NAME ______________________ CLASS __________ DATE __________

Alternative Assessment

Factored Form of a Polynomial, Chapter 6, Form A

TASK: To write the factored form of a polynomial function

HOW YOU WILL BE SCORED: As you work through the task, your teacher will be looking for the following:

- how well you can use the Factor Theorem to write a polynomial in factored form
- whether you can determine the number of complex zeros of a polynomial function
- how well you can use the division algorithm to write a polynomial in factored form with complex component factors

1. Is $x(x - 1)(x + 2)(x - 3)$ the factored form of $f(x) = x^3 - 2x^2 - 5x + 6$. Why or why not?

____

____

2. Describe how you can determine whether $x + 2$ is a factor of $f(x) = 2x^4 - x^3 - 7x^2 + 6x$.

____

____

Write each function in factored form.

3. $f(x) = x^2 - x + 20$ ______________

4. $f(x) = 81x^3 - 16x$ ______________

5. What is the relationship between the linear factors of a polynomial and the zeros of a polynomial?

____

____

6. The graph of the function $f(x) = x^3 - 3x^2 + x - 3$ intersects the x-axis at only one point. How many real zeros does f have? How many complex zeros does f have? What are the zeros of f? Graph f. Then use the division algorithm and the quadratic formula to write f in factored form with complex component factors.

____

____

SELF-ASSESSMENT: If the polynomial P is divided by $x - 2$, the degree of the quotient is 4 and the remainder is 3. What is the degree of P?

NAME ______________________ CLASS ____________ DATE ____________

Alternative Assessment

Exploring Polynomial Function Behavior, Chapter 6, Form B

TASK: To determine the basic shape of polynomial functions of varying degrees

HOW YOU WILL BE SCORED: As you work through the task, your teacher will be looking for the following:

- whether you can determine and classify the behavior of polynomials of varying degree
- how well you can describe the behavior of polynomials of varying degree

Compare the graphs of $f(x) = x^2(x + 2)$ and $g(x) = -x^2(x + 2)$.

1. What are the zeros and the degree of f? What are the zeros and degree of g?

____

____

2. Where is f increasing? Where is f decreasing? What are the turning points? Where is g increasing? Where is g decreasing? What are the turning points?

____

____

3. Describe the effect of the sign of the leading coefficient of g on the shape of the graph of f.

____

Let $f(x) = x^4 - x^3 - 7x^2 + x + 6$.

4. Use the zeros of f to factor f into its linear factors.

____

5. Describe what happens at a turning point. Does f have any zeros that are turning points?

____

6. Describe what happens at a crossing point. How many crossing points does f have?

____

SELF-ASSESSMENT: Write an example of a polynomial in factored form that has degree 3, two crossing point zeros and no turning point zeros.

ANSWERS

Practice & Apply—Chapter 4

Lesson 4.1

1. 3×2 **2.** 2×2

3. $\begin{bmatrix} 11 & -3 \\ 6 & 2 \\ 5 & -5 \end{bmatrix}$ **4.** $\begin{bmatrix} -1 & -1 \\ -8 & -2 \\ 3 & -1 \end{bmatrix}$

5. $\begin{bmatrix} 0 & 7 \\ -6 & 9 \end{bmatrix}$ **6.** $\begin{bmatrix} -2 & 7 \\ 2 & -3 \end{bmatrix}$

7. $\begin{bmatrix} 2x - 3y \\ -2x + 5y \end{bmatrix} = \begin{bmatrix} 6 \\ 11 \end{bmatrix}$

8. $\begin{bmatrix} -3x + 5y \\ 2x + y \\ 6x - y \end{bmatrix} = \begin{bmatrix} 8 \\ -4 \\ 1 \end{bmatrix}$

9. 3×3

10. The amount of subcompact sales for Jose is \$12,000.

11. The amount of compact sales for Joe is \$42,000.

12. The amount of luxury sales for Al is \$111,000.

Lesson 4.2

1. 0 **2.** −1 **3.** 22 **4.** yes **5.** no

6. yes

7. $\begin{bmatrix} 34 & 12 \\ 36 & -12 \\ 16 & 18 \end{bmatrix}$ **8.** $\begin{bmatrix} -4 & 26 \\ -12 & -7 \end{bmatrix}$

9. $\begin{bmatrix} 14 \\ -12 \end{bmatrix}$ **10.** $\begin{bmatrix} -2 & -4 & -6 \\ -3 & -6 & -9 \\ 1 & 2 & 3 \end{bmatrix}$

11. $\begin{bmatrix} 9 & 17 & 15 \\ 89 & 10 & -27 \\ -22 & -22 & -26 \end{bmatrix}$

12. $\begin{bmatrix} -17 & 0.7 & -0.1 \\ 26 & 3.6 & -1.7 \end{bmatrix}$

13. \$11,531 **14.** \$4480 **15.** \$16,011

Lesson 4.3

1. Independent **2.** Inconsistent

3. Dependent **4.** Inconsistent

5. Dependent **6.** Independent **7.** (7, 5)

8. (−3, 1) **9.** (−2, −5) **10.** $\left(4, -\frac{3}{2}\right)$

11. (4, −1) **12.** $\left(-\frac{3}{2}, 3\right)$ **13.** (−7, 8)

14. (6, −4) **15.** (−2, 12)

16. $\begin{cases} x + y = 35 \\ 0.23x + 0.29y = 9.25 \end{cases}$

17. 15 pencils at 23¢; 20 pencils at 29¢

18. $\begin{cases} 2l + 2w = 180 \\ 4l = 5w \end{cases}$

19. $l = 50$ ft, $w = 40$ ft

20. 200 ft^2

Lesson 4.4

1. $\left[\begin{array}{cc|c} 2 & 1 & 2 \\ 5 & -3 & -17 \end{array}\right]$ **2.** $\left[\begin{array}{cc|c} 3 & -2 & 4 \\ 1 & -4 & 15 \end{array}\right]$

3. $\left[\begin{array}{ccc|c} 2 & 1 & 1 & 1 \\ 1 & 3 & -4 & 19 \\ 4 & -2 & 3 & -9 \end{array}\right]$

4. $\left[\begin{array}{ccc|c} 3 & -1 & 2 & 3 \\ 2 & 5 & -3 & -12 \\ 1 & -3 & 4 & 8 \end{array}\right]$

5. (−2, 3)

6. $\left(4, \frac{2}{3}\right)$ **7.** (2, −1, 4) **8.** (5, 0, −2)

9. (4, 1) **10.** (−2, 7) **11.** $\left(\frac{17}{2}, 21\right)$

12. (−2, 1, 4) **13.** (3, 4, −5) **14.** (2, −2, 7)

15. $\begin{cases} x + 2y + 3z = 185 \\ 2x + y + 2z = 150 \\ 3x + 2y + z = 195 \end{cases}$

ANSWERS

16. apple: 30 calories
banana: 40 calories
orange: 25 calories

Lesson 4.5

1. yes 2. no 3. no 4. yes

5. $\begin{bmatrix} -0.14 & -0.29 \\ -0.29 & 0.43 \end{bmatrix}$ 6. $\begin{bmatrix} 1.00 & 1.67 \\ 2.00 & 3.00 \end{bmatrix}$

7. $\begin{bmatrix} -1 & 3 \\ 2 & -5 \end{bmatrix}$ 8. $\begin{bmatrix} 0.09 & 0.18 \\ -0.36 & 0.27 \end{bmatrix}$

9. Does not exist.

10. $\begin{bmatrix} -1.33 & -1.67 \\ -0.33 & -0.67 \end{bmatrix}$

11. $A + 2B + 3C = 125$
$2A + 3B + 4C = 195$
$A + 2B + C = 95$

12. $\begin{bmatrix} 1 & 2 & 3 \\ 2 & 3 & 4 \\ 1 & 2 & 1 \end{bmatrix}$ 13. $\begin{bmatrix} -2.5 & 2 & -0.5 \\ 1 & -1 & 1 \\ 0.5 & 0 & -0.5 \end{bmatrix}$

Lesson 4.6

1. $\begin{bmatrix} 1 & -3 \\ 2 & 9 \end{bmatrix} \begin{bmatrix} x \\ y \end{bmatrix} = \begin{bmatrix} 3 \\ 11 \end{bmatrix}$

2. $\begin{bmatrix} 3 & 1 \\ 2 & -1 \end{bmatrix} \begin{bmatrix} x \\ y \end{bmatrix} = \begin{bmatrix} 13 \\ 2 \end{bmatrix}$

3. $\begin{bmatrix} 3 & 2 & -1 \\ 1 & -4 & 1 \\ 5 & 0 & -3 \end{bmatrix} \begin{bmatrix} x \\ y \\ z \end{bmatrix} = \begin{bmatrix} 7 \\ -9 \\ -1 \end{bmatrix}$

4. $\begin{bmatrix} 2 & -1 & -1 \\ 5 & 2 & 3 \\ 1 & -3 & -7 \end{bmatrix} \begin{bmatrix} x \\ y \\ z \end{bmatrix} = \begin{bmatrix} -8 \\ 2 \\ -30 \end{bmatrix}$

5. $(-5, 3)$ 6. $(2, -3)$ 7. $(2, -1)$

8. $(3, -2, 1)$ 9. $(2, -1, 4)$ 10. $(0, 5, -4)$

11. a. $\begin{bmatrix} 1 & 1 & 0 \\ 0 & 2 & -1 \\ 3 & 0 & -2 \end{bmatrix} \begin{bmatrix} x \\ y \\ z \end{bmatrix} = \begin{bmatrix} -1 \\ -3 \\ 5 \end{bmatrix}$

b. $x = 1, y = -2, z = -1$

12. a. $\begin{bmatrix} 1 & 0 & -2 \\ 0 & 3 & 1 \\ -2 & 1 & 0 \end{bmatrix} \begin{bmatrix} x \\ y \\ z \end{bmatrix} = \begin{bmatrix} 0 \\ 2 \\ 5 \end{bmatrix}$

b. $x = -2, y = 1, z = -1$

13. a. $\begin{bmatrix} 2 & -3 & 1 & -1 \\ 4 & 3 & 2 & 1 \\ 1 & 2 & -3 & -1 \\ 1 & -1 & -1 & 1 \end{bmatrix} \begin{bmatrix} x \\ y \\ z \\ w \end{bmatrix} = \begin{bmatrix} 3 \\ -6 \\ 4 \\ 8 \end{bmatrix}$

b. $x = 1, y = -2, z = -3, w = 2$

14. a. $\begin{bmatrix} 2 & 1 & 2 & 0 \\ 1 & -2 & 3 & -4 \\ 3 & 1 & -1 & 2 \\ -1 & -3 & 0 & 3 \end{bmatrix} \begin{bmatrix} x \\ y \\ z \\ w \end{bmatrix} = \begin{bmatrix} -5 \\ -13 \\ 2 \\ 1 \end{bmatrix}$

b. $x = -1, y = 1, z = -2, w = 1$

Lesson 4.7

1. parallelogram

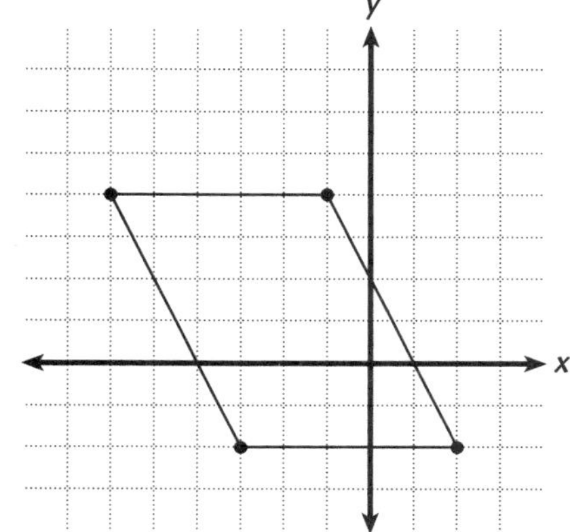

2. isosceles triangle

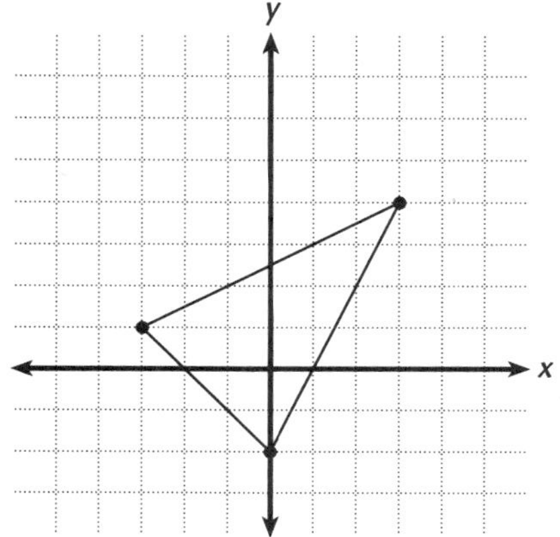

3. $\begin{bmatrix} 1 & 6 & 3 & -2 \\ -4 & -4 & 2 & 2 \end{bmatrix}$; $\begin{bmatrix} -1 & 0 \\ 0 & -1 \end{bmatrix}$

4. $\begin{bmatrix} -4 & -1 & 2 \\ 3 & -3 & 0 \end{bmatrix}$; $\begin{bmatrix} 0 & -1 \\ 1 & 0 \end{bmatrix}$

5. $\begin{bmatrix} -3 & -18 & -9 & 6 \\ 12 & 12 & -6 & -6 \end{bmatrix}$; $\begin{bmatrix} 3 & 0 \\ 0 & 3 \end{bmatrix}$

6. $\begin{bmatrix} 1.5 & -1.5 & 0 \\ 2 & 0.5 & -1 \end{bmatrix}$; $\begin{bmatrix} 0.5 & 0 \\ 0 & 0.5 \end{bmatrix}$

7.

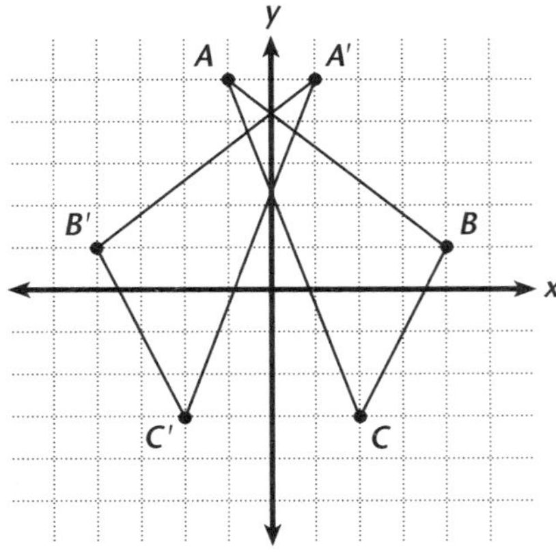

8. The resulting image, $\Delta A'B'C'$, is the reflected image about the y-axis.

Lesson 4.8

1.

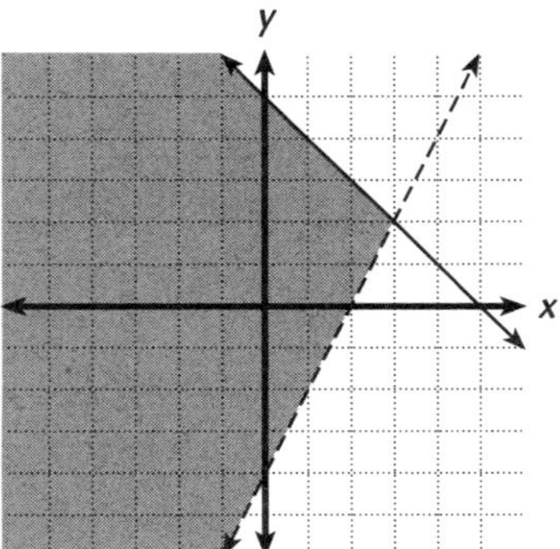

2.

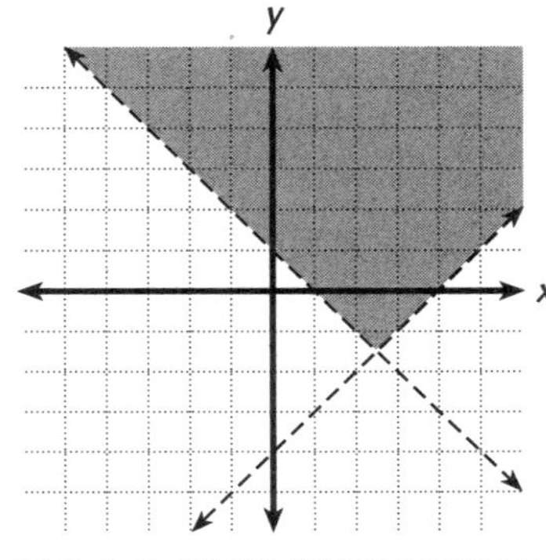

3.

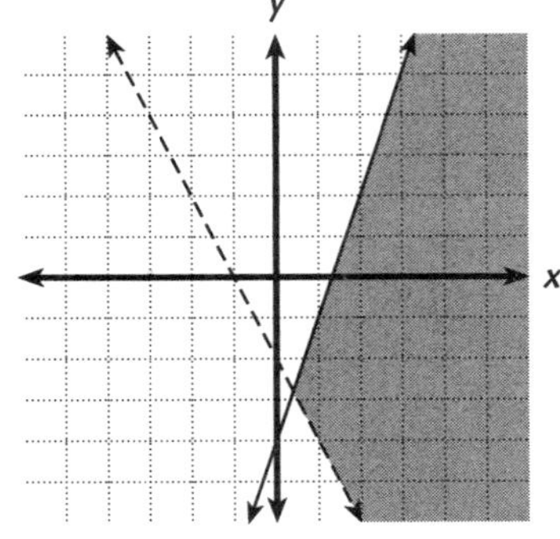

4.

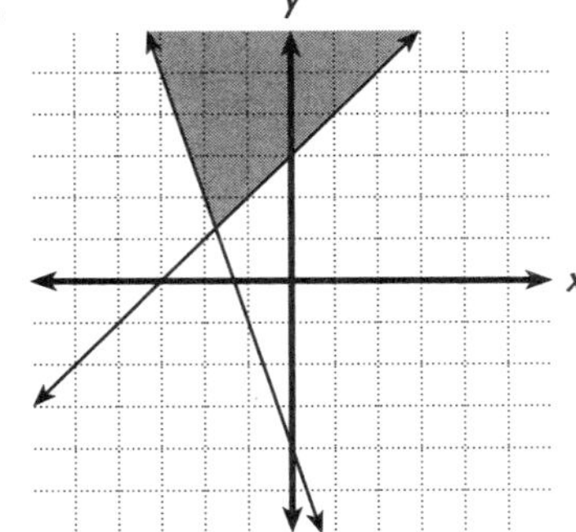

5.

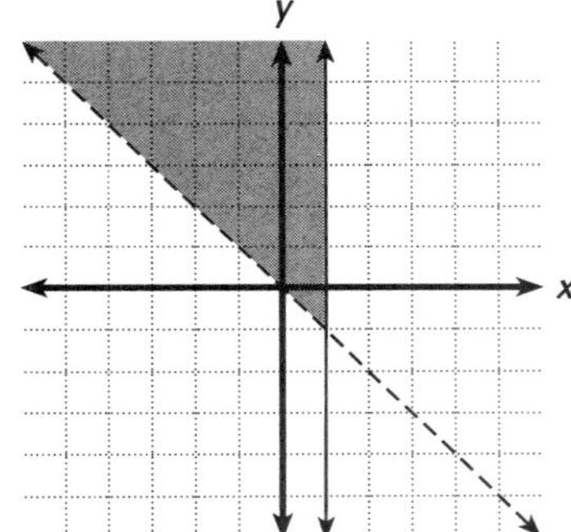

6.

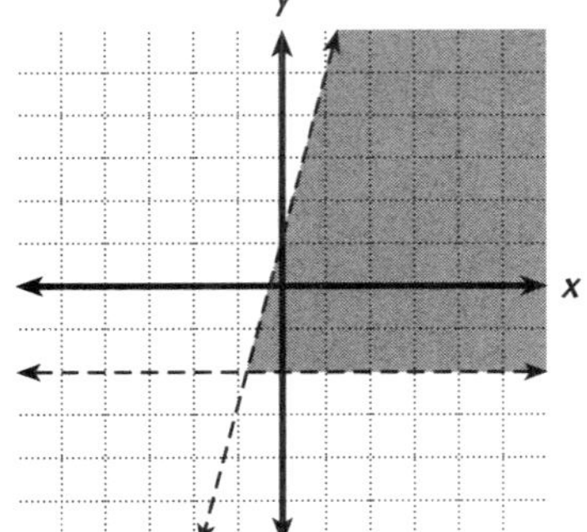

7.

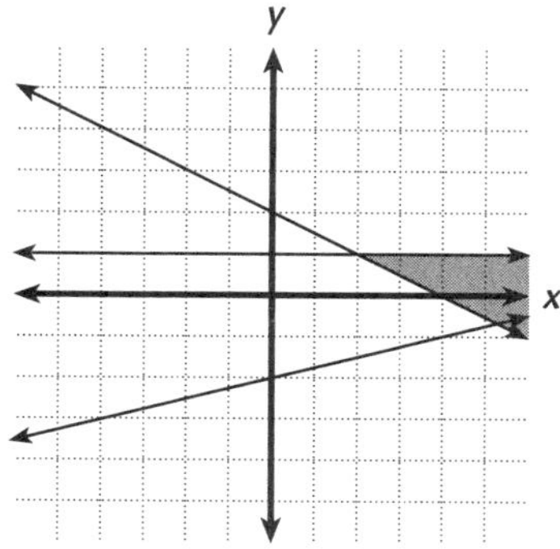

8.

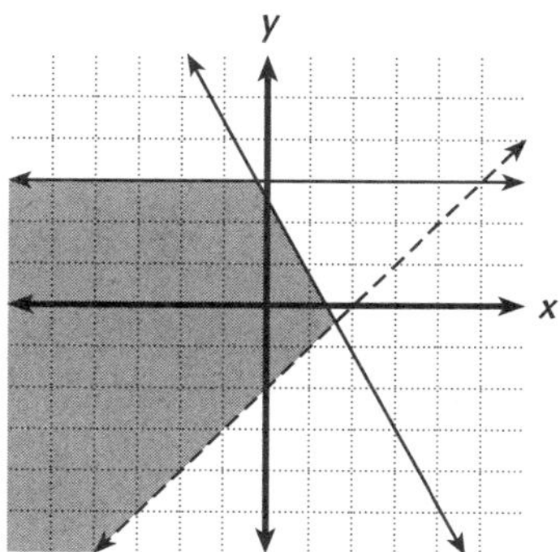

9.

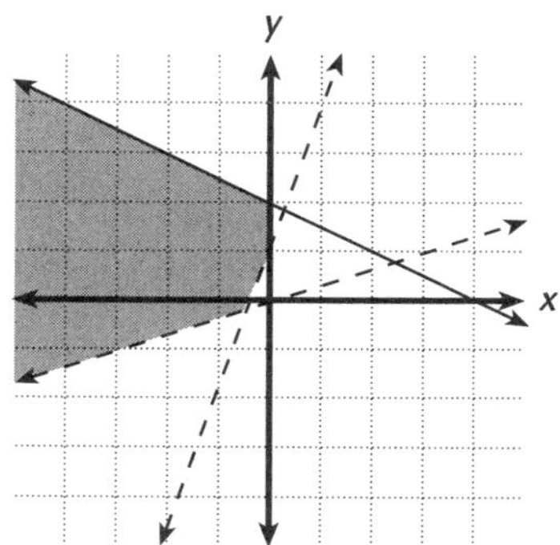

Lesson 4.9

1.

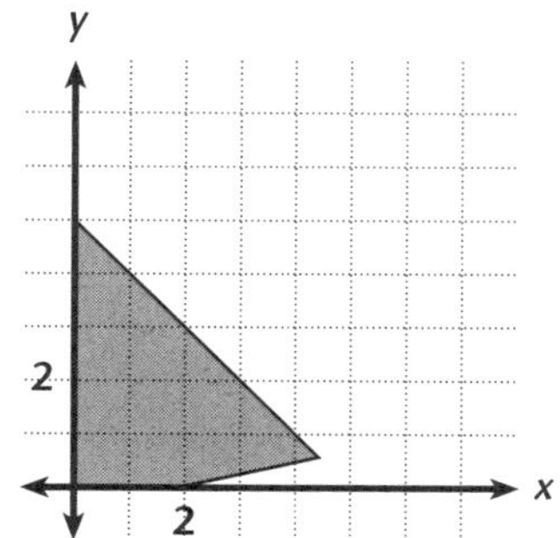

Points and value may vary.

2.

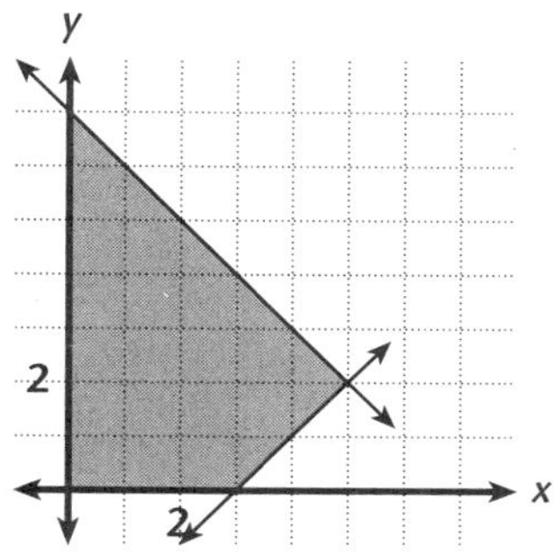

Points and value may vary.

3.

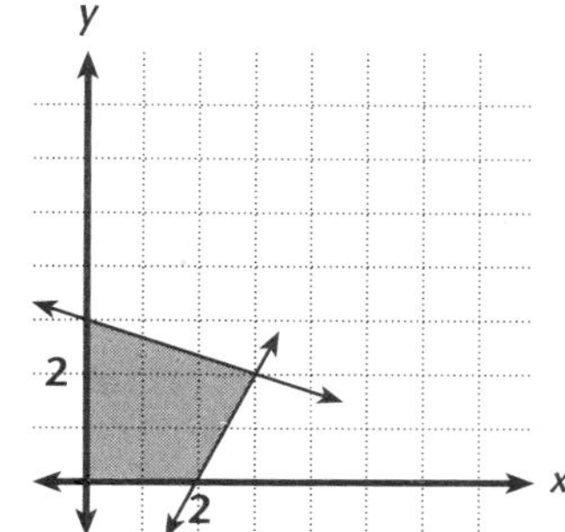

Points and value may vary.

4.

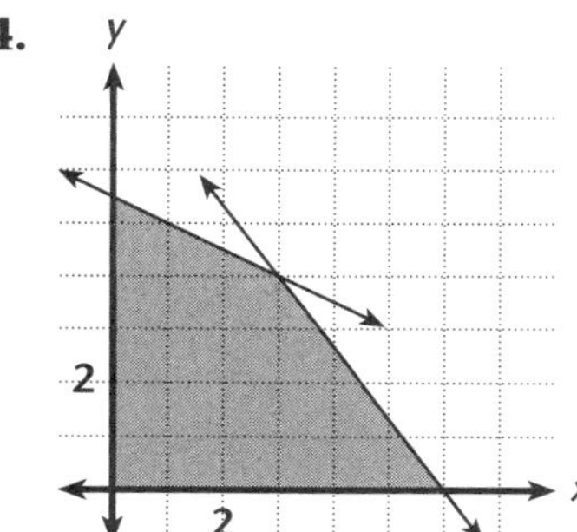

Points and value may vary.

5.

	stage one (h)	stage two (h)	total hours
scientific	2	5	80
graphic	4	2	80

Let x represent the number of scientific calculators and let y represent the number of graphics calculators.

$$\begin{cases} 2x + 4y \leq 80 \\ 5x + 2y \leq 80 \\ x \geq 0 \\ y \geq 0 \end{cases}$$

$P = 25x + 40y$

6.

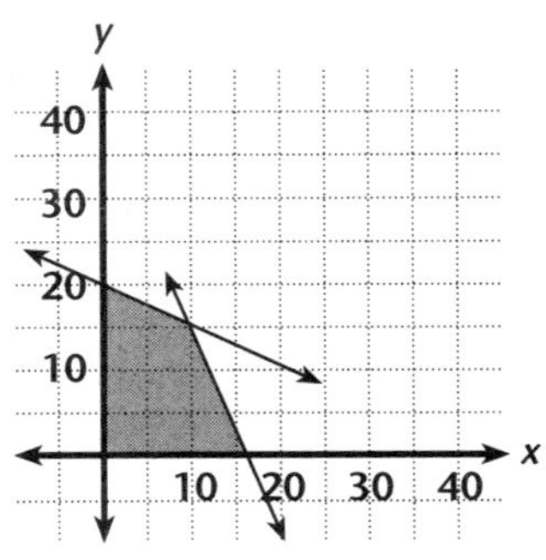

7. Answers may vary.

Lesson 4.10

1. (*a*) and (*b*)

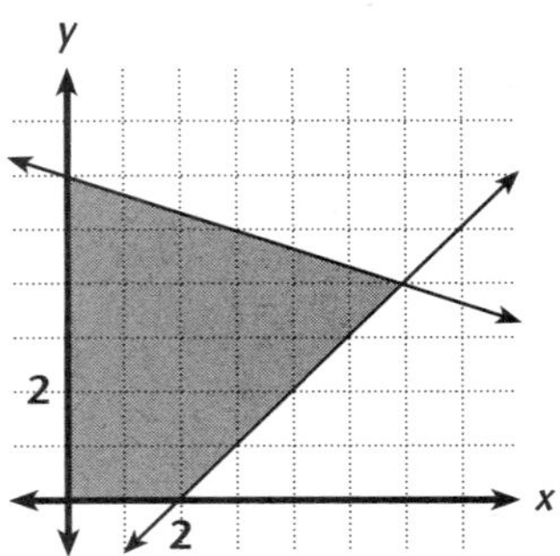

(c) (0, 0), (2, 0), (6, 4), (0, 6)

(d) 0 at (0, 0)

(e) 30 at (0, 6)

2. (*a*) and (*b*)

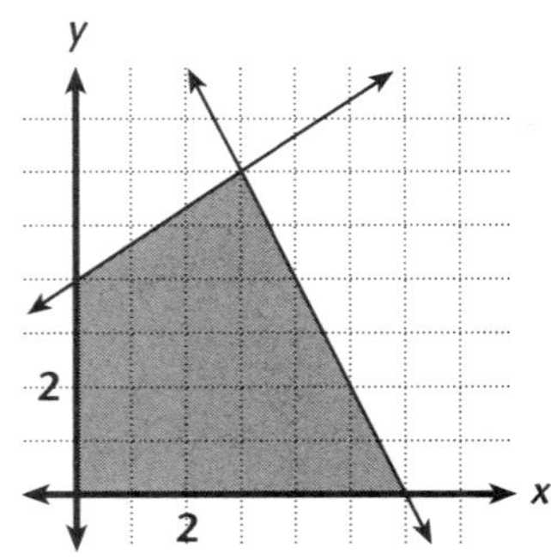

(c) (0, 0), (6, 0), (3, 6), (0, 4)

(d) −4 at (0, 4)

(e) 24 at (6, 0)

3. (*a*) and (*b*)

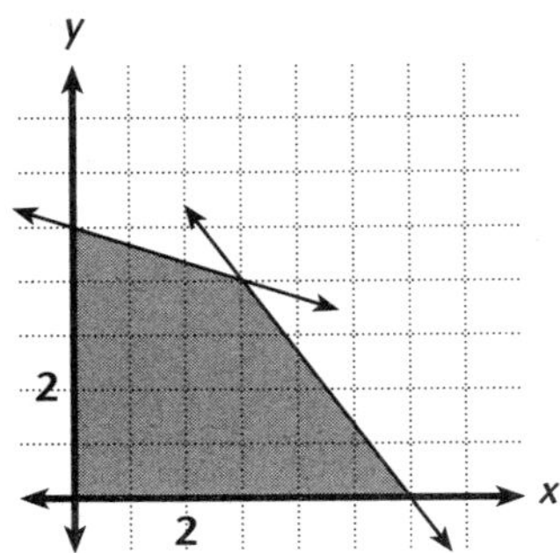

(c) (0, 0), (6, 0) (3, 4), (0, 5)

(d) 0 at (0, 0)

(e) 18 at (3, 4)

4. (*a*) and (*b*)

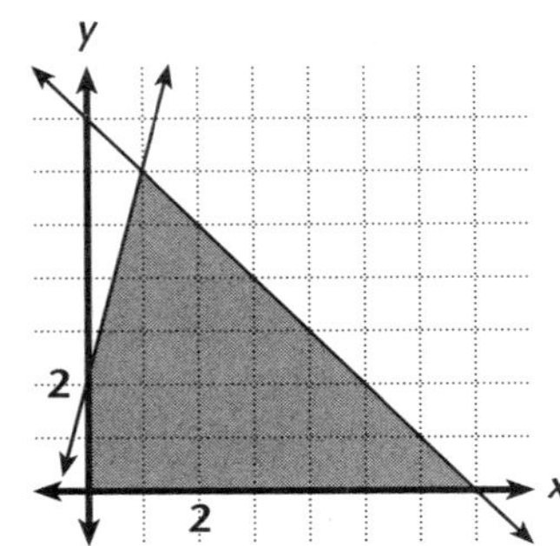

(c) (0, 0), (7, 0), (1,6), (0, 2)

(d) −6 at (1, 6)

(e) 42 at (7, 0)

5.

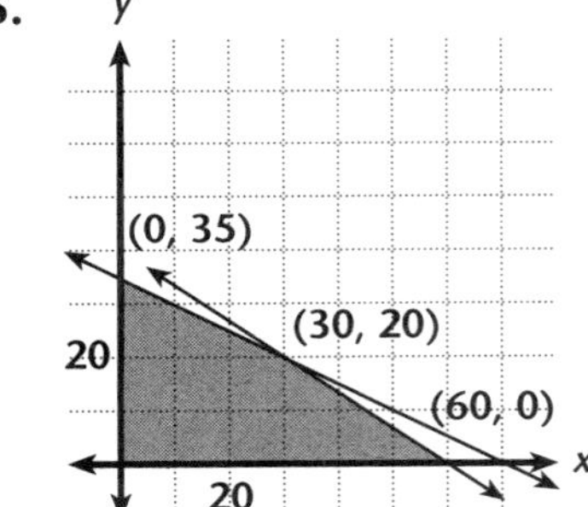

$x \geq 0$; $y \geq 0$; $10x + 15y \leq 600$;
$15x + 30y \leq 1050$

6. $P = 3x + 5y$

7. (0, 0), (0, 35), (30, 20), (60, 0)

8. $190 at (30, 20)

Enrichment — Chapter 4

Lesson 4.1

1. $\begin{bmatrix} 31 & 72 \\ 35 & 55 \end{bmatrix}$

2. cannot be added; matrices do not have the same dimension

3. $\begin{bmatrix} 104 & -36 & 69 \\ -16 & 130 & 44 \end{bmatrix}$

4. $\begin{bmatrix} 7.66 & 9.79 & 11.39 \\ 11.28 & 10.47 & 8.1 \\ 13.67 & 16.33 & 11.34 \end{bmatrix}$

5. $\begin{bmatrix} 10\frac{1}{6} & 13\frac{7}{12} \\ 11\frac{7}{30} & 13\frac{3}{20} \end{bmatrix}$

6. $\begin{bmatrix} 14\frac{10}{21} & 1\frac{3}{5} \\ -2\frac{2}{35} & 1\frac{1}{24} \\ 8\frac{5}{24} & -5\frac{7}{24} \end{bmatrix}$

7. $\begin{bmatrix} 6.5 & 7.08 & 14.77 \\ 7.33 & 10.12 & 10.16 \end{bmatrix}$

8. $\begin{bmatrix} 265 & 71 & 11 \\ 45 & 890 & 710 \\ 265 & 1055 & -327 \end{bmatrix}$

Lesson 4.2

1. 3 **2.** 1 **3.** 4 **4.** 1 **5.** 5 **6.** 9 **7.** 2

8. 6 **9.** 5 **10.** 4

11. The number gives the value of π to 9 decimal places.

Lesson 4.3

1. $(-\frac{1}{3}, 1, -\frac{4}{3})$ **2.** (2, 4, −3) **3.** (−1, 7, 3)

4. (2, 1, 4) **5.** (5, 7, −3) **6.** (4, −6, 9)

7. (3.5, 4, 5.25) **8.** (0.3, 0.1, −0.2)

9. (1.7, 3.2, 4.6)

Lesson 4.4

1. (1, −1, 2) **2.** (4, −1, 2) **3.** (1, −1, 3)

4. (1, 3, −2) **5.** (4, 2, 0) **6.** (0, 4, 1)

7. (1, 1, 1) **8.** (1, −3, 2) **9.** (2, 3, −1)

10. (3, 4, −1) **11.** (4, −1, 2) **12.** (1, 4, 2)

13. (1, −4, 6) **14.** (−3, 2, 5) **15.** (2, −4, 6)

Lesson 4.5

1. $\begin{bmatrix} \frac{3}{5} & \frac{1}{5} \\ -\frac{2}{5} & \frac{1}{5} \end{bmatrix}$

2. $\begin{bmatrix} 3 & -5 \\ -1 & 2 \end{bmatrix}$

3. $\begin{bmatrix} \frac{5}{9} & \frac{2}{9} \\ -\frac{2}{9} & \frac{1}{9} \end{bmatrix}$

4. $\begin{bmatrix} \frac{1}{7} & \frac{1}{14} \\ -\frac{2}{7} & \frac{5}{14} \end{bmatrix}$

5. $\begin{bmatrix} -\frac{1}{13} & \frac{2}{13} \\ \frac{5}{13} & \frac{3}{13} \end{bmatrix}$

6. $\begin{bmatrix} -\frac{1}{4} & \frac{5}{4} \\ \frac{1}{2} & -\frac{3}{2} \end{bmatrix}$

7. $\begin{bmatrix} 1 & 0 \\ \frac{3}{4} & \frac{1}{4} \end{bmatrix}$

8. $\begin{bmatrix} \frac{2}{15} & \frac{1}{6} \\ -\frac{1}{15} & \frac{1}{6} \end{bmatrix}$

9. $\begin{bmatrix} 0 & 1 \\ \frac{1}{6} & \frac{1}{6} \end{bmatrix}$

Lesson 4.6

1. $a = 1, b = 2$ **2.** $a = 0, b = 1$

3. $a = -1, b = 1$ **4.** $a = 2, b = -1$

5. $a = 1, b = -2$ **6.** $a = 1, b = 1$

7. $a = 4, b = -2, c = 4$

8. $a = -2, b = 1, c = 2$

9. $a = 1, b = 0, c = 0$ **10.** $a = 1, b = 2, c = 3$

Lesson 4.7

1. $\begin{bmatrix} -2 & 1 & 0 \\ 4 & 3 & 2 \end{bmatrix}; \begin{bmatrix} 2 & -1 & 0 \\ -4 & -3 & -2 \end{bmatrix}$

2. $\begin{bmatrix} -5 & -3 & 1 \\ -1 & 2 & 4 \end{bmatrix}; \begin{bmatrix} 5 & 3 & -1 \\ 1 & -2 & -4 \end{bmatrix}$

3. $\begin{bmatrix} -3 & -1 & -6 \\ 4 & 5 & 2 \end{bmatrix}; \begin{bmatrix} 3 & 1 & 6 \\ -4 & -5 & -2 \end{bmatrix}$

4. $\begin{bmatrix} -5 & 0 & -6 \\ 2 & 3 & -4 \end{bmatrix}; \begin{bmatrix} 5 & 0 & 6 \\ -2 & -3 & 4 \end{bmatrix}$

5. $\begin{bmatrix} -5 & 1 & -2 \\ 7 & -5 & 3 \end{bmatrix}; \begin{bmatrix} 5 & -1 & 2 \\ -7 & 5 & -3 \end{bmatrix}$

6. $\begin{bmatrix} 0 & 1 & -2 \\ 6 & 4 & -5 \end{bmatrix}; \begin{bmatrix} 0 & -1 & 2 \\ -6 & -4 & 5 \end{bmatrix}$

Lesson 4.8

1.

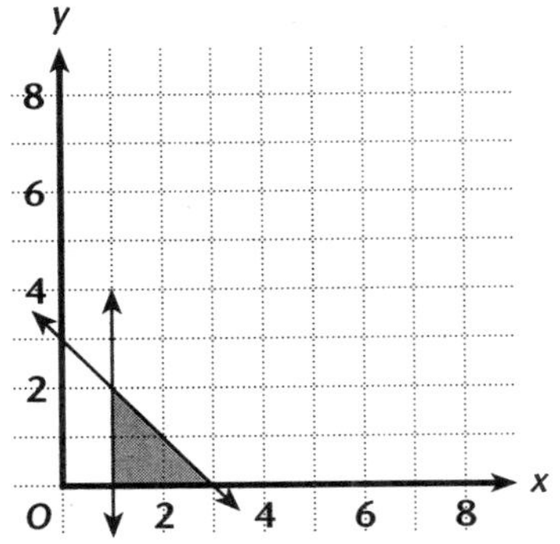

(1,0), (3,0), (1,2)

2.

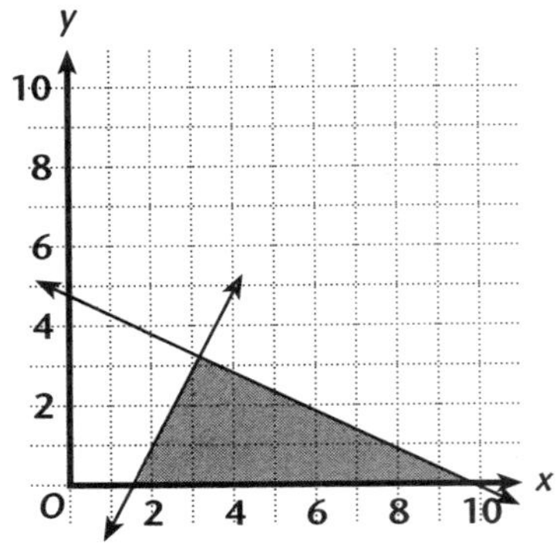

$\left(\frac{3}{2}, 0\right)$, (10, 0), $\left(\frac{16}{5}, \frac{17}{5}\right)$

3.

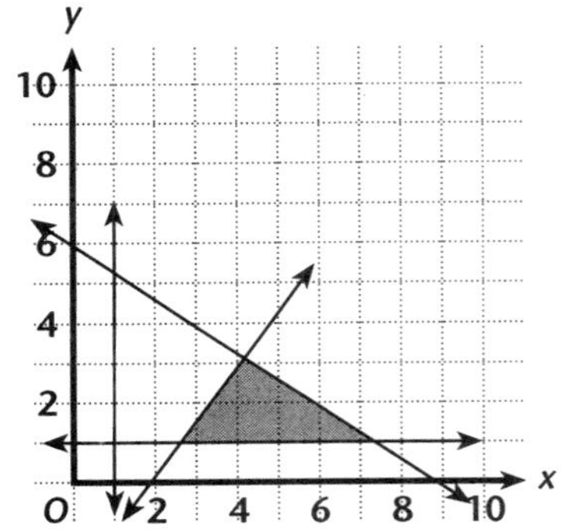

$\left(\frac{8}{3}, 1\right), \left(\frac{15}{2}, 1\right), \left(\frac{54}{13}, \frac{42}{13}\right)$

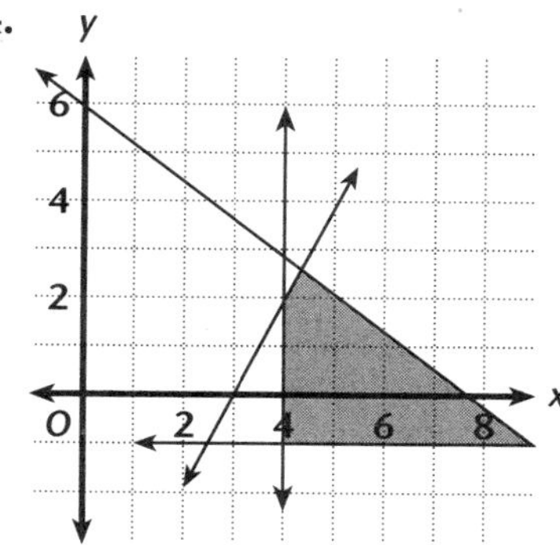

$(4, -1)$, $(4, 2)$, $\left(\frac{48}{11}, \frac{30}{11}\right)$, $\left(\frac{28}{3}, -1\right)$

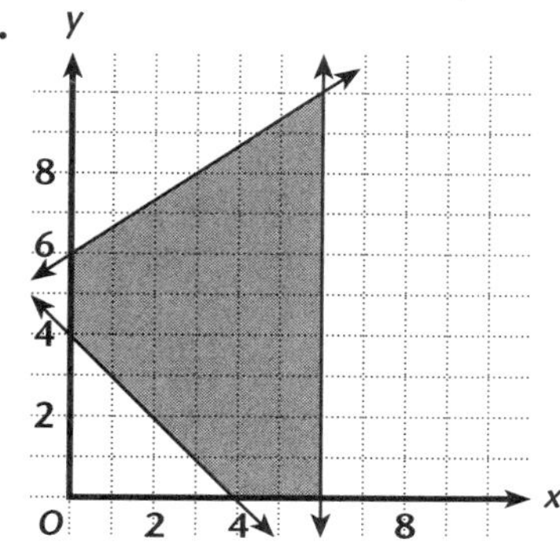

$(0, 4)$, $(0, 6)$, $(6, 10)$, $(6, 0)$, $(4, 0)$

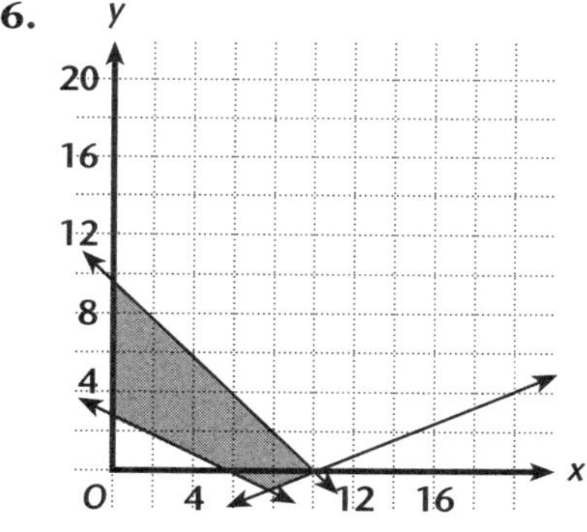

$(0, 10)$, $(0, 3)$, $(10, 0)$, $\left(\frac{70}{9}, -\frac{8}{9}\right)$

Lesson 4.9

1. $5x + 3y \le 480$, $2x + 5y \le 480$, $x \ge 30$, $y \ge 50$, $P = 10x + 8y$; $(30, 50)$, $(66, 50)$, $\left(50\frac{10}{19}, 75\frac{15}{19}\right)$, $(30, 84)$; \$1111.58

2. $10x + 2y \le 500$, $5x + 3y \le 200$, $x \ge 0$, $y \ge 0$, $P = 5x + 8y$; $(0, 0)$, $(40, 0)$, $\left(0, \frac{200}{3}\right)$; \$533.33

3. $20x + 10y \le 480$, $5x + 25y \le 480$, $x \ge 10$, $y \ge 5$, $C = 5x + 15y$; $(10, 5)$, $\left(21\frac{1}{2}, 5\right)$, $(16, 16)$, $\left(10, 17\frac{1}{5}\right)$; \$125

Lesson 4.10

1. $(0, 0)$, $(10, 0)$, $(8, 2)$, $\left(\frac{8}{3}, \frac{14}{3}\right)$, $(0,2)$; 22

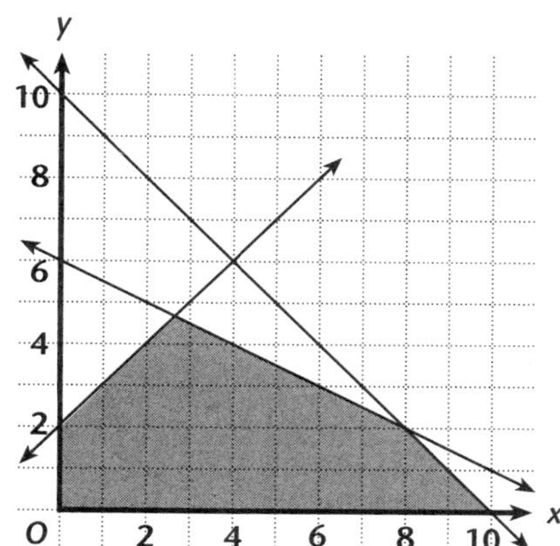

2. $(0, 0)$, $(6, 0)$, $(9, 3)$, $\left(\frac{16}{3}, \frac{20}{3}\right)$, $(0, 10)$; 24

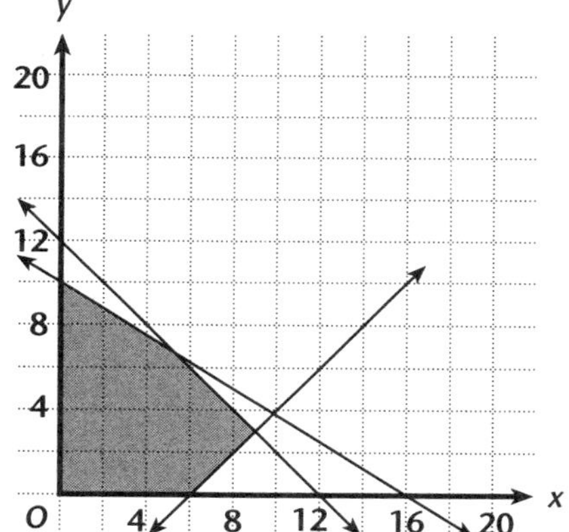

3. $(0, 0)$, $(8, 0)$, $(8, 4)$, $(4, 8)$, $\left(\frac{8}{3}, \frac{26}{3}\right)$, $(0, 6)$; 52

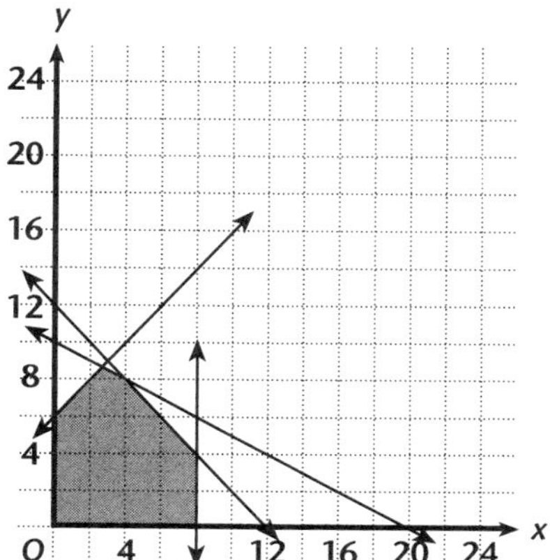

4. $\left(\frac{30}{7}, \frac{24}{7}\right)$, (10, 0), (12, 0), (12, 4), (4, 12), (0, 12); 60

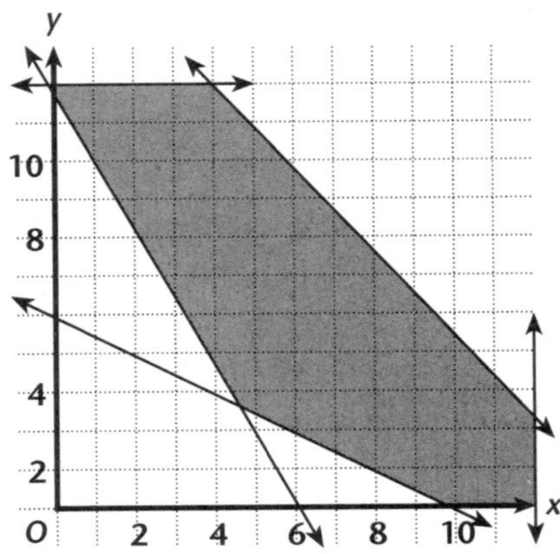

Technology — Chapter 4

Lesson 4.1

1. For I1: A1+E1; For J1: B1+F1; For K1: C1+G1; I2: A2+E2; For J2: B2+F2; For K2: C2+G2; For I3: A3+E3; For J3: B3+F3; For K3: C3+G3

2. For K5: A5−F5; For L5: B5−G5; For M5: C5−H5; For N5: D5−I5; For K6: A6−F6; For L6: B6−G6; For M6: C6−H6; For N6: D6−I6; For K7: A7−F7; For L7: B7−G7; For M7: C7−H7; For N7: D7−I7

3. $\begin{bmatrix} 15 & -4 & 3 \\ 7 & -5 & -1 \\ 6 & 15 & -3 \end{bmatrix}$

4. $\begin{bmatrix} 2 & -6 & -2 & -10 \\ -4 & -2 & -8 & -6 \\ 4 & 5 & 5 & 6 \end{bmatrix}$

5. $\begin{bmatrix} -5 & -4 & 3 \\ -1 & 5 & -3 \\ -6 & 3 & -15 \end{bmatrix}$

6. $\begin{bmatrix} 4 & 0 & 8 & 4 \\ 0 & 6 & 4 & 10 \\ 6 & 5 & 7 & 6 \end{bmatrix}$

7. $\begin{bmatrix} 0 & -2 & -4 \\ 6 & 12 & -16 \\ 20 & 16 & -8 \end{bmatrix}$

8. $\begin{bmatrix} 0 & -3 & -6 \\ 9 & 18 & -24 \\ 30 & 24 & -12 \end{bmatrix}$; $\begin{bmatrix} 0 & -4 & -8 \\ 12 & 24 & -32 \\ 40 & 32 & -16 \end{bmatrix}$

9. Multiply each element of W by 10.

10. Multiply each element of W by −4.

Lesson 4.2

1. A1*E1+B1*E2+C1*E3

2. A2*F1+B2*F2+C2*F3

3. $\begin{bmatrix} 7 & 29 \\ 19 & 65 \end{bmatrix}$

4. $\begin{bmatrix} 8 & 10 & 12 \\ 17 & 25 & 33 \\ 27 & 33 & 39 \end{bmatrix}$

5. Answers will vary. Possible answer:

$$Y = \begin{bmatrix} -1 & 0 \\ 1 & 0 \\ 0 & 0 \end{bmatrix}$$

6. Answers will vary. Possible answer:

$$N = \begin{bmatrix} -18.5 & 1 & 5 \\ -19 & 1 & 5 \end{bmatrix}$$

7. Answers will vary. Possible answer:

$$Y = \begin{bmatrix} 0 & -1 \\ 0 & 2 \\ 0 & -1 \end{bmatrix}$$

8. Answers will vary. Possible answer:

$$N = \begin{bmatrix} 0 & 0 & 0 \\ -19 & 1 & 5 \end{bmatrix}$$

Lesson 4.3

1. $x = 500$; $y = 1500$

2. $x = 0$; $y = 2000$

3. $x = 2250$; $y = 750$

4. $x = 1250$; $y = 1750$

5. $x = 4000$; $y = 0$

6. $x = 1500$; $y = 2500$

7. $x = 2500$; $y = 3500$

8. $x = 312.50$; $y = 4687.50$

9. The graphs of $x + y = 2000$ and $0.09x + 0.05y = 120$ intersect at one point.

Lesson 4.4

1. $x = 2, y = 3, z = -1$

2. $x \approx 0.6667, y \approx 1.4762, z \approx 0.0476$

3. $x = -2, y = 4, z = 5$

4. $x = 1, y = 2, z = 1$

5. $x = -3, y = 5, z = 7$

6. $x = 2, y = 0.5, z = -2$

Lesson 4.5

1. 1; $\begin{bmatrix} 1 & -2 \\ -5 & 11 \end{bmatrix}$

2. 1; $\begin{bmatrix} 4 & -9 \\ -3 & 7 \end{bmatrix}$

3. 1; $\begin{bmatrix} 2 & -5 \\ 7 & -17 \end{bmatrix}$

4. 1; $\begin{bmatrix} -1 & -2 \\ 1 & 1 \end{bmatrix}$

5. 1; $\begin{bmatrix} 10 & -9 \\ -11 & 10 \end{bmatrix}$

6. 1; $\begin{bmatrix} -20 & 13 \\ 3 & -2 \end{bmatrix}$

7. $ad - bc = 1$

8. Switch a and d. Change the sign of b and c.

9. $\begin{bmatrix} a & -a - 1 \\ 1 - a & a \end{bmatrix}$

10. $\begin{bmatrix} a + 1 & -1 \\ -a(a + 2) & a + 1 \end{bmatrix}$

Lesson 4.6

1. $x = 4, y = -5$ 2. $x = -2, y = 7$

3. $x = 1, y = 1$ 4. $x = 3, y = 4$

5. $x = 4, y = 5$ 6. $x = 5, y = 6$

7. $x = 2, y = -3$ 8. $x = 1, y = -4$

9. $x = 0, y = -5$ 10. $x = 10, y = 11$

11. $x = 12, y = 9$ 12. $x = 14, y = 7$

Lesson 4.7

1. $r + s = 2$; $t + u = 1$

2. $r + 4s = 5$; $t + 4u = 4$

3. $5r + 4s = 9$; $5t + 4u = 4$

4. $5r + s = 6$; $5t + u = 1$

5. Answers will vary. Students should give two independent equations involving r and s and two independent equations involving t and u.

6. $\begin{bmatrix} 1 & 1 \\ 0 & 1 \end{bmatrix}$

7. $M = \begin{bmatrix} 1 & 0 \\ 1 & 1 \end{bmatrix}$

8. $M = \begin{bmatrix} 1 & 0 \\ -1 & 1 \end{bmatrix}$

Lesson 4.8

1.

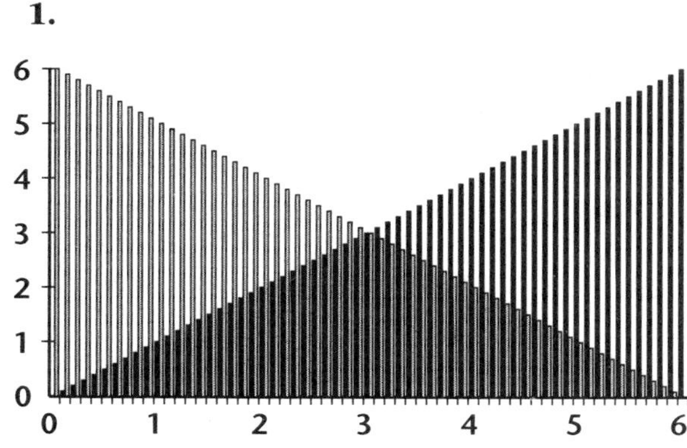

2.

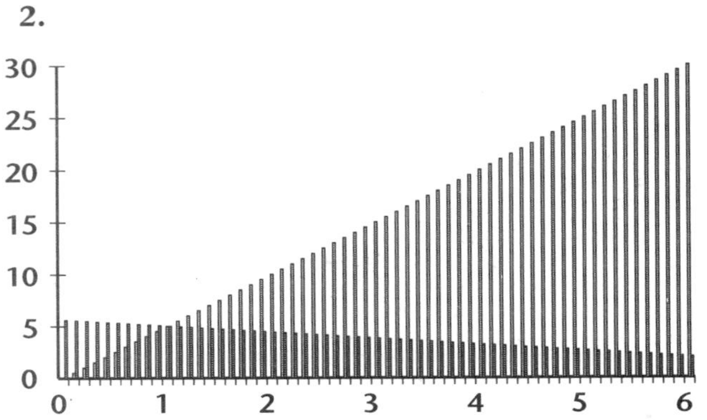

3.

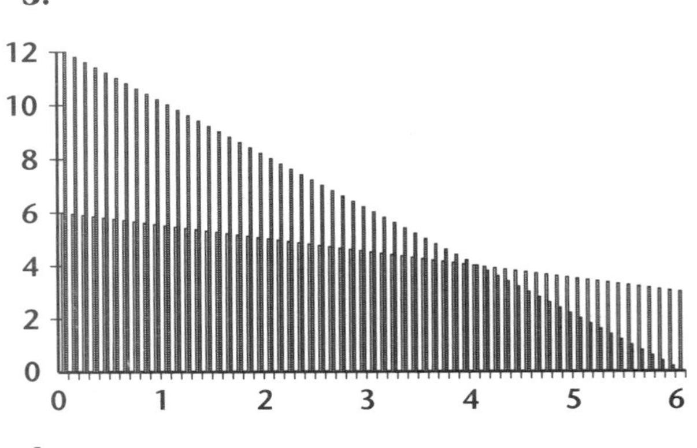

4.

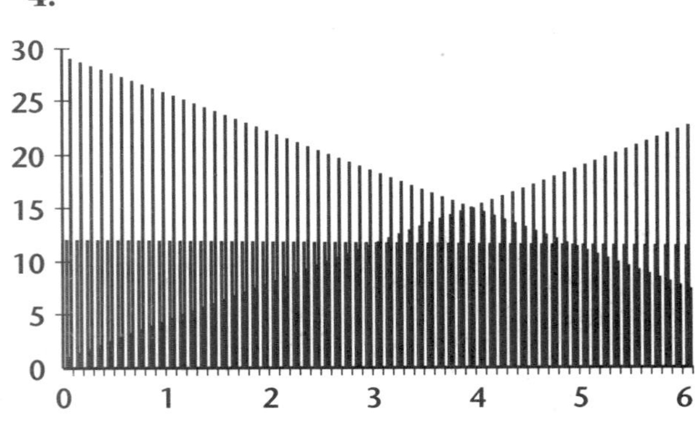

5.

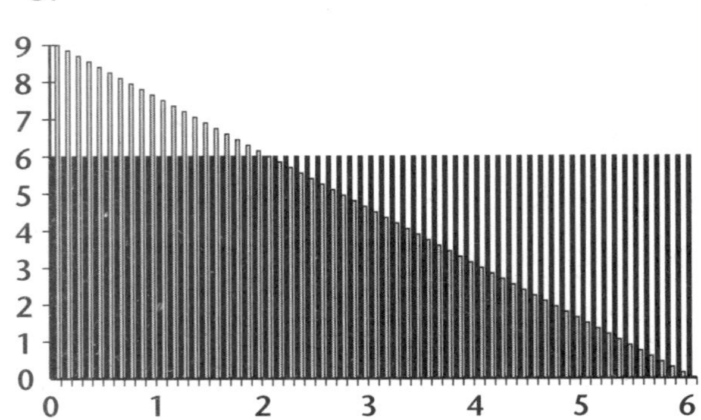

6.

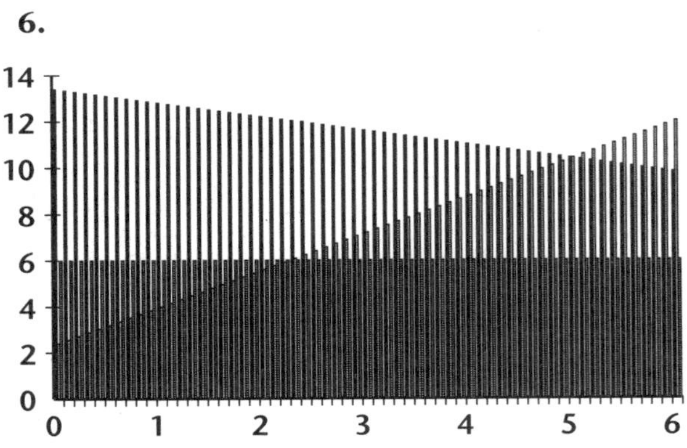

7. Increment the values of x in column A by a smaller amount, such as 0.05.

Lesson 4.9

1. 137.35 **2.** 92 **3.** 43.38 **4.** 99.2

5. 105.1 **6.** 75.973

7. If $A'(r's')$ is on $y = mx$ and in the interior of the feasible region, then $r' < r$ and $s' < s$. If $P = cx + dy$, where c and d are positive, as in Exercises 1–6, then P' obtained from $A'(r's')$ would be less than the value of P' obtained from $A(r, s)$.

Lesson 4.10

1. $0 \le x \le 2.66$ **2.** $2.15 \le x \le 3.00$

3. $0 \le x \le 1.71$ **4.** $1.58 \le x \le 8.53$

5. $x = 4$ **6.** No solution

7. Minimum: 0; Maximum: 62

Lesson Activities — Chapter 4

Lesson 4.1

1. 5

2. $\begin{bmatrix} 0 & 1 & 0 \\ 1 & 0 & 0 \\ 0 & 0 & 1 \end{bmatrix}$, $\begin{bmatrix} 0 & 1 & 0 \\ 0 & 0 & 1 \\ 1 & 0 & 0 \end{bmatrix}$, $\begin{bmatrix} 0 & 0 & 1 \\ 1 & 0 & 0 \\ 0 & 1 & 0 \end{bmatrix}$,

$\begin{bmatrix} 0 & 0 & 1 \\ 0 & 1 & 0 \\ 1 & 0 & 0 \end{bmatrix}$, $\begin{bmatrix} 1 & 0 & 0 \\ 0 & 0 & 1 \\ 0 & 1 & 0 \end{bmatrix}$

3. The maximum score is obtained when Malik is stock clerk, Adelso is cashier, and Paula is deli clerk.

4. The maximum score is $8 + 6 + 9 = 23$.

Lesson 4.2

1. 3 2. *DA*, *DB*, and *DC*. 3. None

4. The vertices are arranged in order according to the number of directed segments that begin at a particular vertex.

5.
$$\begin{array}{c} \\ A \\ B \\ C \\ D \end{array}\begin{array}{c} \begin{array}{cccc} A & B & C & D \end{array} \\ \begin{bmatrix} 0 & 0 & 0 & 0 \\ 1 & 0 & 1 & 1 \\ 1 & 0 & 0 & 0 \\ 1 & 0 & 1 & 0 \end{bmatrix} \end{array}$$

Lesson 4.3

1. Six tours are possible.

2.
$$A = \begin{bmatrix} 0 & 1 & 0 & 0 \\ 0 & 0 & 1 & 0 \\ 0 & 0 & 0 & 1 \\ 1 & 0 & 0 & 0 \end{bmatrix}$$

$$B = \begin{bmatrix} 0 & 0 & 1 & 0 \\ 1 & 0 & 0 & 0 \\ 0 & 0 & 0 & 1 \\ 0 & 1 & 0 & 0 \end{bmatrix}$$

$$C = \begin{bmatrix} 0 & 0 & 0 & 1 \\ 1 & 0 & 0 & 0 \\ 0 & 1 & 0 & 0 \\ 0 & 0 & 1 & 0 \end{bmatrix}$$

$$T = \begin{bmatrix} 0 & 1 & 0 & 0 \\ 0 & 0 & 0 & 1 \\ 1 & 0 & 0 & 0 \\ 0 & 0 & 1 & 0 \end{bmatrix}$$

$$E = \begin{bmatrix} 0 & 0 & 1 & 0 \\ 0 & 0 & 0 & 1 \\ 0 & 1 & 0 & 0 \\ 1 & 0 & 0 & 0 \end{bmatrix}$$

$$F = \begin{bmatrix} 0 & 0 & 0 & 1 \\ 0 & 0 & 1 & 0 \\ 1 & 0 & 0 & 0 \\ 0 & 1 & 0 & 0 \end{bmatrix}$$

(Rows and columns of each matrix are labeled 1, 2, 3, 4.)

3. Both matrices are 4×4.

4.
$$\begin{bmatrix} 17 & 0 & 6 & 12 \\ 15 & 12 & 14 & 0 \\ 0 & 17 & 10 & 15 \\ 10 & 6 & 0 & 14 \end{bmatrix}$$

5. 53

6. The total distance traveled on the tour.

7.

Tour *AD*
$$\begin{bmatrix} 17 & 0 & 6 & 12 \\ 10 & 6 & 0 & 14 \\ 15 & 12 & 14 & 0 \\ 0 & 17 & 10 & 15 \end{bmatrix}$$

Tour *BD*
$$\begin{bmatrix} 10 & 6 & 0 & 14 \\ 0 & 17 & 10 & 15 \\ 15 & 12 & 14 & 0 \\ 17 & 0 & 6 & 12 \end{bmatrix}$$

Tour *CD*
$$\begin{bmatrix} 15 & 12 & 14 & 0 \\ 0 & 17 & 10 & 15 \\ 17 & 0 & 6 & 12 \\ 10 & 6 & 0 & 14 \end{bmatrix}$$

Tour *ED*
$$\begin{bmatrix} 10 & 6 & 0 & 14 \\ 15 & 12 & 14 & 0 \\ 17 & 0 & 6 & 12 \\ 0 & 17 & 10 & 15 \end{bmatrix}$$

Tour *FD*
$$\begin{bmatrix} 15 & 12 & 14 & 0 \\ 10 & 6 & 0 & 14 \\ 0 & 17 & 10 & 15 \\ 17 & 0 & 6 & 12 \end{bmatrix}$$

8. Tours *F* (1-4-2-3-1) and *E* (1-3-2-4-1) minimize the distance.

Lesson 4.4

1. 2 2. 5

3. There is no direct route from a city to itself.

4.
$$\begin{bmatrix} 9 & 0 & 0 & 0 \\ 0 & 4 & 2 & 4 \\ 0 & 2 & 1 & 2 \\ 0 & 4 & 2 & 4 \end{bmatrix}$$

5. The number of ways to travel from one city to another city by passing through exactly one other city.

6. 2

7. The number of ways to travel from one city to another by passing through exactly two other cities.

Lesson 4.5

1. By multiplying C^{-1} (the inverse of matrix C) by the matrix $\begin{bmatrix} 1035 \\ 388 \\ 154 \end{bmatrix}$.

2. Possible answer (using the coding matrices in the example and the matrix $\begin{bmatrix} 3 \\ 1 \\ 20 \end{bmatrix}$ for "cat"): $\begin{bmatrix} 218 \\ 84 \\ 37 \end{bmatrix}$

3. Possible answer: Without knowing what matrix was used to determine the coding matrix (C), it would take a long time to guess the correct entries or find them by trial and error.

4–5. Check students' matrices.

Lesson 4.6

1. Rachel's payoff matrix:

$$\begin{array}{cc|ccc} & & & \text{Carl} & \\ & & \mathbf{0} & \mathbf{1} & \mathbf{2} \\ & \mathbf{0} & 0 & 0 & 0 \\ \text{Rachel} & \mathbf{1} & -1 & 1 & -1 \\ & \mathbf{2} & -2 & -2 & 2 \end{array}$$

2. Carl's payoff matrix:.

$$\begin{array}{cc|ccc} & & & \text{Rachel} & \\ & & \mathbf{0} & \mathbf{1} & \mathbf{2} \\ & \mathbf{0} & 0 & 1 & 2 \\ \text{Carl} & \mathbf{1} & 0 & -1 & 2 \\ & \mathbf{2} & 0 & 1 & -2 \end{array}$$

3. No. The sum of the matrices R and $C \neq \begin{bmatrix} 0 & 0 & 0 \\ 0 & 0 & 0 \\ 0 & 0 & 0 \end{bmatrix}$.

Lesson 4.7

1. $AR = \begin{bmatrix} -2 & 0 & 2 & 0 \\ 0 & -4 & 0 & 4 \end{bmatrix}$

rhombus is reflected through x-axis

2. $BR = \begin{bmatrix} 2 & 0 & -2 & 0 \\ 0 & 4 & 0 & -4 \end{bmatrix}$

rhombus is reflected through y-axis

3. $CR = \begin{bmatrix} 0 & 4 & 0 & -4 \\ -2 & 0 & 2 & 0 \end{bmatrix}$

rhombus if reflected through line $y = x$

4. $DR = \begin{bmatrix} 0 & -4 & 0 & 4 \\ 2 & 0 & -2 & 0 \end{bmatrix}$

rhombus is reflected through line $y = x$

5. $ER = \begin{bmatrix} 0 & -4 & 0 & 4 \\ -2 & 0 & 2 & 0 \end{bmatrix}$

rhombus is rotated 90° counterclockwise

6. $FR = \begin{bmatrix} 2 & 0 & -2 & 0 \\ 0 & -4 & 0 & 4 \end{bmatrix}$

rhombus is rotated 180°

7. $GR = \begin{bmatrix} 0 & 4 & 0 & -4 \\ 2 & 0 & -2 & 0 \end{bmatrix}$

rhombus is rotated 90° clockwise

8. $HR = \begin{bmatrix} -2 & 4 & 2 & -4 \\ 2 & 4 & -2 & -4 \end{bmatrix}$

rhombus changes size and orientation

9. All except HR are isometries.

10. Same as BR: rhombus is reflected through y-axis

11. Same as AR: rhombus is reflected through x-axis

12. Same as AR: rhombus is reflected through x-axis

13. Same as BR: rhombus is reflected through y-axis

Lesson 4.8

1. $\begin{bmatrix} 5 & 20 \\ 10 & 15 \end{bmatrix}$ 2. $\begin{bmatrix} 400 \\ 450 \end{bmatrix}$

3. $5c + 20t \leq 400$ or $c + 4t \leq 80$
$10c + 15t \leq 450$ or $2c + 3t \leq 90$

4.

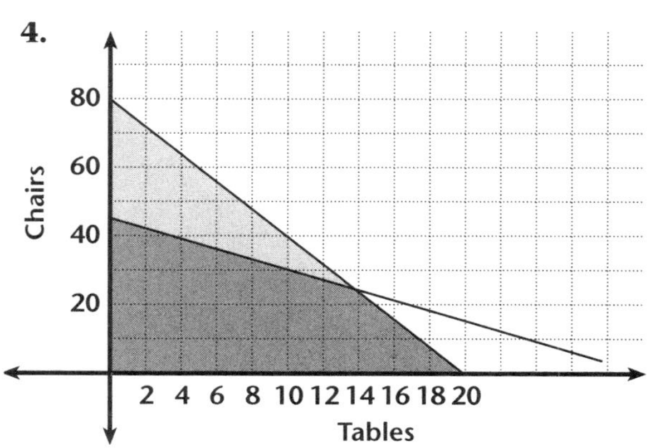

5. $c \geq 0$
$t \geq 0$

6. $P = 45c + 80t$ 7. 24 chairs and 14 tables

Lesson 4.9

1. That the probability of playing each coin is greater than 0.

2. That the sum of the probabilities is 1.

3. $\left(\frac{3}{5}, \frac{2}{5}, \frac{1}{5}\right)$

4. If Rebecca plays the dime $\frac{3}{5}$ of the time and the quarter $\frac{2}{5}$ of the time, she will win $\frac{1}{5}$ of the time.

5. $$\begin{aligned} a - b &\leq v \\ -2a + b &\leq v \\ a + b &= 1 \\ a &> 0 \\ b &> 0 \end{aligned}$$

6. $\left(\frac{2}{5}, \frac{3}{5}, \frac{1}{5}\right)$

Lesson 4.10

1. [0.526 0.474] 2. [0.497 0.503]

3. [0.452 0.548]

4. The probabilities of living in the suburbs appears to be increasing.

Assessment — Chapter 4

Assessing Prior Knowledge 4.1

	$x = 1$	$x = 2$	$x = 3$
$y = 2x + 6$	8	10	12
$y = 12x - 18$	−6	6	18

Quiz 4.1

1. 8; 8 freshmen were absent on Wednesday.

2. 4th row; 4th column

3. 4×5 4. 0 5. 2

6. $\begin{bmatrix} -4x + 7y \\ -3x + 6y \\ 5x - y \end{bmatrix} = \begin{bmatrix} 12 \\ 5 \\ -2 \end{bmatrix}$ 7. $\begin{bmatrix} 2 & -8 & 8 \\ -4 & 0 & 3 \end{bmatrix}$

8. $\begin{bmatrix} -2 & -14 & 10 \\ -4 & 8 & -7 \end{bmatrix}$

Assessing Prior Knowledge 4.2

1. 2×4 2. 3×2 3. 3×3 4. 1×5

Quiz 4.2

1. $\begin{bmatrix} -5 & 20 & -15 \\ -1 & 4 & -3 \\ 0 & 0 & 0 \end{bmatrix}$

2. not possible; inner dimensions are not equal

3. $\begin{bmatrix} 2 \\ -2 \\ 7 \end{bmatrix}$ 4. $\begin{bmatrix} 2 \\ 3 \end{bmatrix}$

5. **Krazy Kats** **Choco Mania** **Wonder Wafers** **Price**

$[14 \quad 23 \quad 18] \begin{bmatrix} 0.35 \\ 0.50 \\ 0.40 \end{bmatrix}$

= [(14(0.35) + (23)(0.50) + (18)(0.40)
= \$23.60

Assessing Prior Knowledge 4.3

1. $y = -\frac{3}{4}x + 3$ **2.** $y = 4x + 32$

3. $y = 10x - 2$ **4.** $y = \frac{5}{4}x$

Quiz 4.3

1. independent **2.** dependent

3. inconsistent **4.** inconsistent

5. $x = -\frac{3}{10}; y = -\frac{9}{4}$ **6.** no solution

7. $x = 9; y = \frac{14}{3}$ **8.** $x = \frac{27}{11}; y = \frac{26}{11}$

Assessing Prior Knowledge 4.4

1. 10 **2.** −9.5

Quiz 4.4

1. $x = 6; y = 10, z = 14$

2. $x = 27; y = 13; z = -2$

3. $\left[\begin{array}{ccc|c} 3 & -2 & -1 & -24 \\ -2 & 5 & 3 & 15 \\ 1 & 6 & -4 & -5 \end{array}\right]$

4. $x = -9; y = 9; z = \frac{11}{4}$

5. $x = 0; y = \frac{3}{2}, z = -\frac{7}{4}$

Assessing Prior Knowledge 4.5

1. $-\frac{1}{3}$ **2.** −7 **3.** 0.4 **4.** $-\frac{5}{6}$

Quiz 4.5

1. not inverses **2.** $\begin{bmatrix} \frac{7}{18} & -\frac{5}{18} \\ \frac{1}{9} & -\frac{2}{9} \end{bmatrix}$

3. $\begin{bmatrix} \frac{1}{13} & -\frac{3}{26} \\ \frac{3}{13} & \frac{2}{13} \end{bmatrix}$ **4.** $\begin{bmatrix} 3 & 2 \\ 8 & 5 \end{bmatrix}$ **5.** $\begin{bmatrix} \frac{1}{2} & 0 \\ -\frac{3}{10} & \frac{1}{5} \end{bmatrix}$

6. $\begin{bmatrix} -5 & 7 \\ 3 & -4 \end{bmatrix}$ **7.** does not exist

Mid-Chapter Assessment

1. d **2.** c **3.** c **4.** c **5.** $\begin{bmatrix} 7 & 39 \\ -5 & -11 \\ -6 & -30 \end{bmatrix}$

6. $x = 2; y = 2$ **7.** $x = 2; y = -1; z = -3$

Assessing Prior Knowledge 4.6

1. dependent **2.** inconsistent **3.** independent

Quiz 4.6

1. $\begin{bmatrix} 1 & 2 \\ 2 & 3 \end{bmatrix} \begin{bmatrix} x \\ y \end{bmatrix} = \begin{bmatrix} 11 \\ 18 \end{bmatrix}; \begin{bmatrix} x \\ y \end{bmatrix} = \begin{bmatrix} 3 \\ 4 \end{bmatrix}$

2. $\begin{bmatrix} 0 & 9 & 2 \\ 3 & 2 & 1 \\ 1 & -1 & 0 \end{bmatrix} \begin{bmatrix} a \\ b \\ c \end{bmatrix} = \begin{bmatrix} 14 \\ 5 \\ -1 \end{bmatrix}; \begin{bmatrix} a \\ b \\ c \end{bmatrix} = \begin{bmatrix} 1 \\ 2 \\ -2 \end{bmatrix}$

3. $\begin{bmatrix} -1 & 0 & 1 \\ 0 & 2 & -1 \\ 1 & 1 & 1 \end{bmatrix} \begin{bmatrix} l \\ m \\ n \end{bmatrix} = \begin{bmatrix} -4 \\ -1 \\ 6 \end{bmatrix}; \begin{bmatrix} l \\ m \\ n \end{bmatrix} = \begin{bmatrix} 5 \\ 0 \\ 1 \end{bmatrix}$

4. $\begin{bmatrix} 1 & 9 & 2 \\ 3 & 2 & 1 \\ 1 & -1 & 2 \end{bmatrix} \begin{bmatrix} x \\ y \\ x \end{bmatrix} = \begin{bmatrix} 14 \\ 5 \\ -5 \end{bmatrix}; \begin{bmatrix} x \\ y \\ z \end{bmatrix} = \begin{bmatrix} \frac{11}{10} \\ \frac{19}{10} \\ -\frac{21}{10} \end{bmatrix}$

Assessing Prior Knowledge 4.7

1. triangle **2.** square **3.** rectangle

4. parallelogram

Quiz 4.7

1. $\begin{bmatrix} 4 & 8 & 5 & 1 \\ -1 & -5 & -8 & -4 \end{bmatrix}$ **2.** $\begin{bmatrix} 6 & 6 & 12 \\ 6 & 12 & 6 \end{bmatrix}$

3. $\begin{bmatrix} 3 & -3 & -1 & 1 \\ 6 & 6 & 8 & 8 \end{bmatrix}$

Assessing Prior Knowledge 4.8

1.

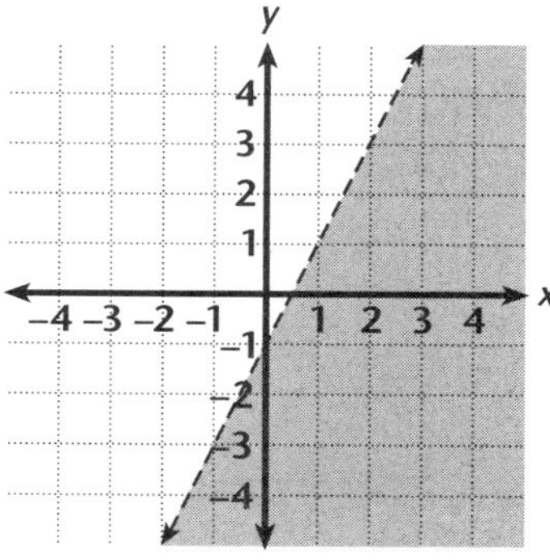

2.

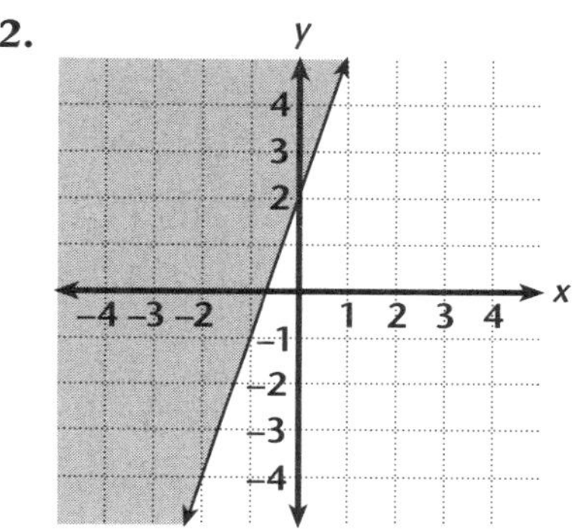

Quiz 4.8

1.

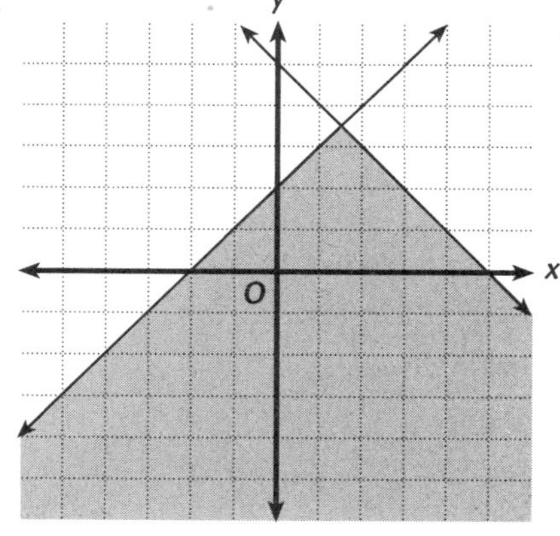

2.

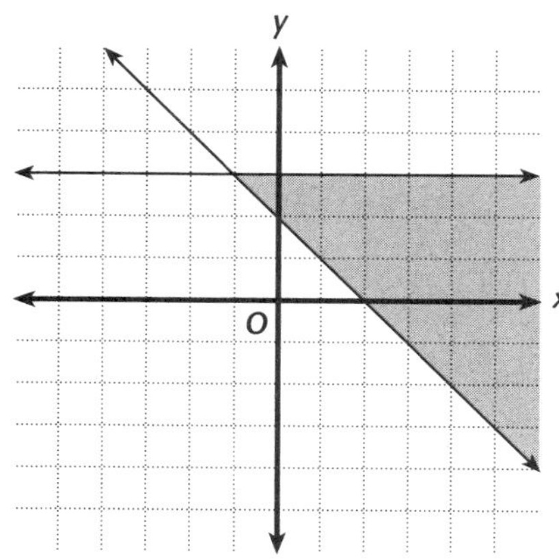

3.

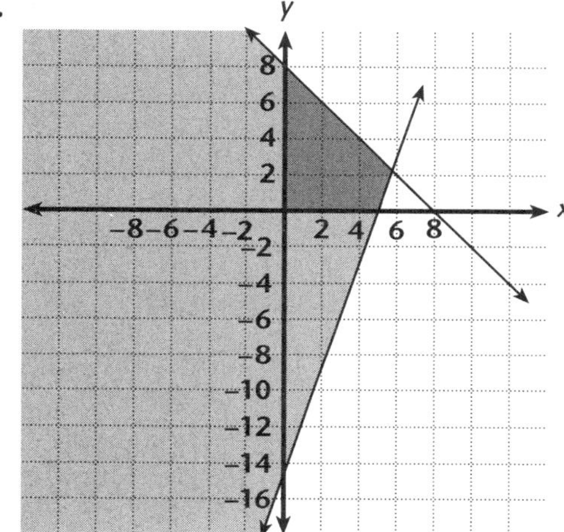

4.

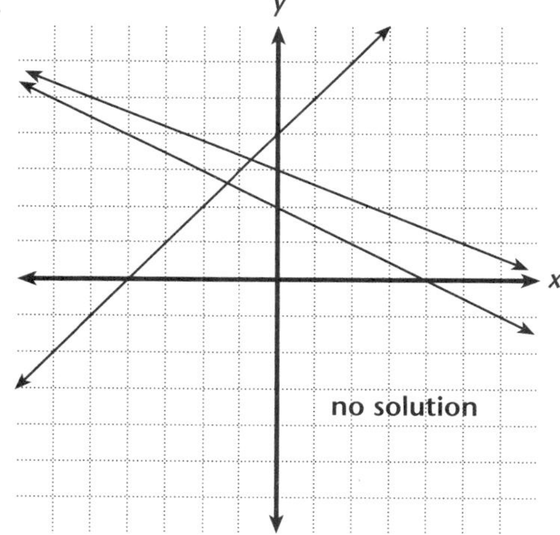

ANSWERS

Assessing Prior Knowledge 4.9

1. $-\frac{8}{3}$; 40 **2.** $-\frac{1}{3}$; $-\frac{1}{12}$ **3.** $\frac{7}{40}$; 0

Quiz 4.9

1.

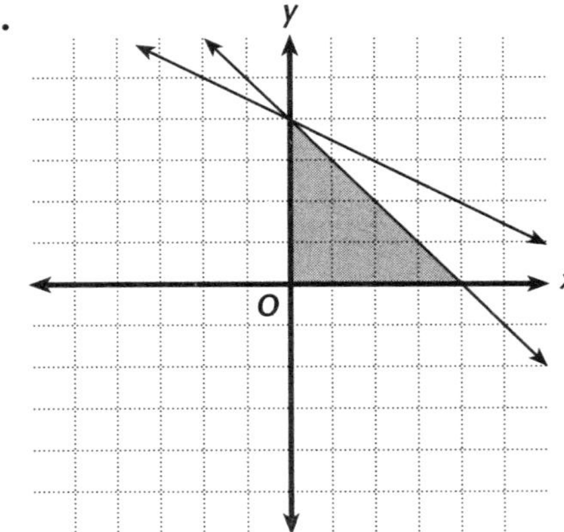

Answers will vary. Possible answers: (1, 1), (1, 2), (2, 1)

2.

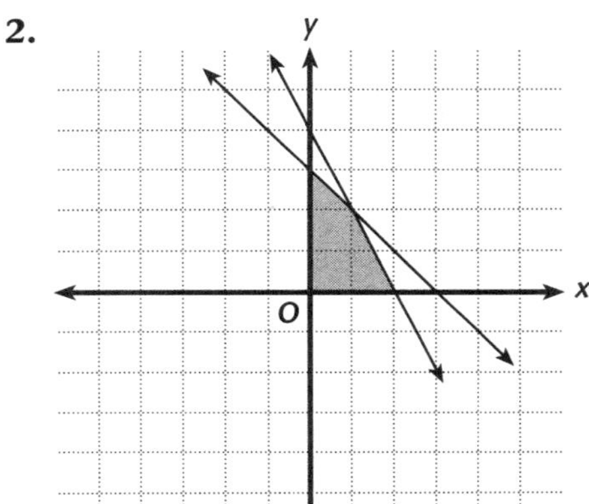

Answers will vary. Possible answers: (1, 1), $(\frac{1}{2}, 1)$, $(\frac{3}{2}, \frac{1}{12})$

3. Answers will vary. Possible answer: $10 = 4x + 3y$

4. Answers will vary. Possible answer: $4 = 3x + y$

Assessing Prior Knowledge 4.10

1. (4, 6) **2.** (−3, 7) **3.** (−1, −5)

Quiz 4.10

1. $(5\frac{1}{4}, 1\frac{1}{2})$; $P = 13\frac{1}{2}$ **2.** (0, 0); C = 0

3. (0, 2); P = 8 **4.** $\left(\frac{84}{11}, \frac{23}{11}\right)$; $C = 21.6$

Chapter Assessmen, Form A

1. d **2.** d **3.** b **4.** d **5.** d **6.** c **7.** c

8. c **9.** c **10.** b **11.** a

Chapter Assessment, Form B

1. $\begin{bmatrix} -3 & 4 & 1 \\ 2 & -5 & -5 \\ 1 & -2 & 5 \end{bmatrix} \begin{bmatrix} x \\ y \\ z \end{bmatrix} = \begin{bmatrix} 5 \\ -7 \\ 12 \end{bmatrix}$

2. $\begin{bmatrix} 12 & -12 & 12 \\ -4 & 8 & 0 \end{bmatrix}$ **3.** inconsistent

4. $\left[\begin{array}{ccc|c} -1 & 3 & -4 & 2 \\ 2 & 5 & -3 & 1 \\ 0 & -9 & 18 & -13 \end{array}\right]$

5. $x = -1$; $y = 6$; $z = 4$

6. $\begin{bmatrix} \frac{7}{3} & \frac{4}{3} & -\frac{2}{3} \\ \frac{2}{3} & \frac{2}{3} & -\frac{1}{3} \\ -2 & -1 & 1 \end{bmatrix}$

7. $\begin{bmatrix} 2 & 1 & -1 \\ 4 & -1 & 4 \\ 0 & -3 & 2 \end{bmatrix} \begin{bmatrix} x \\ y \\ z \end{bmatrix} = \begin{bmatrix} 3 \\ 0 \\ 6 \end{bmatrix}$

8. 144 pennies, 72 nickels, 53 dimes

9. $\begin{bmatrix} -6 & -6 & 2 & 2 \\ 2 & 4 & 4 & 2 \end{bmatrix}$

10.

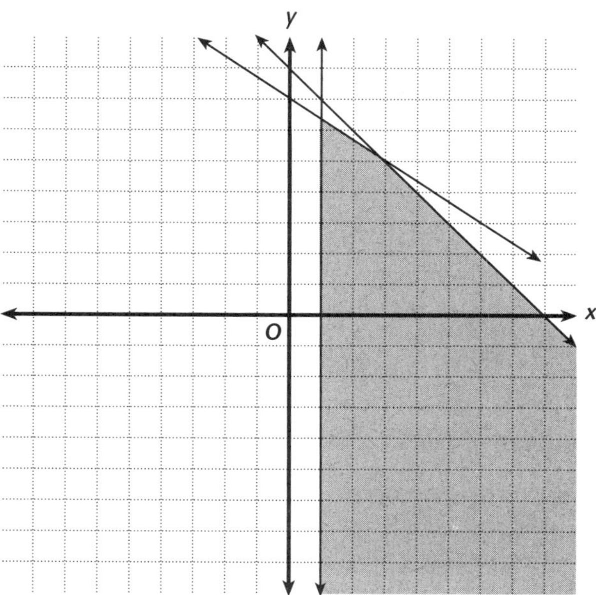

11.

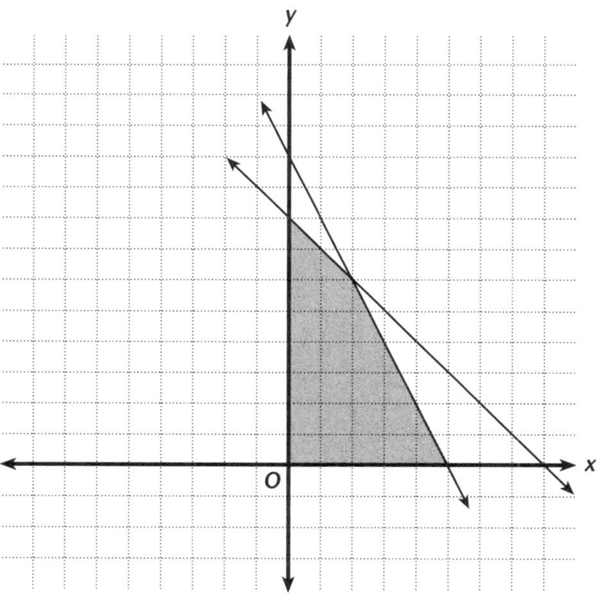

12. (0, 8)

Alternative Assessment — Chapter 4

Form A

1. $A = 100t + 800$; $A = 60t + 1200$

2. Answers may vary. Possible answer:

$$\begin{bmatrix} -100 & 1 \\ -60 & 1 \end{bmatrix} \begin{bmatrix} t \\ A \end{bmatrix} = \begin{bmatrix} 800 \\ 1200 \end{bmatrix}$$

3. $t = 10$ and $A = \$1800$

4. Since the solution of a system must satisfy each equation in the system, check the values for A and t in each original equation.

5. After 10 years, both debts would be $1800.

Score Point 4: Distinguished

The student demonstrates a comprehensive understanding of applying matrix algebra to solve a real-world problem. The student uses perceptive, creative, and complex mathematical reasoning throughout the task. He or she is able to use sophisticated, precise, and appropriate mathematical language throughout the task. Theoretical knowledge is apparent and applied to concrete situations as the student successfully demonstrates a comprehensive understanding of core concepts throughout the task.

Score Point 3: Proficient

The student demonstrates a broad understanding of applying matrix algebra to solve a real-world problem. The student uses perceptive mathematical reasoning most of the time. He or she is able to use precise and appropriate mathematical language most of the time. Theoretical knowledge is apparent and applied to concrete situations as the student attempts to draw conclusions based on his or her investigations.

Score Point 2: Apprentice

The student demonstrates an understanding of applying matrix algebra to solve a real-world problem. He or she uses mathematical reasoning at times during the task. He or she uses appropriate mathematical language some of the time. Student attempts to apply theoretical knowledge to the task but may be able to draw conclusions from his or her investigation.

Score Point 1: Novice

The student demonstrates a basic understanding of applying matrix algebra to solve a real-world problem. He or she uses appropriate mathematical language some of the time. Theoretical knowledge is extremely weak and many responses are irrelevant or illogical. He or she may fail to follow directions and has great difficulty in communicating his or her responses.

Score Point 0: Unsatisfactory

Student fails to make an attempt to complete the task and his or her responses are just an attempt to fill the page or restate the problem.

Form B

1. The region is polygonal in shape.
2. The solutions are located in a region that has boundary lines and does not extend beyond its boundary lines.
3. Answers may vary. Substitute the x- and y-values of each point into each inequality. If it makes the inequality true, the point is a solution to the system.
4. All points in the solution region of a system of linear inequalities indicate the feasible region.
5. $A(3, 2)$, $B(6, 2)$, $C(9, 3)$, $D(3, 9)$. The coordinates represent the maximum or minimum values of P that are feasible.
6. The maximum value of the objective function occurs at $C(9, 3)$.

Score Point 4: Distinguished

The student demonstrates a comprehensive understanding of solving systems of linear inequalities. The student uses perceptive, creative, and complex mathematical reasoning throughout the task. He or she is able to use sophisticated, precise, and appropriate mathematical language throughout the task. Theoretical knowledge is apparent and applied to concrete situations as the student successfully demonstrates a comprehensive understanding of core concepts throughout the task.

Score Point 3: Proficient

The student demonstrates a broad understanding of solving systems of linear inequalities. The student uses perceptive mathematical reasoning most of the time. He or she is able to use precise and appropriate mathematical language most of the time. Theoretical knowledge is apparent and applied to concrete situations as the student attempts to draw conclusions based on his or her investigations.

Score Point 2: Apprentice

The student demonstrates an understanding of solving systems of linear inequalities. He or she uses mathematical reasoning at times during the task. He or she uses appropriate mathematical language some of the time. Student attempts to apply theoretical knowledge to the task but may be able to draw conclusions from his or her investigation.

Score Point 1: Novice

The student demonstrates a basic understanding of solving systems of linear inequalities. He or she uses appropriate mathematical language some of the time. Theoretical knowledge is extremely weak and many responses are irrelevant or illogical. He or she may fail to follow directions and has great difficulty in communicating his or her responses.

Score Point 0: Unsatisfactory

Student fails to make an attempt to complete the task and his or her responses are just an attempt to fill the page or restate the problem.

Practice & Apply — Chapter 5

Lesson 5.1

1. $y = 2x^2 - 14x - 5$; $a = 2$, $b = -14$, $c = -5$
2. x-intercepts: $(-3, 0)$, $(3, 0)$; vertex: $(0, -18)$
3. x-intercepts: $(-3, 0)$, $(3, 0)$; vertex: $(0, 18)$

4.

Number of hamburgers	50	100	150	200
Profit ($)	112	199.5	262	299.5

Number of hamburgers	250	300	350	400
Profit ($)	312	299.5	262	199.5

5.

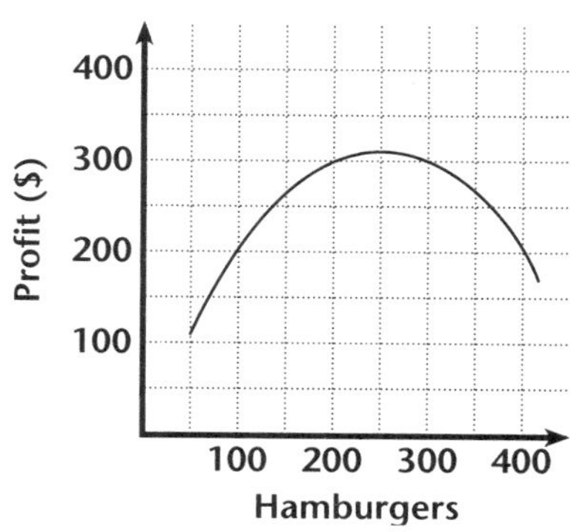

6. quadratic

7. When 250 hamburgers are produced.

8. 250 9. $312

Lesson 5.2

1. one 2. two 3. 6 or -6 4. 8 or -4

5. 5 or -1 6. $\frac{1}{5}$ or $-\frac{1}{5}$ 7. 3 or -3

8. 8 or -8 9. 5 or -5 10. 3 or -3

11. 7 or -7 12. 4 or 2 13. $-\frac{10}{3}$ or $-\frac{14}{3}$

14. 28 15. 159.25 16. 13 17. 8 18. 6

19. 10 20. 11.40 21. 4.24 22. 12.04

Lesson 5.3

1.

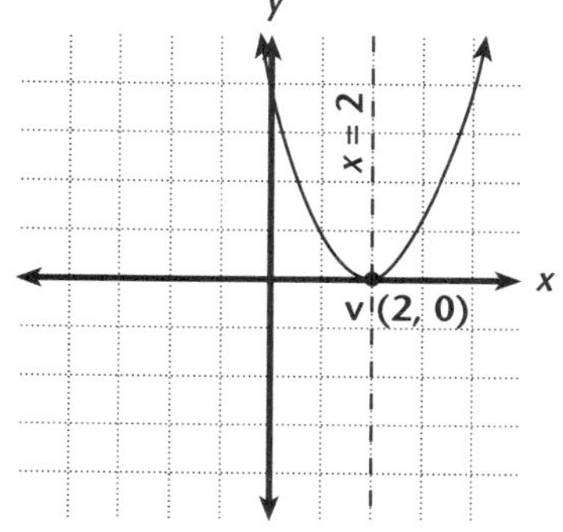

2.

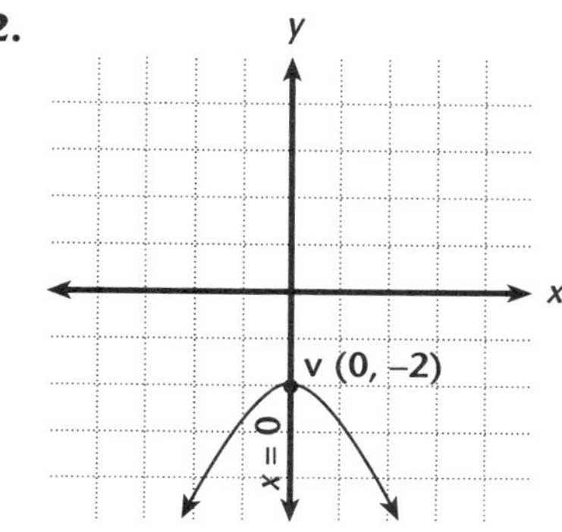

3. $(-3, -2)$; $x = -3$

4. $(-4, 0)$; $x = -4$

5. decreasing for $x < 1$; increasing for $x > 1$

6. decreasing for $x > -1$; increasing for $x < -1$

7. $(0, 0)$, $(-2, 0)$

8. $(1, 0)$

9. 6

10. $y^2 = -x^2 + 9$

11. 1.5 s

Lesson 5.4

1. 16; $x^2 + 8x + 16$ 2. 64; $x^2 + 16x + 64$

3. $22x$ 4. $30x$ 5. x 6. 36 7. $\frac{81}{4}$

8. $\frac{1}{9}$ 9. $x = 3$ or $x = 21$

10. $x = -7$ or $x = 4$ **11.** $x = 0$ or $x = 9$

12. $x = 0$ or $x = -1$ **13.** $x = 9$ or $x = 2$

14. $x = 3$ or $x = -3$ **15.** $x = -0.5$ or $x = 0.7$

16. $x = -1.5$ or $x = -5$ **17.** $x = 3$ or $x = -1.3$

18. $x = 2$ or $x = -0.5$

19. 1 second or 4 seconds

20. 31.25 m **21.** 11.25 m

Lesson 5.5

1. $a = 1, b = -3, c = -4$

2. $a = -2, b = 1, c = 1$

3. $2x^2 - x + 1 = 0$ **4.** $\frac{2}{3}; -\frac{1}{3}$

5. -1.3 or -4.7 **6.** 1.7 or -1.7 **7.** 0 or 0.8

8. -3 or 0.7 **9.** -0.7 or 0.3

10. 1.9 or -1.9 **11.** none, one, or two

12. $\frac{2}{3}$ and -1 **13.** $-\frac{1}{2}$ **14.** 1 **15.** 6

Lesson 5.6

1. less than 0 **2.** greater than 0

3. equal to 0 **4.** one **5.** two **6.** two

7. none **8.** $13i$ **9.** $7i\sqrt{2}$ **10.** $3i\sqrt{3}$

11. -9 **12.** -25 **13.** $-\sqrt{6}$ **14.** $32i$

15. $-10i$ **16.** $-4i$ **17.** *d*

18. Answers will vary. Possible answer: $x^2 + 49 = 0$

19. $\pm 3i\sqrt{6}$ **20.** $1 \pm 3i\sqrt{6}$ **21.** $\pm 4i$

22. $\pm 2i$

Lesson 5.7

1. $4; -2i$ **2.** $3; 2i$ **3.** $0; -2i$ **4.** $5; 0$

5. $5 + 3i$ **6.** $2 - 7i$ **7.** $-8 - 12i$

8. $21 + 3i$ **9.** $10 + 10i$ **10.** $27 - 21i$

11. $-30 + 15i$ **12.** $-6 - 10i$

13. $x^2 - 10x + 28 = 0$ **14.** $2 - 3i$ **15.** $-5i$

16. $-7 + 2i$ **17.** $-3 \pm i\sqrt{6}$ **18.** $-\frac{1}{5} \pm \frac{i\sqrt{6}}{10}$

19. $\frac{1}{2} \pm \left(\frac{\sqrt{3}}{2}\right)i$ **20.** $-7 \pm i\sqrt{11}$

21. $\frac{5}{4} \pm \left(\frac{\sqrt{23}}{4}\right)i$ **22.** $-\frac{1}{2} \pm \left(\frac{\sqrt{3}}{2}\right)i$ **23.** B

24. A **25.** D **26.** C

Lesson 5.8

1. $a = 4, b = -3, c = 4$

2. $a = 3, b = -2, c = -3$ **3.** $y = \frac{3}{4}(x - 2)^2$

4. $19 \stackrel{?}{=} (-2)^2 - 6(-2) + 3 \stackrel{?}{=} 4 + 12 + 3 = 19$

5. $y = 4x^2 - 2$ **6.** $y = -2x^2 + 6x + 5$

7. $y = 4x^2 - 20x + 25$ **8.** $y = x^2 + 1$

9. $d = 0.07\,s^2 + 0.02\,s + 26$

10. 89.6 ft **11.** 40 mi/h

Lesson 5.9

1. $y \geq x^2 - 4$ **2.** $-1 \leq x \leq 5$

3. $x < -2$ or $x > 3$ **4.** no solution

5. $x \geq 2$ or $x \leq 1$

6.

Fertilizer	1	2	3	4	5	6
Area	80	92	100	104	104	100

Fertilizer	7	8	9
Area	92	80	64

7.

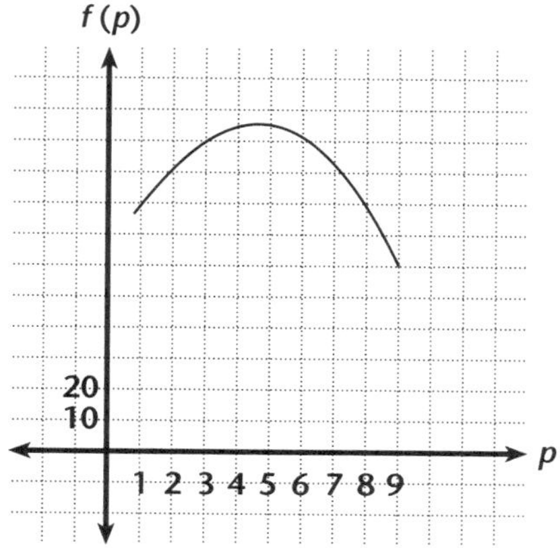

8. 4.5 pounds

9. $f(p) \leq 104.5 \text{ lb/ft}^2$

10. $p > 4.5$ lb

Enrichment—Chapter 5

Lesson 5.1

1. b; 5 sets; 93,750 apples
2. c; \$6.50; \$7290.00
3. a; \$40; \$6400
4. c; \$31.48; \$9908.33
5. c; \$14.50; \$10,512.50
6. a; \$35; \$3062.50

Lesson 5.2

1. ±11 = V **2.** 5, −1 = Q **3.** ±16 = R

4. ±7 = A **5.** 2 = H **6.** ±19 = T

7. 30 = N **8.** 27 = L **9.** ±14 = H

10. ±22 = H **11.** 25 = I **12.** ±12 = S

13. ±17 = O **14.** 24 = W **15.** 3 = E

16. ±13 = T **17.** ±18 = O **18.** 28 = W

19. ±21 = W **20.** ±6 = U **21.** 10 = S

22. 4, −2 = S **23.** 29 = I **24.** 8 = R

25. ±15 = E **26.** 20 = S **27.** ±26 = L

28. 23 = O **29.** ±9 = E **30.** 1 = T

THE SQUARES VS. THE ROOTS: WHO WILL WIN?

Lesson 5.3

$y = 2x^2 - 16x + 30$	$y = x^2 - 6x + 10$	$y = 2x^2 - 3x + 4$	$y = -x^2 + 6x - 9$
$y = 3x^2 + 2x - 8$	$y = \frac{1}{2}x^2 - 3x + \frac{19}{2}$	$y = 4x^2 - 8x + 15$	$y = -2x^2 - 16x - 20$
$y = \frac{1}{4}x^2 + \frac{11}{2}x - \frac{7}{4}$	$y = x^2 + 6$	$y = x^2 - 5x + 11.75$	$y = (x - 4)^2$
$y = 7x^2 - 8x + 17$	$y = 3x^2 - 54x + 242$	$y = \frac{1}{2}x^2 - 3x + \frac{1}{4}$	$y = \frac{1}{4}x^2 - 6x + 32$
$y = x^2 + 18$	$y = -x^2 - 8x + 5$	$y = -\frac{1}{3}x^2 + 4x^2 - \frac{5}{3}$	$y = -x^2 - \frac{1}{3}x + 5$
$y = \frac{1}{2}x^2 - 8x + 1$	$y = -x^2 + 10x + 38$	$y = -\frac{2}{3}x^2 + 6x - 7$	$y = -4x^2 + 40x - 97$
$y = 3x^2 - 14x + 11$	$y = -x^2 - 14x - 34$	$y = 3x^2 - 5x + 12$	$y = 4x^2 + 40x + 113$
$y = x^2 + 5x - 18$	$y = x^2 - 8x + 13$	$y = \frac{1}{3}x^2 - \frac{8}{3}x + \frac{28}{3}$	$y = 4x^2 - 12x + 9$
$y = x^2 - 8x + 19$	$y = -x^2 - 11x + 20$	$y = \frac{1}{4}x^2 - 4x + 16$	$y = x^2 + 18x - 20$
$y = x^2 + 16x + 16$	$y = -\frac{1}{2}x^2 + 6x - 6$	$y = x^2 + 8$	$y = x^2 + 4x + 4$

ANSWERS

Lesson 5.4

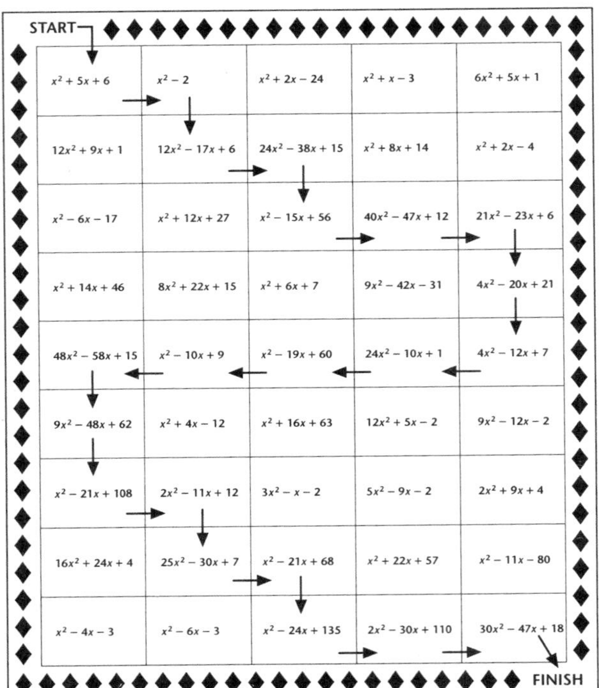

START

$x^2 + 5x + 6$	$x^2 - 2$	$x^2 + 2x - 24$	$x^2 + x - 3$	$6x^2 + 5x + 1$
$12x^2 + 9x + 1$	$12x^2 - 17x + 6$	$24x^2 - 38x + 15$	$x^2 + 8x + 14$	$x^2 + 2x - 4$
$x^2 - 6x - 17$	$x^2 + 12x + 27$	$x^2 - 15x + 56$	$40x^2 - 47x + 12$	$21x^2 - 23x + 6$
$x^2 + 14x + 46$	$8x^2 + 22x + 15$	$x^2 + 6x + 7$	$9x^2 - 42x - 31$	$4x^2 - 20x + 21$
$48x^2 - 58x + 15$	$x^2 - 10x + 9$	$x^2 - 19x + 60$	$24x^2 - 10x + 1$	$4x^2 - 12x + 7$
$9x^2 - 48x + 62$	$x^2 + 4x - 12$	$x^2 + 16x + 63$	$12x^2 + 5x - 2$	$9x^2 - 12x - 2$
$x^2 - 21x + 108$	$2x^2 - 11x + 12$	$3x^2 - x - 2$	$5x^2 - 9x - 2$	$2x^2 + 9x + 4$
$16x^2 + 24x + 4$	$25x^2 - 30x + 7$	$x^2 - 21x + 68$	$x^2 + 22x + 57$	$x^2 - 11x - 80$
$x^2 - 4x - 3$	$x^2 - 6x - 3$	$x^2 - 24x + 135$	$2x^2 - 30x + 110$	$30x^2 - 47x + 18$

FINISH

Lesson 5.7

START

$(4 + 3i) + (6 - 2i) = 10 + i$	$(2 - 3i) - (4 + i) = 2 - 4i$	$2i(3 + 5i) = 6i - 10$	$7(2i + 1) = 14 - 7i$	$(2 + i)(3 - i) = 7 + i$
$4i(3 - 5i) = 20 + 12i$	$2(3 - 4i) + 6i = 6 - 2i$	$3i(2 - i) = 6i - 3$	$(5 + 3i) - 2(1 - i) = 3 + 5i$	$(4 + 2i) - i(3) = 4 + i$
$(5 - 3i) - (2 + 4i) = -3 + i$	$4(-3 + i) = -12 + 4i$	$2i(i + 6) = -2 + 12i$	$(2 + i)(2 - i) = 5$	$(5 + 6i)(3 - i) = 21 + 13i$
$(5i + 3)i = 5 - 3i$	$(8 - 5i) - (3 - 4i) = 5 - i$	$6i(3 - i) = -6 + 18i$	$-5i(3 + 7i) = -35 - 15i$	$(6 + 4i) - (2 + 7i) = 4 - 3i$
$(8 - 3i) - (9 + i) = 1 - 4i$	$3(2 + 4i) - (1 + i) = 5 + 11i$	$5(7 + 3i) = 35 + 15i$	$(8i + 3)(2i - 4) = -28 - 26i$	$3i(51) = 153i$
$-7i(8 - 15i) = 105 - 56i$	$(3 - 5i) - (-4 + 6i) = 7 - 11i$	$5i(3i - 6) = -15 - 30i$	$(9 - 11i)(2 + i) = 29 - 13i$	$2i(3i) = 6$
$15(3 - 4i) = 60 - 45i$	$(8 + 3i)(8 - 3i) = 55$	$(4 - 3i) - (7 + 6i) = 3 - 9i$	$(2 + 3i) + 3(6 - i) = 20$	$78i(4 - 9i) = 63 + 28i$
$-4i(9 - 3i) = 12 - 36i$	$14(3 - 6i) = 42 - 84i$	$8(9 - 2i) = 72 - 16i$	$(9 + 3i)(4 - 2i) = 42 - 6i$	$-4i(18 + i) = -4 - 72i$
$3(2 + 4i) + 2(3 - i) = 12 + 10i$	$3(4 - 2i) - (2 + i) = 12 - 7i$	$(3 - 2i)(3 + 2i) = 13$	$(4 - 3i)(5 + i) = 23 - 11i$	$(4 + 7i) + (-1 + 2i) = 3 + 9i$
$(5 + 3i) - (8 - 4i) = -3 - i$	$3(6 - i) + 4i = 6 + i$	$(5 - 3i)(-4i) = 12 - 20i$	$(8 + 3i) - (-4 - i) = 12 + 4i$	$(7 - 5i) - (-2 - 3i) = 9 - 8i$
$(12 - 3i)(-5i) = 15 - 60i$	$(5 - 9i) + (4 - 6i) = 9 - 15i$	$-7i(4 - 5i) = 35 - 28i$	$(6i - 3) - 4(-2 + i) = 5 + 2i$	$2(3 + 4i) + (-5 + i) = 1 + 9i$
$(2 + 4i)(3 - 5i) = 26 - 2i$	$3(9i - 1) + 12i = -3 + 24i$	$-12i(4 - 5i) = 60 - 48i$	$(9 - 5i)(-8 + 2i) = 62 + 58i$	$17i(3 - i) = 17 + 51i$

FINISH

Lesson 5.5

1. 8 items; \$1468 **2.** 2.5 s; 106 ft

3. 25 items; \$825 **4.** 2 s; 64 ft

5. 10 h; 310 mg **6.** 15 items; \$875

7. 250 paintings; \$62,700

Lesson 5.6

1. I **2.** M **3.** A **4.** G **5.** I **6.** N

7. A **8.** R **9.** Y **10.** N **11.** U **12.** M

13. B **14.** E **15.** R **16.** S **17.** E

18. X **19.** I **20.** S **21.** T

Lesson 5.8

1. yes; $y = 2x^2 - 3x + 1$ **2.** no **3.** no

4. yes; $y = -x^2 + 7x + 16$

5. yes; $y = 3x^2 - x + 8$

6. yes; $y = -2x^2 + 9x - 10$

7. no **8.** yes; $y = 3x^2 - 4x + 5$

9. yes; $y = \frac{1}{2}x^2 + 7x - 9$

10. yes; $y = x^2 - 20x + 16$

Lesson 5.9

1. $y \geq x^2 - 8x + 12$ **2.** $y > -x^2 - 10x - 16$

3. $y > x^2 - 3x - 4$ **4.** $y > x^2 - 4x - 21$

5. $y \leq x^2 - 7x + 10$ **6.** $y \leq -x^2 + 5x + 6$

ANSWERS

Technology — Chapter 5

Lesson 5.1

1.

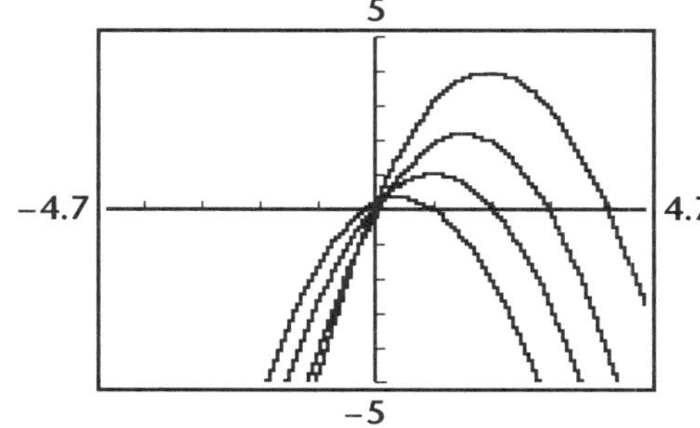

2.

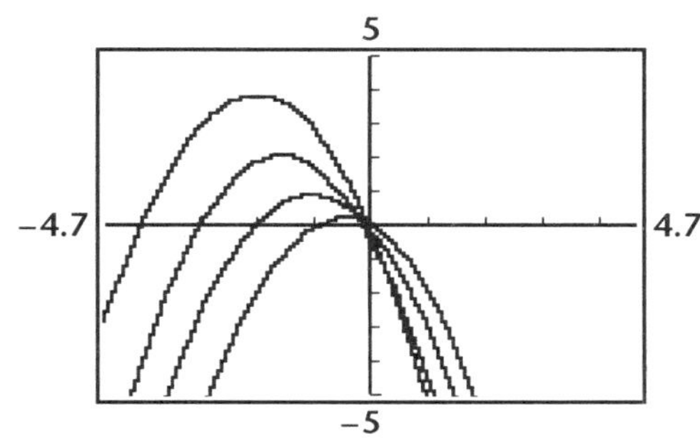

3. As the x-intercepts move father apart (b increases), the distance between the x-axis and the maximum point increases.

4. As the x-intercepts move farther apart (b decreases), the distance between the x-axis and the maximum point increases.

5.

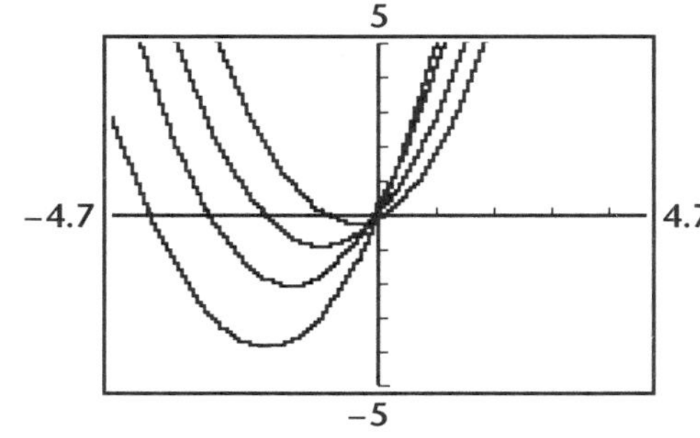

6.

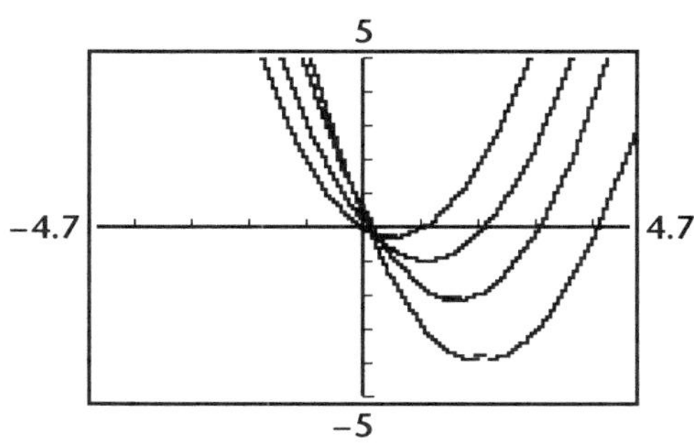

7. As the x-intercepts move farther apart (b increases), the distances between the x-axis and the minimum point increases.

8. As the x-intercepts move farther apart (b decreases), the distance between the x-axis and the minimum point increases.

9. The x-value of the highest point is one half the distance between the x-intercepts or $\frac{b}{2}$.

10. The x-value of the lowest point is one half the distance between the x-intercepts or $-\frac{b}{2}$.

Lesson 5.2

1. 19.1445351 2. 17.2783844

3. 24.0000000 4. 27.8241136

5. 33.0044231 6. 19.2425976

7. 33.7989899 8. 22.6274170

9. 16.9442719 10. $y = 3$ or 7

Lesson 5.3

1.

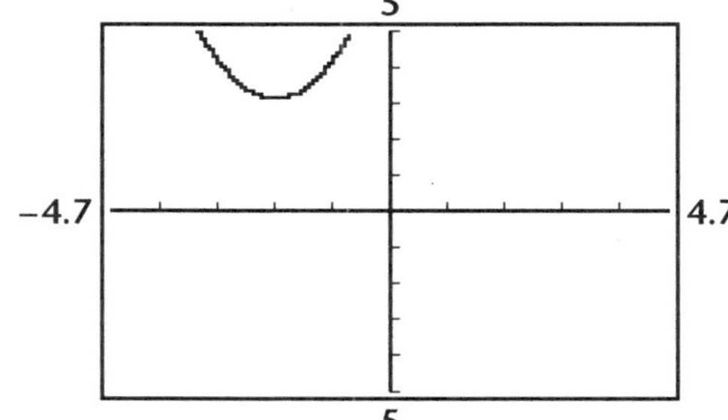

2.

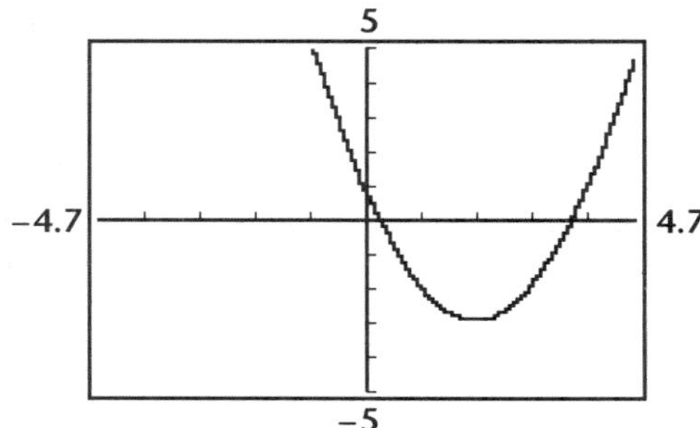

3.

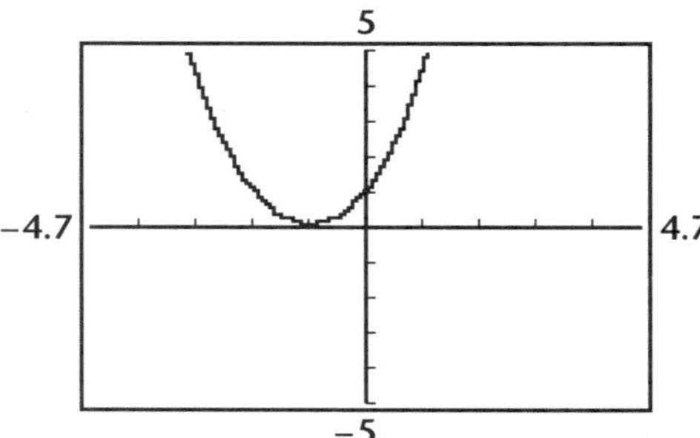

4.

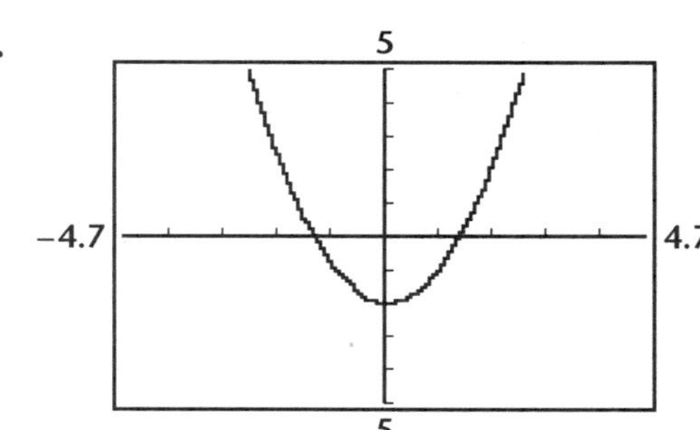

5.

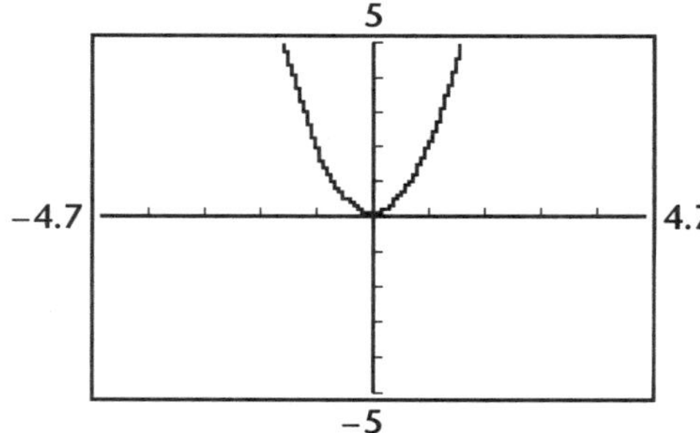

6.

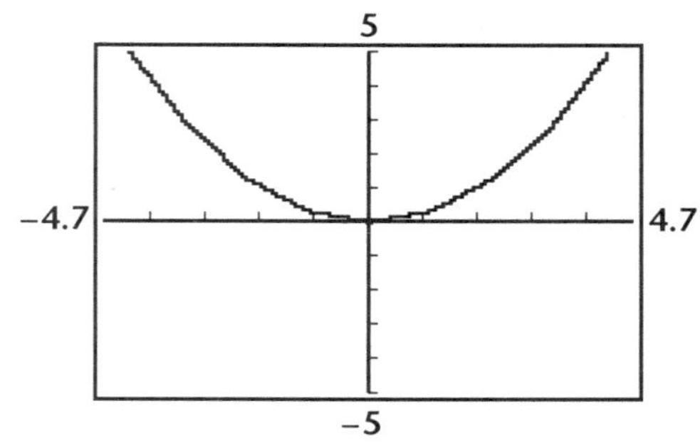

7.

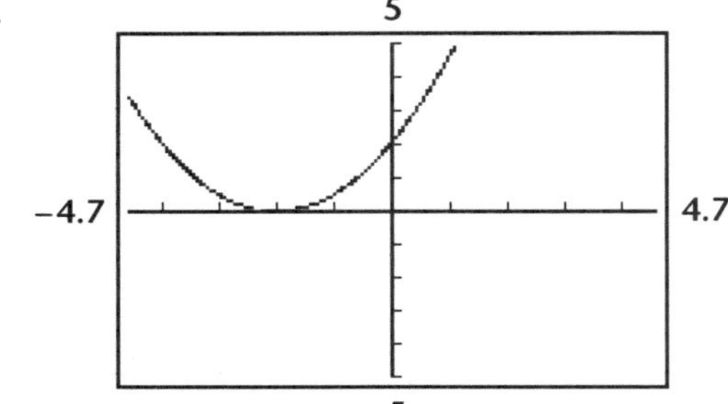

8.

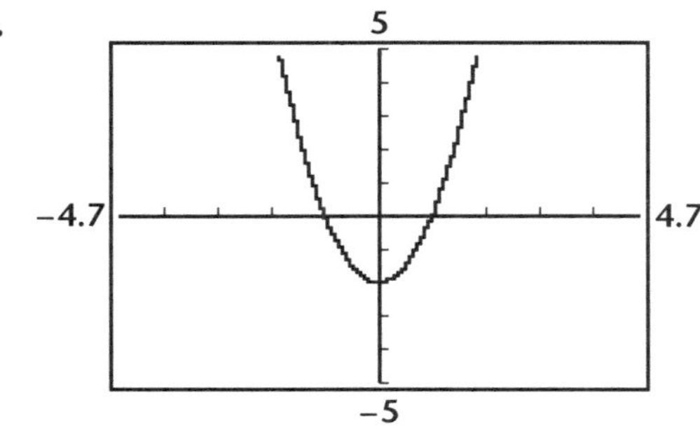

9. The graph of the parent function is made deeper, shifted to the left, and shifted up.

10. The graph of the parent function is made deeper, shifted to the right, and shifted down.

Lesson 5.4

1.

2.

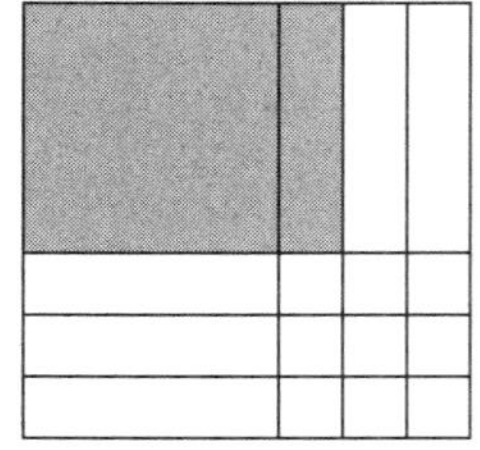

3. 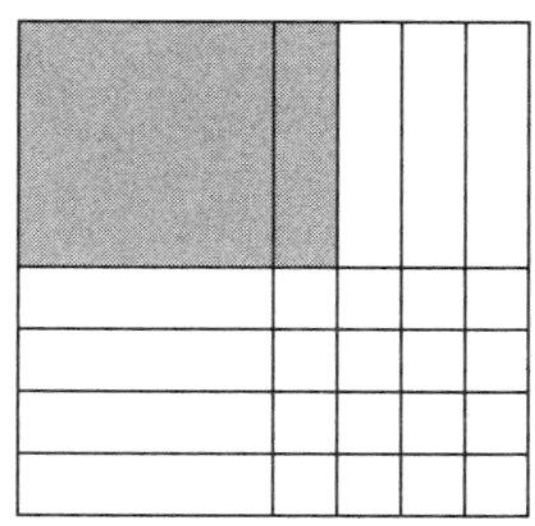

4. 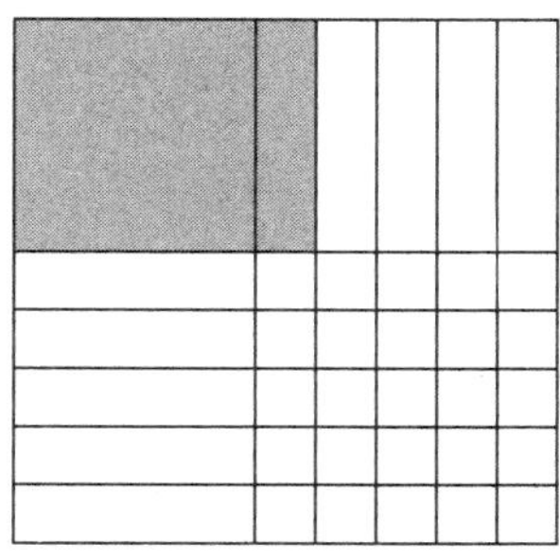

5. To the original diagram, add $2n - 1$ x tiles and n^2 unit tiles, where $n = 1, 2, 3, \ldots$

6. To the original diagram, add $2n - 2$ x tiles and n^2 unit tiles, where $n = 1, 2, 3, \ldots$

7. To the original diagram, add $2n - 3$ x tiles and n^2 unit tiles, where $n = 1, 2, 3, \ldots$

8. To the original diagram, add $2n$ x tiles and $n^2 - 1$ unit tiles, where $n = 1, 2, 3, \ldots$

9. To the original diagram, add $2n$ x tiles and $n^2 - 2$ unit tiles, where $n = 2, 3, 4, \ldots$

Lesson 5.5

For Exercises 1–8, check student reports against the graphs shown.

1. 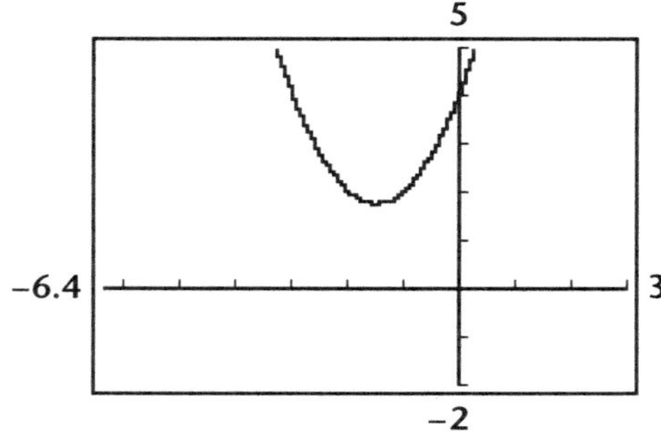

2. 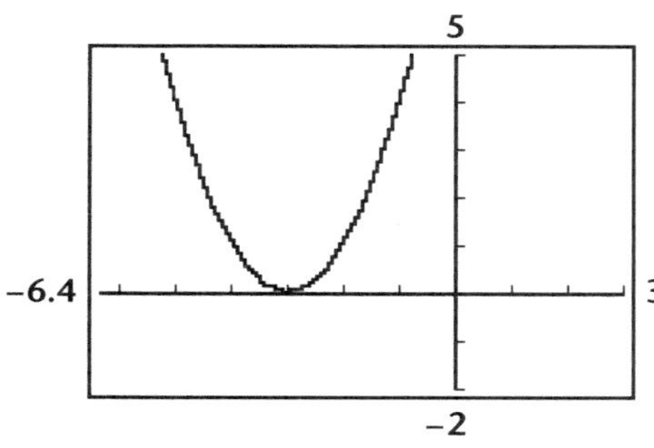

3. 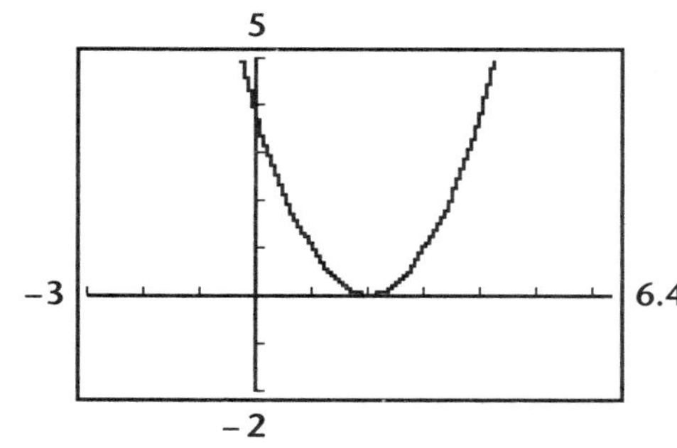

4. 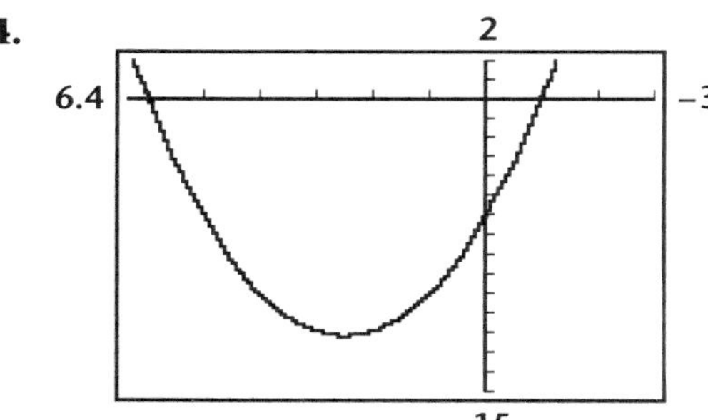

5. 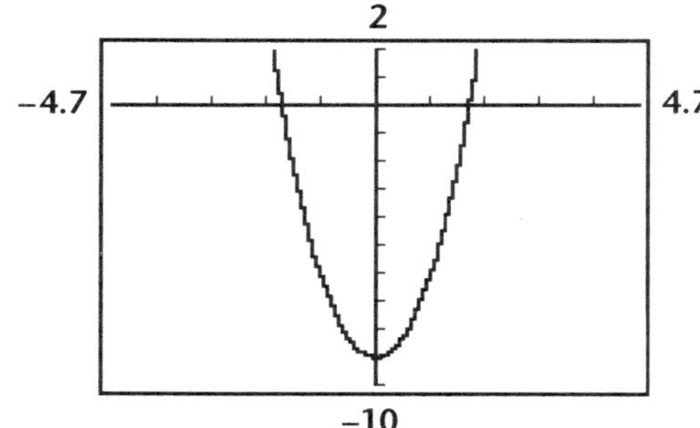

6. 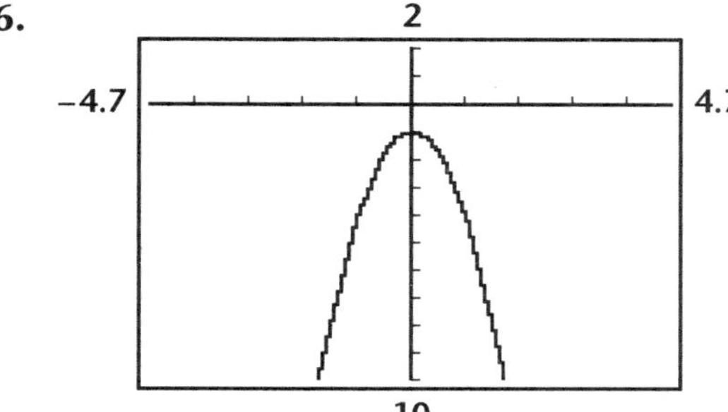

7.

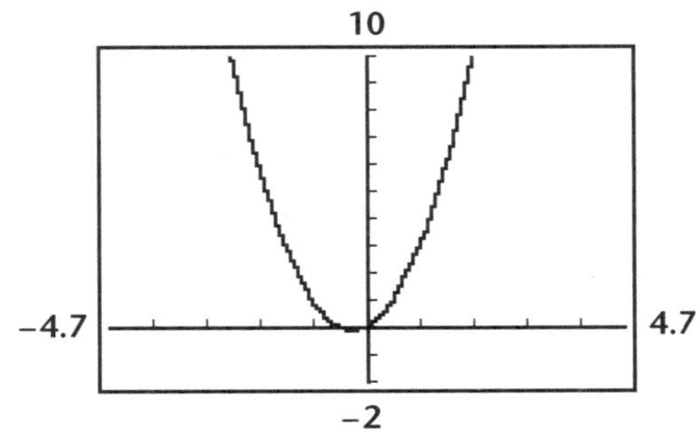

8.

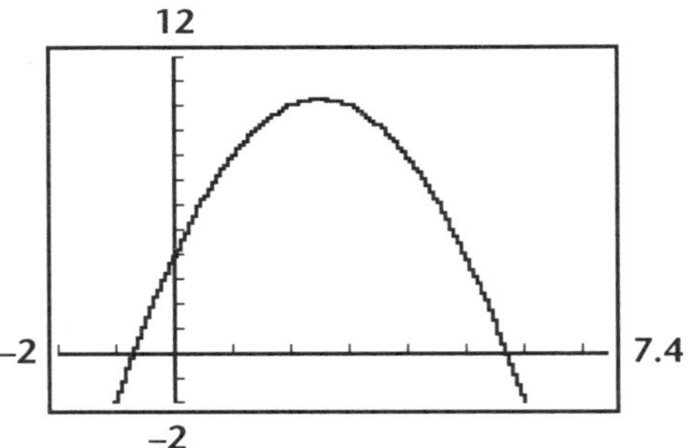

9. Reports will vary. The y-intercept is 0. The x-intercepts are 0 and $-\frac{b}{a}$. If $a > 0$, the graph opens upward. If $a < 0$, the graph opens downward.

10. Reports will vary. The y-intercept is c. The x-intercept is 0. If $a > 0$, the graph opens upward. If $a < 0$, the graph opens downward.

Lesson 5.6

1. IF(D2>=0,(−B2−SQRT(D2))/(2*A2), " ")
2. IF(D2>=0,0, " ")
3. IF(D2>=0, " ",−B2/(2*A2))
4. IF(D2>=0," ",SQRT(−D2)/(2*A2))
5. $-3 \pm 1.41421356i$
6. -4.7912878 or -0.2087122
7. $0.83333333 \pm 1.90758719i$
8. -2
9. $-6, -2, 1,$ or 11

Lesson 5.7

1.

	A	B	C	D
1	REAL	IMAG	REAL	IMAG
2	1	0	1	0
3	1	0	1	0
4	1	0	1	0

2.

	A	B	C	D
1	REAL	IMAG	REAL	IMAG
2	2	0	4	0
3	4	0	16	0
4	16	0	256	0

3.

	A	B	C	D
1	REAL	IMAG	REAL	IMAG
2	0	0.5	−0.25	0
3	−0.25	0	0.0625	0
4	0.0625	0	0.00390625	0
5	0.00390625	0	1.5259E-05	0

4.

	A	B	C	D
1	REAL	IMAG	REAL	IMAG
2	0	−2.5	−6.25	0
3	−6.25	0	39.0625	0
4	39.0625	0	1525.87891	0
5	1525.87891	0	2328306.44	0
6	2328306.44	0	5.421E+12	0

5.

	A	B	C	D
1	REAL	IMAG	REAL	IMAG
2	0.45	0.2	0.165	0.18
3	0.1625	0.18	−0.0059938	0.0585
4	−0.0059938	0.0585	−0.0033863	0.0007013
5	−0.0033863	−0.0007013	1.0975E-05	4.7494E-06

6.

	A	B	C	D
1	REAL	IMAG	REAL	IMAG
2	1.01	1.15	−0.3024	2.323
3	−0.3024	2.323	−5.3048832	−1.4049504
4	−5.3048832	−1.4049504	26.1679006	14.9061957
5	26.1679006	14.9061957	462.564351	780.127692
6	462.564351	780.127692	−394633.44	721718.519

7.

	A	B	C	D
1	REAL	IMAG	REAL	IMAG
2	−0.5	−0.24	0.1924	0.24
3	0.1924	0.24	−0.0205822	0.092352
4	−0.0205822	0.092352	−0.0081053	−0.0038016
5	−0.0081053	−0.0038016	5.1243E-05	6.1626E-05
6	5.1243E-05	6.1626E-05	−1.172E-09	6.315E-09

8.

	A	B	C	D
1	REAL	IMAG	REAL	IMAG
2	−1.2	1.3	−0.25	−3.12
3	−0.25	−3.12	−9.6719	1.56
4	−9.6719	1.56	91.1120496	−30.176328
5	91.1120496	−30.176328	7390.79481	−5498.8542
6	7390.79481	−5498.8542	24386450.6	−81281806

9. Given $z = r + si$: If either part is 0 and the other part is not equal to 1, z^n tends to 0 or increases without bound. If $z = 1$, $z^n = z$.

10. Given $z = r + si$: If both parts are greater than 1 in absolute value, z^n tends to "increase" without bound. If both parts are less than 1 in absolute value, then z^n tends to go to 0.

Lesson 5.8

1. Let $c = 1$. $f(x) = 2x^2 - 3x + 1$

2. $f(x) = x^2 + x - 6$ 3. $f(x) = x^2 - 4x - 5$

4. $f(x) = 4x^2 - 1$ 5. $f(x) = 9x^2 + 6x + 1$

6. Let $c = 3$; $f(x) = -x^2 + 5x + 3$

7. Let $c = 6$; $f(x) = 2x^2 - 5x + 6$

8. There is no quadratic function whose graph contains all four points.

9. Slope of $\overline{AB}$: $\frac{2-0}{3-0} = \frac{2}{3}$; Slope of $\overline{BC}$: $\frac{4-2}{6-3} = \frac{2}{3}$, so the slopes of the line segments are equal.

10. You will find that $a = 0$. If $a = 0$, the function is not quadratic.

Lesson 5.9

1. $-2.2 \le x \le 2.2$

2. $-2.2 < x < -1$ or $1 < x < 2.2$

3. $0.4 \le x < 1.2$

4. $-2.4 \le x < -1.3$ or $2.3 < x \le 2.4$

5. $x < -0.6$ or $x > 1.6$

6. $-0.6 \le x \le 1.6$

7. $0.2 < x \le 0.3$ or $3.3 \le x < 4.3$

8. no solution

9. $0.5x + 1 < x^2 + x - 2 < 4$. The symbol $\le$ may replace $<$.

10. Answers may vary. Possible answer: $2x - 3 < x^2 - 4x + 1 < -2.5x$

Lesson Activities — Chapter 5

Lesson 5.1

1.

x	0	5	10	15	20
$P(x)$	−780	−505	−280	−105	20

x	25	30	35	40	45
$P(x)$	95	120	95	20	−105

2. As the prices increase from $0 to $45, the profit increases and then decreases.

3. Check students' graphs. Possible answer: The graph suggests that at first, profit increases rapidly, then more slowly, until a maximum point is reached. Thereafter, profit decreases slowly, then more rapidly.

4. $P(x) = 0$ between \$40 and \$41 and between \$19 and \$20.

5. The break-even points occur at prices of about \$20 and about \$40.

6. Christen should charge \$30 to maximize her profit, because (30, 120) is the vertex of the parabola.

Lesson 5.2

1. $x = 5$ and 1 **2.** two

3.

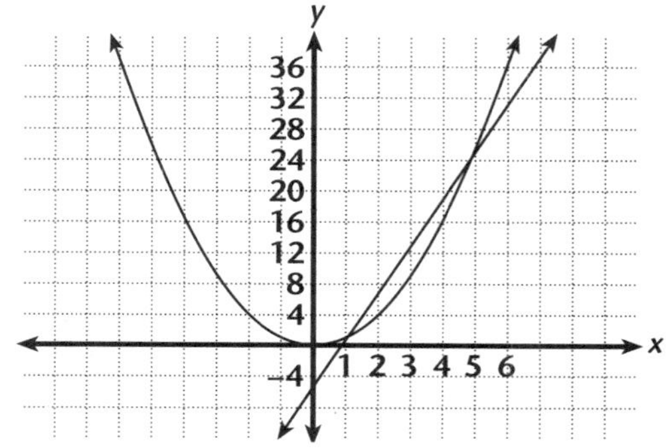

two; (1, 1), (5, 25)

4. $x = -2$ **5.** one

6.

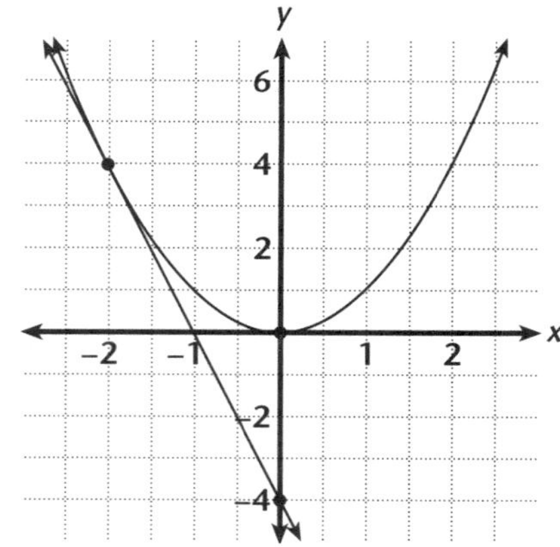

one; (−2, 4)

7. no real-number solutions **8.** none

9.

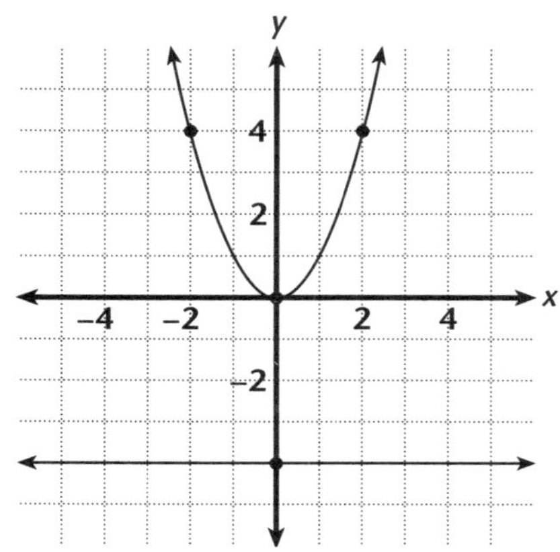

does not intersect

10. The number of points of intersection is the same as the number of real solutions.

Lesson 5.3

1. Vertical shift: 3; horizontal shift: 1 **2.** 2

3. $f(x) = 2(x - 1) + 3$ **4.** $g(x) = 3x^2 - 2$

5. $h(x) = -0.5(x - 2)^2 + 3$

Lesson 5.4

1. $a^2 + 2ab + b^2$ **2.** $(a + b)^2$

3. The 3^2 area is subtracted twice so it must be added once to obtain a single subtraction of 3^2.

4. $(a - b)^2 = a^2 - 2ab + b^2$

5. missing parts are: x, x, and 8; $(x + 8)^2 = x^2 + 16x + 64$

6. missing parts are: $x - 8$, $x - 8$, and x; $(x - 8)^2 = x^2 - 16x + 64$

Lesson 5.5

1. $x_1 = -0.2192235936$; $x_2 = -2.280776406$

2. $x_1 = 2.280776406$; $x_2 = 0.2192235936$

3. $x_1 = -1\text{E-}15$ or $-1(10)^{-15}$; $x_2 = -1$

4. $x_1 = -1\text{E-}15$ or $-1(10)^{-15}$; $x_2 = -1$

5. Depending on the calculator students use, the solutions may or may not be the same. On some calculators, the solution obtained using the quadratic formula may round the value −1E-15.

6. A calculator produces only a finite number of digits and must round off answers at some point.

Lesson 5.6

1.

R	F	L	B
F	L	B	R
L	B	R	F

2. F 3. B

4.

	i	-1	$-i$	1
i	-1	$-i$	1	i
-1	$-i$	1	i	-1
$-i$	1	i	-1	$-i$
1	i	-1	$-i$	1

5. A multiplication table for powers of i.

Lesson 5.7

1. $-\frac{5}{7}$ 2. $-\frac{3}{2}$ 3. $-\frac{5}{7}$ 4. $-\frac{3}{2}$

5. Since the opposite sides of $PQRS$ have the same slope, they are parallel.

6.

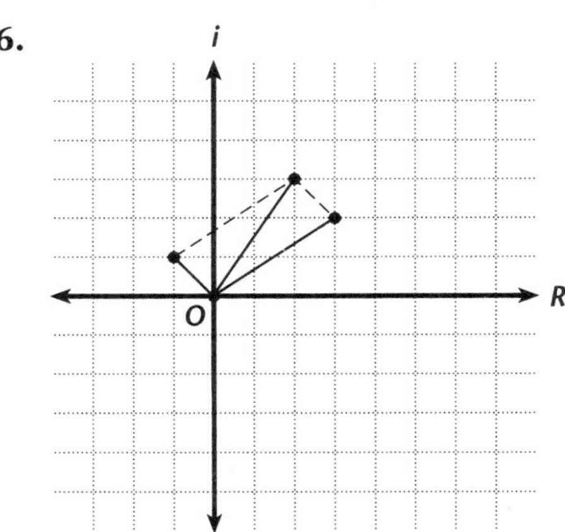

7. Plot the ordered pair $(-2, i)$ for the complex number $-(2 - i)$; $2 + 4i$

8. The resultant of two vectors acting at any angle may be represented by the diagonal of a parallelogram. The two vectors are drawn as the sides of the parallelogram and the resultant (sum) is its diagonal.

Lesson 5.8

1.

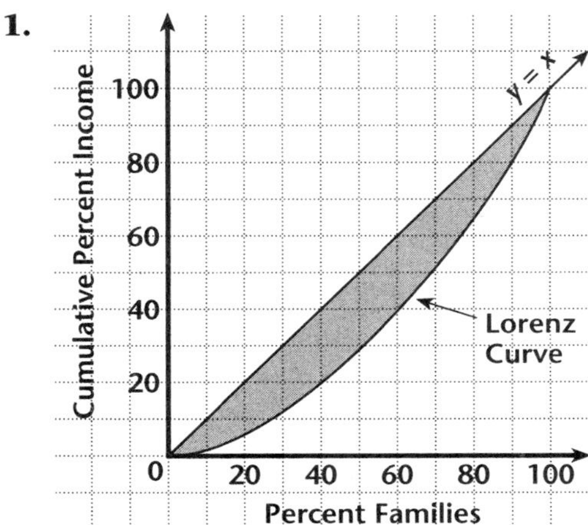

2. 20% 3. about 80%

4. $y = 0.008x^2 + 0.21x - 2$

5. See answer to Exercise 1.

6. The distribution of income is not equal amongst all families.

7. The farther the Lorenz Curve is from the line of perfect equality, the less equal the distribution income.

ANSWERS

Lesson 5.9

1.

Number of tickets in excess of 30	Number of tickets sold	Price of ticket	Bus company income
1	31	27.25	844.75
2	32	26.50	848.00
3	33	25.75	849.75
4	34	25.00	850.00
5	35	24.25	848.75
6	36	23.50	846.00
7	37	22.75	841.75
8	38	22.00	836.00
9	39	21.25	828.75
10	40	20.50	820.00
11	41	19.75	809.75
12	42	19.00	798.00

2. 34 tickets **3.** a quadratic function

4. $y = (30 + x)(28 - 0.75x)$

Assessment — Chapter 5

Assessing Prior Knowledge 5.1

1. $3x^2 - 13x + 4$ **2.** $-4x^2 - 17x + 15$

Quiz 5.1

1. $f(x) = 2x^2 - 6x - 80$; $a = 2$, $b = -6$, $c = -80$

2. 900 ft^2 **3.** -5 and $\frac{4}{3}$

4. $\left(-\frac{11}{6}, \frac{-361}{12}\right)$; minimum

5. increasing: $x > -\frac{11}{6}$; decreasing: $x < -\frac{11}{6}$

6. x-intercepts: $-\frac{3}{2}, \frac{4}{3}$

vertex: $\left(-\frac{1}{12}, \frac{-289}{24}\right)$

Assessing Prior Knowledge 5.2

1. 9 **2.** -12 **3.** $3\sqrt{11}$ **4.** $-2\sqrt{17}$

Quiz 5.2

1. $x = \pm 13$ **2.** $x = 96$ **3.** $x = \pm\frac{\sqrt{15}}{2}$

4. no real solution **5.** $\sqrt{73}$ **6.** $2\sqrt{34}$

7. $\sqrt{145}$ **8.** $\sqrt{149}$ **9.** $9 + \sqrt{74} + \sqrt{65}$

Assessing Prior Knowledge 5.3

1.

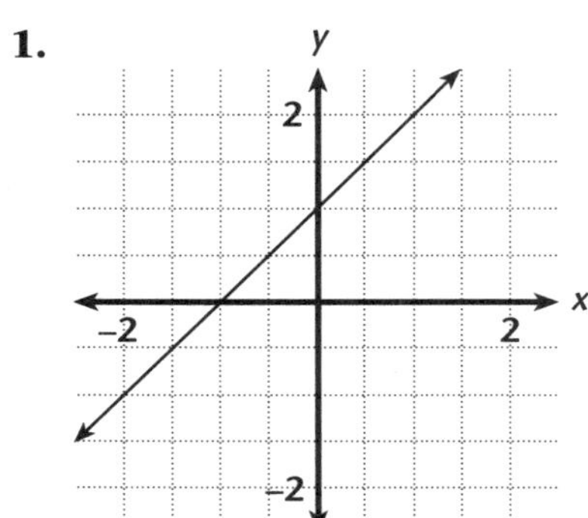

2.

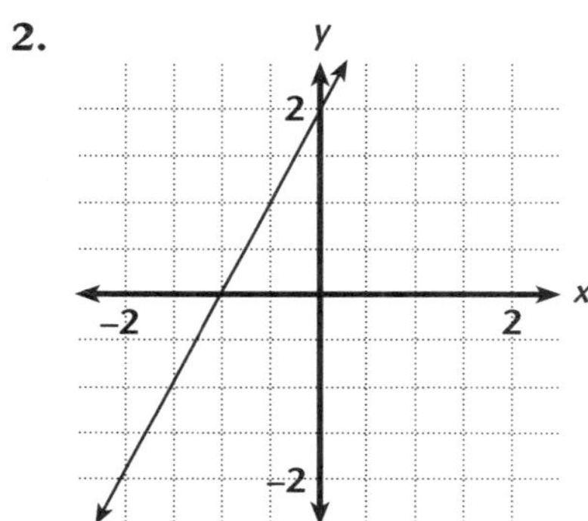

Quiz 5.3

1. $(3, 5)$ **2.** $x = 3$

3. Answers may vary, but should be in the form $y = -a(x + 3)^2 + 2$.

4. Increasing: $x < 4$; Decreasing: $x > 4$

5. no x-intercepts

Assessing Prior Knowledge 5.4

1. ± 10 **2.** ± 12 **3.** $-2, 4$

ANSWERS

Quiz 5.4

1. $x^2 + 5x + 4$

2. Answers will vary.

3. 25; $x^2 + 10x + 25$ **4.** $+4$ **5.** $+\frac{25}{4}$

6. 1.8, −6.8 **7.** 0.4, −4.4

Assessing Prior Knowledge 5.5

1. ± 7 **2.** ± 2 **3.** $\pm\frac{5}{3}$ **4.** −4

Quiz 5.5

1. 0, 8 **2.** no real solutions **3.** 4.3, −3.3

4. 1.9, −1.4

5. no *x*-intercepts
vertex: (1, 4)

6. no *x*-intercepts

vertex: $\left(-\frac{3}{10}, \frac{155}{100}\right)$

7. *x*-intercepts: −6.1, 1.1

vertex: $\left(-\frac{5}{2}, -\frac{47}{4}\right)$

8. *x*-intercepts: 2.4, 3.6
vertex: (3,1)

Mid-Chapter Assessment

1. d **2.** c **3.** a **4.** b **5.** 14

6. $\left(-\frac{9}{2}, -\frac{49}{4}\right)$ **7.** 3.4, 0.6 **8.** 3.6, −7.6

Assessing Prior Knowledge 5.6

1. 40 **2.** 1 **3.** 44 **4.** −20

Quiz 5.6

1. 0 **2.** 2 **3.** −25 **4.** $-i$ **5.** $5\sqrt{3}$

6. −3 **7.** $t = \pm 2i\sqrt{2}$ **8.** $y = \pm 5i$

Assessing Prior Knowledge 5.7

1. $5 + 9a$ **2.** $-3 - x$ **3.** $-12y - 9y^2$

4. $-15 + 26m - 8m^2$

Quiz 5.7

1. $x = \frac{1}{2} \pm \frac{i\sqrt{51}}{6}$ **2.** $x = \frac{5}{2} \pm \frac{i\sqrt{7}}{2}$

3. $x = \frac{7}{2} \pm \frac{i\sqrt{3}}{2}$ **4.** $x = \frac{3}{10} \pm \frac{i\sqrt{131}}{10}$

5. Real: −3
Imaginary: 7

6. Real: 0
Imaginary: 2

7. $5 + 2i$ **8.** $-3 - i$

9.

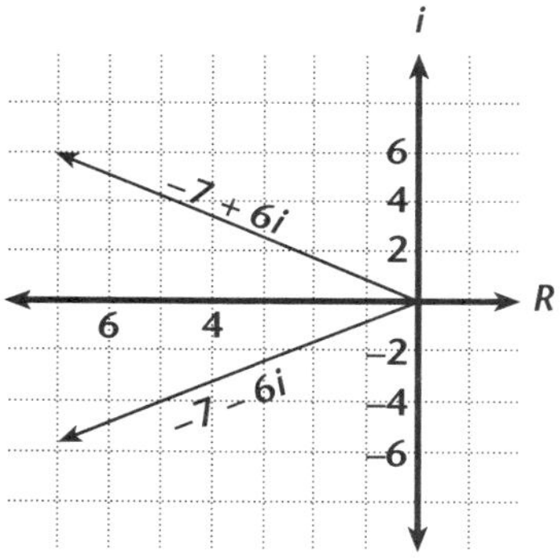

i
6
4
2
−7 + 6i
R
6
4
−2
−4
−6
−7 − 6i

10.

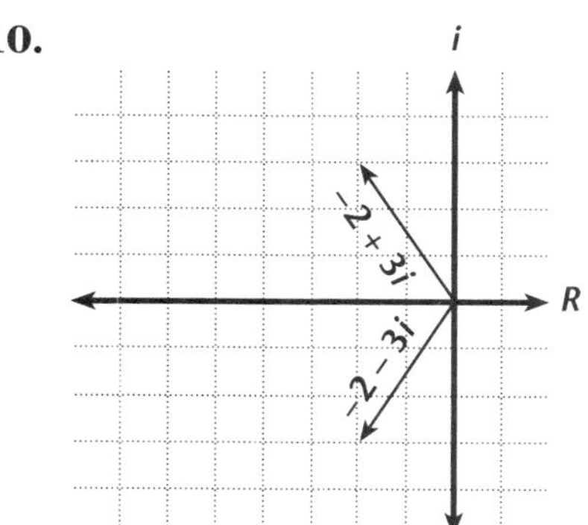

Assessing Prior Knowledge 5.8

1. $\begin{bmatrix} 2 & -5 \\ -1 & 3 \end{bmatrix}$ **2.** $\begin{bmatrix} -1 & 0 \\ 0 & 1 \end{bmatrix}$

3. $\begin{bmatrix} 0.2 & -0.2 & 0.2 \\ 1 & -0.5 & -1 \\ 0.4 & 0.1 & -0.6 \end{bmatrix}$

Quiz 5.8

1. $y = 2x^2 + 5x - 4$ **2.** $y = x^2 - 7x + 4$

3. $y = 11.5x^2 + 43.5x - 18$ **4.** ≈ 23.1 ft

5. ≈ 1.9 s **6.** ≈ 18.9 ft **7.** ≈ 1.3 s and 2.5 s

Assessing Prior Knowledge 5.9

1. $-1, \frac{5}{2}$ **2.** ± 3 **3.** 1 **4.** $-1, 3$

Quiz 5.9

1. 6 and -1 **2.** $-1 < x < 6$

3. $x < -1$ and $x > 6$

4.

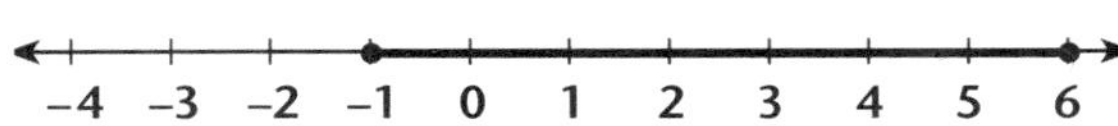

5. $x < \frac{5 - \sqrt{57}}{8}$ and $x > \frac{5 + \sqrt{57}}{8}$

6.

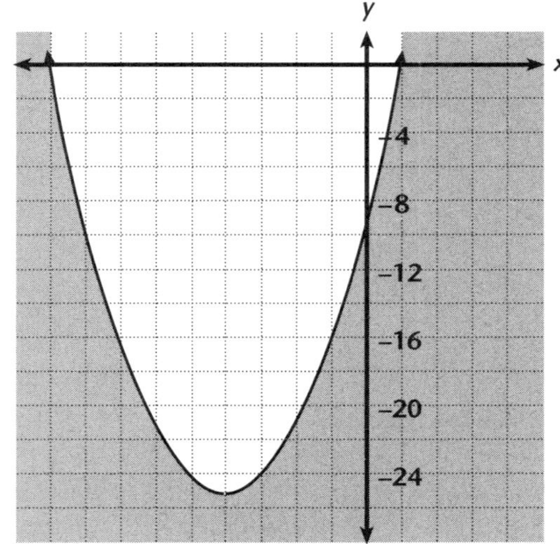

Chapter Assessment, Form A

1. b **2.** c **3.** a **4.** c **5.** d **6.** c **7.** b

8. a **9.** a **10.** d **11.** c **12.** b **13.** b

14. b **15.** d **16.** c

Chapter Assessment, Form B

1. $x = -3, \frac{5}{2}$ **2.** (2, 25) **3.** $x = 11$

4. $5\sqrt{5}$ **5.** $6\sqrt{2}$ **6.** $x = 6$

7.

8. $7, -1$ **9.** $6.7, -2.7$ **10.** 0

11. $y = 2x^2 - 5x + 7$ **12.** $2 \leq x \leq 10$

13. $-5 + i$

14.

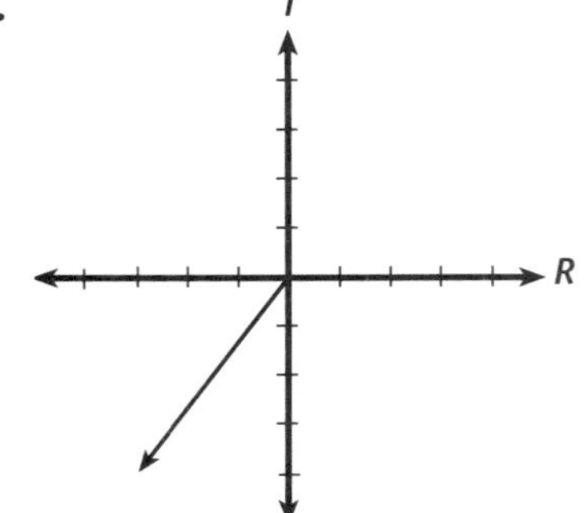

15. $x > \frac{5}{2}$ **16.** $-3i$ **17.** $m = \pm 3i\sqrt{6}$

18. $x > \frac{5}{6}$ **19.** $4\sqrt{3}$

Alternative Assessment — Chapter 5

Form A

1. Vertex: (2, 1); Axis of Symmetry: $x = 2$. x-intercepts: 1, 3; Answers may vary. Solve the equation $(x - 2)^2 = 1$.

2. Translate g 2 units to the right and 1 unit up.

3. The graph of f is narrower than the graph of g.

4. f is increasing when $x > -1$ and f is decreasing when $x < -1$. g is increasing when $x < -1$ and g is decreasing when $x > -1$. The negative coefficient of x^2 changes the direction of the opening of the quadratic function.

5. $f(x) = 9x^2$

Score Point 4: Distinguished

The student demonstrates a comprehensive understanding of analyzing graphs of quadratic functions. The student uses perceptive, creative, and complex mathematical reasoning throughout the task. He or she is able to use sophisticated, precise, and appropriate mathematical language throughout the task. Theoretical knowledge is apparent and applied to concrete situations as the student successfully demonstrates a comprehensive understanding of core concepts throughout the task.

Score Point 3: Proficient

The student demonstrates a broad understanding of analyzing graphs of quadratic functions. The student uses perceptive mathematical reasoning most of the time. He or she is able to use precise and appropriate mathematical language most of the time. Theoretical knowledge is apparent and applied to concrete situations as the student attempts to draw conclusions based on his or her investigations.

Score Point 2: Apprentice

The student demonstrates an understanding of analyzing graphs of quadratic functions. He or she uses mathematical reasoning at times during the task. He or she uses appropriate mathematical language some of the time. Student attempts to apply theoretical knowledge to the task but may be able to draw conclusions from his or her investigation.

Score Point 1: Novice

The student demonstrates a basic understanding of analyzing graphs of quadratic functions. He or she uses appropriate mathematical language some of the time. Theoretical knowledge is extremely weak and many responses are irrelevant or illogical. He or she may fail to follow directions and has great difficulty in communicating his or her responses.

Score Point 0: Unsatisfactory

Student fails to make an attempt to complete the task and his or her responses are just an attempt to fill the page or restate the problem.

Form B

1. When the three points are plotted on a coordinate plane, a quadratic function can be found with a graph that fits the three points.

2. $$\begin{bmatrix} 0.0625 & 0.25 & 1 \\ 0.25 & 0.5 & 1 \\ 0.3906 & 0.625 & 1 \end{bmatrix} \begin{bmatrix} a \\ b \\ c \end{bmatrix} = \begin{bmatrix} 30 \\ 100 \\ 120 \end{bmatrix}$$

   $a = -320, b = 520, c = -80$
   $f(x) = -320x^2 + 520x - 80$

3. About 100 sheets

4. The maximum size for the staple is $\frac{13}{16}$-inch. The staple will hold about 130 sheets of paper.

5. Answers will vary. A larger staple would be cumbersome to handle and require a great deal of pressure.

Score Point 4: Distinguished

The student demonstrates a comprehensive understanding of finding a quadratic model for data. The student uses perceptive, creative, and complex mathematical reasoning throughout the task. He or she is able to use sophisticated, precise, and appropriate mathematical language throughout the task. Theoretical knowledge is apparent and applied to concrete situations as the student successfully demonstrates a comprehensive understanding of core concepts throughout the task.

Score Point 3: Proficient

The student demonstrates a broad understanding of finding a quadratic model for data. The student uses perceptive mathematical reasoning most of the time. He or she is able to use precise and appropriate mathematical language most of the time. Theoretical knowledge is apparent and applied to concrete situations as the student attempts to draw conclusions based on his or her investigations.

Score Point 2: Apprentice

The student demonstrates an understanding of finding a quadratic model for data. He or she uses mathematical reasoning at times during the task. He or she uses appropriate mathematical language some of the time. Student attempts to apply theoretical knowledge to the task but may be able to draw conclusions from his or her investigation.

Score Point 1: Novice

The student demonstrates a basic understanding of finding a quadratic model for data. He or she uses appropriate mathematical language some of the time. Theoretical knowledge is extremely weak and many responses are irrelevant or illogical. He or she may fail to follow directions and has great difficulty in communicating his or her responses.

Score Point 0: Unsatisfactory

Student fails to make an attempt to complete the task and his or her responses are just an attempt to fill the page or restate the problem.

Practice & Apply—Chapter 6

Lesson 6.1

1. polynomial **2.** neither **3.** linear

4. neither **5.** third **6.** -2, 1, and 3

7. $f(x) = -x^3 + 2x^2 + 5x - 6$

8. -1 **9.** -6 **10.** 4 **11.** 3 **12.** 5

13. 6

14. 2; Possible answer: $f(x) = (x - 2)(x + 3)$

15. 4; Possible answer: $f(x) = (x + 3)(x + 2)(x - 1)(x - 2)$

16. 1; Possible answer: $f(x) = -2(x - \frac{1}{2})$

Lesson 6.2

1. $9x^2 - 25$ **2.** $x^3 - 49x$ **3.** $0.16x^2 - 1$

4. $\frac{1}{9}x^2 - \frac{1}{64}$ **5.** $16x^2 + 72x + 81$

6. $10x^2 - 21x - 10$ **7.** $(2x - 5)(x - 3)$

8. $(3x + 2)(2x - 3)$ **9.** $(2x - 3)(x - 1)$

10. $(3x + 2)(x + 3)$ **11.** $(3x - 1)(3x - 1)$

12. $(12x + 5)(12x - 5)$ **13.** a

14. Possible answer: $f(x) = 2x(x - 1)(x - 4)$ or $f(x) = 2x^3 - 10x^2 + 8x$

15. $x(x - 4)(x - 1)$

16. 0, 4, and 1

Lesson 6.3

1. c

2.
$$
\begin{array}{r}
2x^2 + 5x + 2 \\
3x - 2\,\overline{)\,6x^3 + 11x^2 - 4x - 4} \\
\underline{6x^3 - 4x^2} \quad\quad\quad \\
15x^2 - 4x \quad\; \\
\underline{15x^2 - 10x} \quad\; \\
6x - 4 \\
\underline{6x - 4} \\
0
\end{array}
$$

3. $2x^2 + 11x + 32$ R 102 **4.** $x^2 + x + 1$ R 0

5. $x^3 - 7x^2 + 8x - 11$ R 18 **6.** $x - 9$ R 30

7. no **8.** $x^2 - 4x + 6$ **9.** -1, 2, and 4

10. $(x + 1)(x - 2)(x - 4)$

11. $(x - 1)(x - 3)(x + 1)$

ANSWERS

Lesson 6.4

1. C **2.** A **3.** B

4. $f(x) = x(2x - 1)(x + 1)^2$

5.

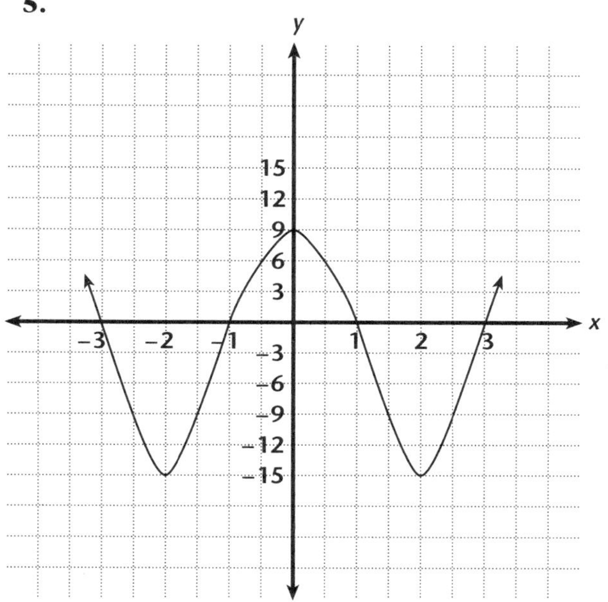

6. 3 **7.** at $x = -2$, $x = 0$, and $x = 2$

8. $-3, -1, 1,$ and 3

9. $x < -3$, $-1 < x < 1$, $x > 3$

10. $-2 < x < 0$, $x > 2$

Lesson 6.5

1. $V(x) = x(12 - 2x)(16 - 2x)$ **2.** $0 < x < 6$

3. 194 in.^{2}

4. $600(1.07)^3 + 700(1.07)^2 + 800(1.07) + 900 = \$3292.46$

5. 5% **6.** $21.87

7. $f(x) = 30(1 - 0.01x)^3$ where x = % discount

8. The function is always decreasing because the successive discounts lower the invoice price.

Enrichment — Chapter 6

Lesson 6.1

1. $f(x) = (x - 1)(x - 3)^2(x - 5)$

2. $f(x) = (x + 5)^2(x + 1)x(x - 2)^3$

3. $f(x) = (x + 2)^6\left(x + \frac{1}{2}\right)(x - 4)^4$

4. $f(x) = (x + 7)^{11}(x + 3)^5x^6(x - 5)$

5. $f(x) = \left(x + \frac{2}{3}\right)^2\left(x + \frac{1}{2}\right)x^3(x - 2)^5$

6. $f(x) = (x + 10)^2(x + 8)(x + 4)^7(x + 2)^5$

7. $f(x) = (x + 16)(x + 13)(x + 12)^2x(x - 5)^3(x - 8)$

8. $f(x) = (x + 43)(x + 27)(x + 19)^3x^6(x - 7)^2(x - 15)$

9. $f(x) = (x - 4)^8(x - 9)^2(x - 15)^8$

Lesson 6.2

1. $x^2 - 9 = (x + 3)(x - 3)$	**2.** $x^2 + x - 6 = (x + 5)(x - 1)$	**3.** $x^2 + x - 12 = (x - 4)(x + 3)$	**4.** $3x^2 - 9x + 6 = 3(x - 1)(x - 2)$
5. $4x^2 - 4x - 3 = (4x - 3)(x - 1)$	**6.** $6x^2 - 5x - 4 = (2x + 1)(3x - 4)$	**7.** $4x^3 - x = x(2x + 1)(2x - 1)$	**8.** $5x^2 - 15x + 10 = (5x - 5)(x - 2)$
9. $14x^2 + 49x - 28 = (2x - 1)(7x + 28)$	**10.** $3x^3 + 6x^3 - 105x = 3x(x - 5)(x + 7)$	**11.** $6x^2 - 7x - 20 = (3x + 4)(2x - 5)$	**12.** $12x^2 + x - 63 = (3x - 7)(4x + 9)$
13. $4x^2 - 1 = (2x - 1)(2x + 1)$	**14.** $4x^2 + x = (2x + 1)(2x + 1)$	**15.** $6x^2 - 15x = 3(2x^2 - 5x)$	**16.** $3x^2 - 26x - 9 = (3x + 1)(x - 9)$
17. $4x^2 + 5x - 6 = (2x - 1)(2x + 1)$	**18.** $6x^3 - 11x^2 - 72x = x(3x + 8)(2x - 9)$	**19.** $6x^2 + 7x - 55 = (3x + 11)(2x - 5)$	**20.** $24x^2 + 29x - 4 = (6x - 2)(4x + 2)$
21. $x^2 + 3x - 108 = (x - 9)(x + 12)$	**22.** $9x^2 - 3x - 2 = (3x - 1)(3x + 2)$	**23.** $5x^2 + 30x - 80 = (5x - 10)(x + 8)$	**24.** $9x^3 - x = x(3x + 1)(3x - 1)$
25. $14x^2 + 29x - 15 = (12x - 3)(7x + 5)$	**26.** $9x^2 - 16 = (3x + 4)(3x - 4)$	**27.** $18x^2 + 33x - 40 = (3x + 8)(6x - 5)$	**28.** $24x^2 + 34x - 45 = (8x - 50)(3x + 9)$
29. $6x^2 - x - 40 = (3x + 8)(2x - 5)$	**30.** $x^3 - 10x^2 + 25x = x(x - 5)^2$	**31.** $9x^2 - 24x + 16 = (3x - 4)^2$	**32.** $25x^2 - 70x + 49 = (5x + 7)^2$
33. $8x^2 + 26x - 99 = (4x - 9)(2x + 11)$	**34.** $42x^2 + 31x - 21 = (3x - 7)(12x + 3)$	**35.** $64x^2 + 48x + 9 = (8x + 3)(8x - 3)$	**36.** $56x^2 + 3x - 20 = (7x - 4)(8x + 5)$
37. $18x^2 + 24x - 40 = (8x - 8)(2x + 5)$	**38.** $24x^2 + 34x - 45 = (6x - 5)(4x + 9)$	**39.** $10x^2 - 3x - 27 = (5x - 9)(2x + 3)$	**40.** $2x^3 - 4x^2 - 38x = 2(x^2 + 4x)(x - 6)$
41. $12x^2 + 12x - 24 = 12(x - 1)(x + 2)$	**42.** $16x^2 - 64x + 48 = 4(x - 1)(4x - 12)$	**43.** $50x^2 - 300x + 450 = 50(x^2 - 6x + 9)$	**44.** $6x^2 + 35x - 209 = (2x + 19)(3x - 11)$

2. $(x + 3)(x - 2)$ **3.** $(x + 4)(x - 3)$

5. $(2x + 1)(2x - 3)$ **8.** $5(x - 1)(x - 2)$

9. $7(2x - 1)(x + 4)$ **12.** $(3x + 7)(4x - 9)$

14. $x(4x + 1)$ **15.** $3x(2x - 5)$

17. $(4x - 3)(x + 2)$ **20.** $(8x - 1)(3x + 4)$

22. $(3x + 1)(3x - 2)$ **23.** $5(x - 2)(x + 8)$

25. $(7x - 3)(2x + 5)$ **28.** $(4x + 9)(6x - 5)$

29. $(3x - 8)(2x + 5)$ **32.** $(5x - 7)^2$

34. $(7x - 3)(6x + 7)$ **35.** $(8x + 3)^2$

37. $8(x - 1)(2x + 5)$ **40.** $2x(x + 4)(x - 6)$

42. $16(x - 1)(x - 3)$ **43.** $50(x - 3)^2$

Lesson 6.3

1. 30,001 **2.** −56 **3.** 358,986 **4.** 33,891

5. 33,058,244 **6.** 54,190 **7.** 89,994

8. 7,005 **9.** 34,373 **10.** 25,945

11. 2,971,640 **12.** 299,736

13. −6,605,610 **14.** 1,174,152

Lesson 6.4

Lesson 6.5

1. about 655 cubic units **2.** \$4817.46

3. \$432.62 **4.** \$15,823.79 **5.** about \$700

Technology — Chapter 6

Lesson 6.1

1.

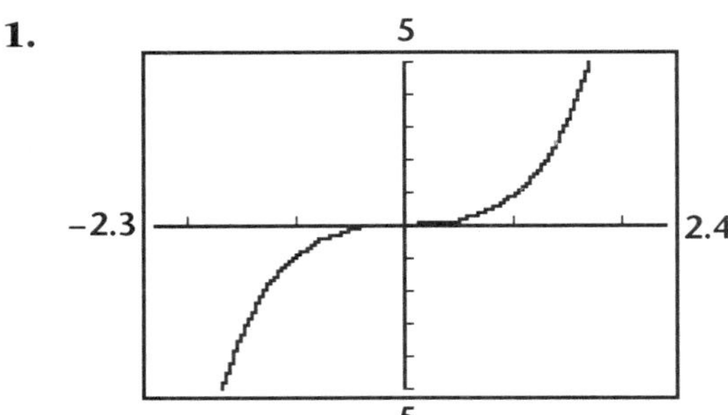

2.

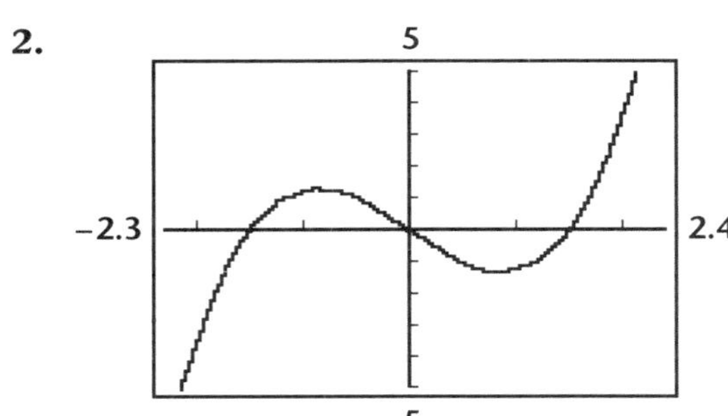

3.

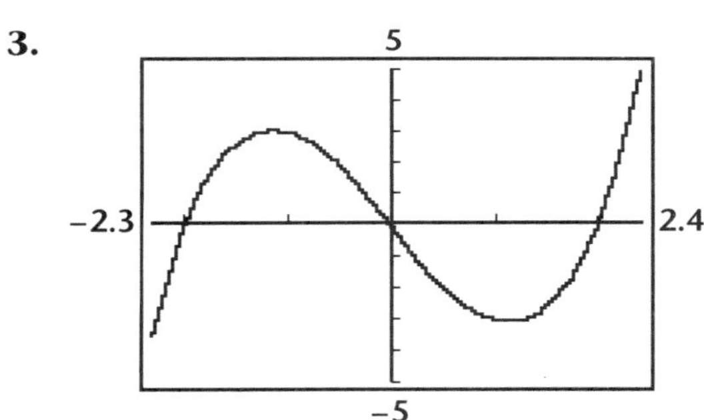

4. The graph is symmetric about the origin for any value of a and the expanded form of the polynomial contains only odd powers of x. If $a = 0$, there is one root counted three times and the graph passes through the third and first quadrants. If a is not 0, there are three distinct real roots, two of which are the same distance from the origin. As a increases, the roots move farther apart, the peaks get higher, and the valleys get lower.

5.

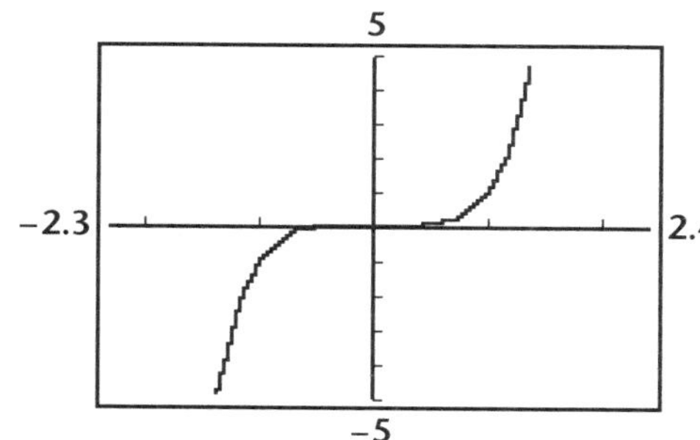

6.

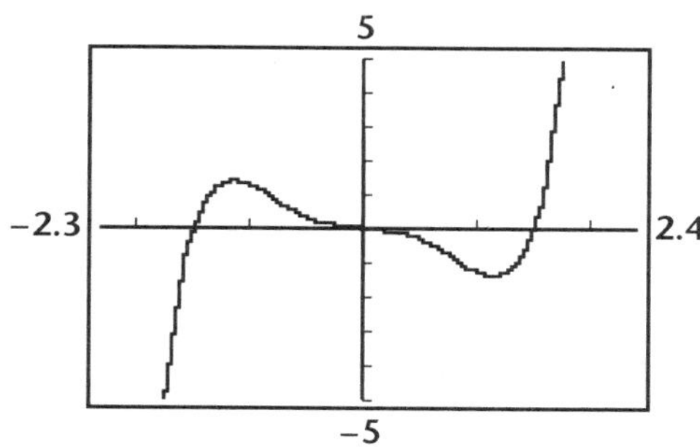

7.

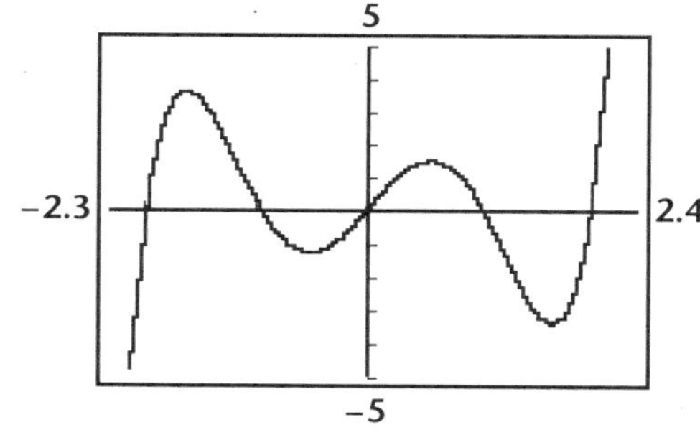

8. When the polynomial is expanded, it contains only odd powers of x and the graph is symmetric about the origin. If $a \neq 0$, $b \neq 0$, and $a = b$, the graph has two peaks, two valleys, and no turning points between them. The graph touches the x-axis twice and crosses it once.

9. When the polynomial is expanded, it contains only odd powers of x and the graph is symmetric about the origin. If $a \neq 0$, $b \neq 0$, $a \neq b$, the graph has two peaks, two valleys, and no turning points between them. The graph crosses the x-axis 5 times.

10. If $a = b = 0$, the graph is a U-shape opening upward with a minimum point at the origin. If $a = 0$ and $b \neq 0$, the graph has one peak at the origin and two valleys an equal distance from the origin. If $a \neq 0$, $b \neq 0$, $a \neq b$, the graph has one peak and two valleys an equal distance from the origin. The valleys do not have minimums on the x-axis. If $a \neq 0$, $b \neq 0$, and $a = b$, the graph has one peak and two valleys an equal distance from the origin. The valleys do have minimums on the x-axis. In each case, the expanded polynomial contains only even powers of x and the graph is symmetric about the y-axis.

Lesson 6.2

1. all n **2.** $n = 0$ **3.** $n = 0$, or 1

4. $n = 0$, or 2 **5.** no n **6.** $n = 1, 4, 9, 16, \ldots$

7. $n = 0$, or 4 **8.** $n = 0$, or 4

9.
$$x = \frac{-2n \pm \sqrt{4n^2 - 4(1)(n)}}{2}$$
$$= \frac{-2n \pm 2\sqrt{n^2 - n}}{2}$$
$$= -n \pm \sqrt{n^2 - n}$$

10. The graph suggests that the only horizontal lines ($y = k$, k an integer) that intersect the graph when n is an integer are the horizontal lines $y = 0$ and $y = 1$.

Lesson 6.3

1.

	A	B	C	D	E
1	2	1	−5	10	−3
2		1	−3	4	5

2.

	A	B	C	D	E
1	−2	1	−3	−9	2
2		1	−5	1	0

3.

	A	B	C	D	E
1	−4	1	−3	−26	9
2		1	−7	2	1

4.

	A	B	C	D	E
1	3	2	−1	−12	−9
2		2	5	3	0

5. Quotient: $x^2 - 3x + 4$
Remainder: 5

6. Quotient: $x^2 - 5x + 1$
Remainder: 0

7. Quotient: $x^2 - 7x + 2$
Remainder: 1

8. Quotient: $2x^2 + 5x + 3$
Remainder: 0

9. The elements in row 2 (except for the last element) give the coefficients of the quotient and the last element gives the remainder. If the remainder is zero, then $D(x)$ is a factor of $P(x)$.

Lesson 6.4

In Exercises 1–9, answers will vary. A sample answer is given.

1. $f(x) = (x + 3)(x)(x - 1)$
2. $f(x) = (x + 2)^2x^2(x - 1)^2$
3. $f(x) = -(x + 2)^2x^2(x - 1)^2$
4. $f(x) = x(x + 1)^2$
5. $f(x) = x^2(x - 1)$
6. $f(x) = x^2(x + 1)^2$
7. $f(x) = (x + 2)^2x^2(x - 2)^2$
8. $f(x) = (x + 2)(x)(x - 1)^2(x - 3)^2$
9. $f(x) = (x + 2)x^3(x - 1)^2(x - 3)^2$

Lesson 6.5

1. BJ*(1+5/100)^(30−AJ) where AJ represents the *j*th row of column A and BJ represents the *j*th row of column B.
2. CJ−BJ where BJ represents the *j*th row of column B and CJ represents the *j*th row of column C.
3. Amount: \$6643.88; Interest: \$3643.88
4. Amount: \$41,151.90; Interest: \$17,151.90
5. Amount: \$8525.33; Interest: \$4775.33
6. Amount: \$43,962.58; Interest: \$18,962.58
7. $100(1.05)^{29} + 100(1.05)^{28} + \ldots 100(1.05)^2 + 100(1.05)^1 + 100(1.05)^0$
8. $1000(1.044)^{23} + 1000(1.044)^{22} + \ldots 1000(1.044)^2 + 1000(1.044)^1 + 1000(1.044)^0$
9. \$212

Lesson Activities — Chapter 6

Lesson 6.1

1. Possible answer: Count the number of 1×1, 2×2, 3×3, and so on, squares contained in the given figure.

2.

1	4	9	16	25	36
0	1	4	9	16	25
0	0	1	4	9	16
0	0	0	1	4	9
0	0	0	0	1	4
0	0	0	0	0	1
1	5	14	30	55	91

3. 91

4.

1		5		14		30		55		91
	4		9		16		25		36	
		5		7		9		11		
			2		2		2			

5. Possible answer: The finite difference is constant at the third difference. The polynomial is third degree.

Lesson 6.2

1. Check students' graphs.

2. The zero is $n = 2$.

3. The domain must be greater than or equal to zero, where the graph touches the x-axis.

4. The function is a cubic polynomial, whose graph crosses the x-axis.

5. Check students' graphs. The zero of E is $n = 2$.

6. The multiplicity of 2 is two. The factor is repeated twice, therefore it produces multiple roots.

Lesson 6.3

1. $(((((x + 1)x - 2)x - 5)x + 2)x - 8)x + 3$

2. $4x^3 + 3x^2 - x + 1$

3. $P(x) = (((x - 5)x + 1)x - 2)x + 1$; The quotient is $x^3 - 6x^2 + 7x - 9$ and the remainder is 10.

4. $f(x) = (((x + 1)x - 11)x - 31)x - 20$

5. $f(4) = (((4 + 1)4 - 11)4 - 31)4 - 20 = 0$
$f(-1) = (((-1) + 1)(-1) - 11(-1) - 31(-1) - 20 = 0$

Lesson 6.4

1. 4 **2.** 10

3. (1)(2)(6) + (1)(3)(6) + (1)(4)(6) + (1)(5)(6) + (2)(3)(6) + (2)(4)(6) + (2)(5)(6) + (3)(4)(6) + (3)(5)(6) + (4)(5)(6)

4. 1960 **5.** 11 = (1)(2) + (1)(3) + (2)(3)

6. 35 = [(1)(2) + (1)(3) + (2)(3)] + (1)(4) + (2)(4) + (3)(4)

7. 85 = [(1)(2) + (1)(3) + (2)(3)] + (1)(4) + (2)(4) + (3)(4)] + (1)(5) + (2)(5) + (3)(5) + (4)(5)

8. 175 = [(1)(2) + (1)(3) + (2)(3)] + [(1)(4) + (2)(4) + (3)(4)] + [(1)(5) + (2)(5) + (3)(5) + (4)(5)] + (1)(6) + (2)(6) + (3)(6) + (4)(6) + (5)(6)

9. 2

Lesson 6.5

1. \$597.79

2. $(((((((1000r - 88.85)r - 88.85)r - 88.85)r - 88.85)r - 88.85)r - 88.85)r - 88.85)r - 88.85$

3. \$346.67

4. $1000r^8 - 88.85r^7 - 88.85r^6 - 88.85r^5 - 88.85r^4 - 88.85r^3 - 88.85r^2 - 88.85r - 88.85$

5. $1000r^{12} - 88.85r^{11} - 88.85r^{10} - 88.85r^9 - 88.85r^8 - 88.85r^7 - 88.85r^6 - 88.85r^5 - 88.85r^4 - 88.85r^3 - 88.85r^2 - 88.85r - 88.85$

6. $(((((2000x - 177.70)x - 177.70)x - 177.70)x - 177.70)x - 177.70)x - 177.70)$, where $x = 1.01$

7. \$1029.83

8. $2000x^{12} - 177.70x^{11} - 177.50x^{10} - 177.70x^9 - 177.70x^8 - 177.70x^7 - 177.70x^6 - 177.60x^5 - 177.70x^4 - 177.70x^3 - 177.70x^2 - 177.70x - 177.70$

9. The degree of the polynomial is the number of installments.

Assessment — Chapter 6

Assessing Prior Knowledge 6.1

1. 26 ft **2.** 40 ft^2 **3.** 120 ft^3

Quiz 6.1

1. degree three **2.** degree five

3. $a_1 = -1, a_3 = 5, a_4 = -2$

4. $x = 3, x = -\frac{5}{2}, x = -2, x = \frac{1}{3}$

5. $x = \frac{2}{5}, x = -3, x = \frac{1}{2}$

6. $x = -3, x = -2$ 7. $x = -3$ 8. $x = -2$

9. $D: x > -3$ 10. 0.147

Assessing Prior Knowledge 6.2

1. −3, 7 2. −2, 4 3. −2, 0 4. −5, 5

Quiz 6.2

1. $(x - 3)(x - 2)$ 2. $(x + 2)(x + 4)$

3. $(x - 7)(x + 3)$ 4. $(x + 9)(x - 2)$

5. $(2x - 3)(x + 4)$ 6. $(3x - 2)(3x + 2)$

7. $x = -0.33, x = 2; f(x) = (3x + 1)(x - 2)$

8. $x = 0.5, x = -1; f(x) = (2x - 1)(x + 1)$

9. $x = 0, x = -0.5, x = 3;$
$f(x) = x(2x + 1)(x - 3)$

10. $x = 0, x = 0.67, x = -3;$
$f(x) = x(3x - 2)(x + 3)$

11. If you set the linear factors equal to zero and solve for x, the solutions are the zeros of the function.

12. $f(x) = x^2 - 2x - 3$

Assessing Prior Knowledge 6.3

1. $x^4 + 0x^3 + x^2 - 4x + 3$

2 $x^3 + 0x^2 + 0x - 1000$

3. $2x^3 + 0x^2 - 3x + 6$

4. $3x^4 + 0x^3 + x^2 + 0x + 2$

Quiz 6.3

1. four 2. $x = -3$

3. $x = -\frac{2}{3}, x = \frac{1}{2}$ 4. $x = -2, x = 3$

5. yes; degree 4

6. $(x + 3)(x + 1)(x - 2)(x - 4)$

7. $(x - 3)$ and $(x + 1)$

Mid-Chapter Assessment

1. c 2. c 3. d 4. b 5. true 6. false

7. $(x + 5)(x + 7)$ 8. $(x + 7)(x - 3)$

9. $(x + 3)$ and $(x + 4)$

10. $f(x) = x(7 - x)(x + 2)$

11.

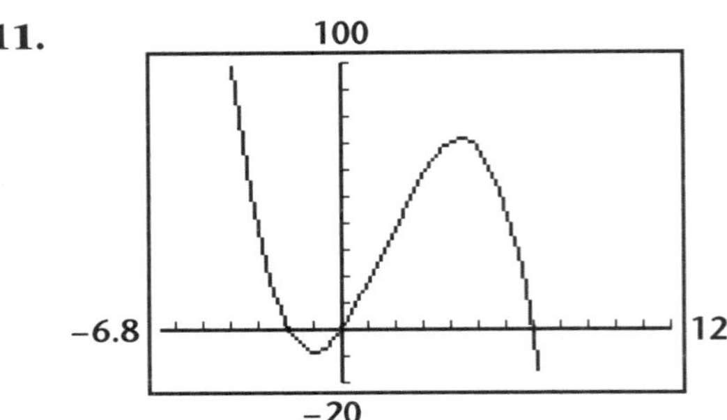

12. 1 through 6

13. width = 4.4 m, length = 2.6 m,
height = 6.4 m

14. 73.2 cubic meters

Assessing Prior Knowledge 6.4

1. 0 2. −1 3. −2, 2 4. 3

Quiz 6.4

1. $x = -1, x = 3, x = -2, x = 1$; degree 5

2. no; f has 3 crossing points and 1 turning point

3. The left end turns upward, the right end turns downward.

4. $x = 1$ 5. f has 3 crossing points.

6. Answers will vary. Possible answer: $f(x) = (x + 2)(x + 3)(x + 4)^2$.

7. $x = -2.2, x = -1.5, x = 0.2$

8. $(x + 2.2)(x + 1.5)(x - 0.2)$

9. $x < -1.9 < x < -0.45$

10. $-1.9 < x < -0.45$

11. $x = -1.9$ and $x = -0.4$

Assessing Prior Knowledge 6.5

1. ≈ 15.6 seconds **2.** ≈ 3916 feet

Quiz 6.5

1. $4x(x - 5)(x - 6)$

2.

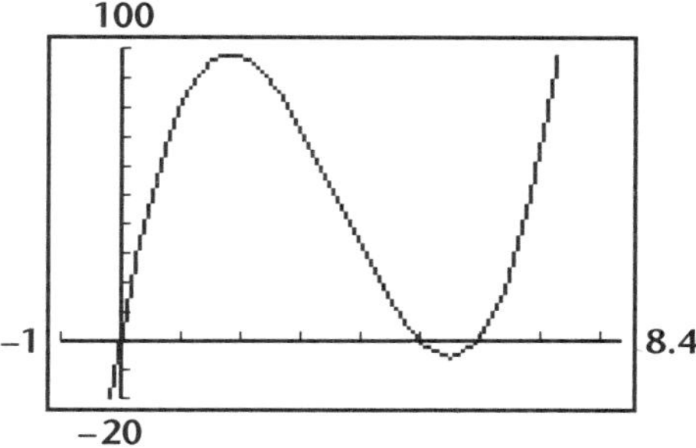

3. 96 ft^2 **4.** $0 < x < 5$ **5.** \$1775.98

6. \$175.98

7. no; he will still be \$199.70 short

Chapter 6 Assessment, Form A

1. c **2** b **3.** a **4.** b **5.** d **6.** a **7.** c

8. d **9.** a **10.** b **11.** c **12.** b **13.** b

14. a **15.** c **16** d **17.** b

Chapter 6 Assessment, Form B

1. $x = -3, x = 2$ **2.** $x = -3$ **3.** $x = 2$

4. $x \geq -3$ **5.** $y = 18.5$

6. $V(x) = x(5 - x)(x - 2)$ **7.** $(3x + 2)(x + 1)$

8. $(3x - 4)(3x + 4)$ **9.** $f(x) = x^2 - 4x - 5$

10. 7 **11.** $(x + 5)(x + 6)(x - 10)$

12. $(x + 2)$ and $(x + 3)^2$

13. $f(x) = 3(x - 2)(x + 4)$

14. 4 **15.** $x = -2$ **16.** $x = 1$ and $x = 3$

17. $x = -6.7, x = -3.5, x = 1.6, x = 3.6$

18. $x = -5.4; x = -1.2, x = 2.6$

19. $-5.4 < x < -1.1$ and $x > 2.7$

20. $x < -5.4, -1.1 < x < 2.7$

21. $V(x) = 4x(x - 6)(x - 4)$

22. $0 < x < 4$ **23.** above 68 cubic in.

24. \$1936.74 **25.** \$736.74 **26.** at least \$350

Alternative Assessment — Chapter 6

Form A

1. There is one more linear factor than the degree of the polynomial function. Therefore, the linear factors are not the in the factored form of *f*.

2. Use the Factor Theorem to determine whether $x + 2$ divides *f* exactly.

3. $f(x) = (x - 5)(x + 4)$

4. $f(x) = x(9x + 4)(9x - 4)$

5. Every polynomial $P_n(x)$ of degree n can be written as a factored product of exactly n linear factors corresponding to exactly n zeros (counting multiplicities).

6. *f* has 1 real zero and 2 complex zeros. The real zero is 3. The complex zeros are i and $-i$. $f(x) = (x - 3)(x + i)(x - i)$.

Score Point 4: Distinguished

The student demonstrates a comprehensive understanding of the factored form of a polynomial. The student uses perceptive, creative, and complex mathematical reasoning throughout the task. He or she is able to use sophisticated, precise, and appropriate mathematical language throughout the task. Theoretical knowledge is apparent and applied to concrete situations as the student successfully demonstrates a comprehensive understanding of core concepts throughout the task.

Score Point 3: Proficient

The student demonstrates a broad understanding of the factored form of a polynomial. The student uses perceptive mathematical reasoning most of the time. He or she is able to use precise and appropriate mathematical language most of the time. Theoretical knowledge is apparent and applied to concrete situations as the student attempts to draw conclusions based on his or her investigations.

Score Point 2: Apprentice

The student demonstrates an understanding of the factored form of a polynomial. He or she uses mathematical reasoning at times during the task. He or she uses appropriate mathematical language some of the time. Student attempts to apply theoretical knowledge to the task but may be able to draw conclusions from his or her investigation.

Score Point 1: Novice

The student demonstrates a basic understanding of the factored form of a polynomial. He or she uses appropriate mathematical language some of the time. Theoretical knowledge is extremely weak and many responses are irrelevant or illogical. He or she may fail to follow directions and has great difficulty in communicating his or her responses.

Score Point 0: Unsatisfactory

Student fails to make an attempt to complete the task and his or her responses are just an attempt to fill the page or restate the problem.

Form B

1. The zeros of f and 0 (multiplicity 2), and -2. The degree of f is 3. The zeros of g are 0 (multiplicity 2), and -2. The degree of f is 3.

2. f is increasing when $x < -\frac{4}{3}$ or $x > 0$.

f is decreasing when $-\frac{4}{3} < x < 0$.

g is increasing when $-\frac{4}{3} < x < 0$.

g is decreasing when $x < -\frac{4}{3}$ of $x > 0$.

3. The negative coefficient inverts the shape of the graph.

4. $f(x) = (x - 1)(x + 1)(x - 2)(x + 3)$

5. At a turning point, the polynomial changes from increasing to decreasing or from decreasing to increasing. f has 3 turning points.

6. At a crossing point, the graph changes signs. f has 4 crossing points.

Score Point 4: Distinguished

The student demonstrates a comprehensive understanding of polynomial function behavior. The student uses perceptive, creative, and complex mathematical reasoning throughout the task. He or she is able to use sophisticated, precise, and appropriate mathematical language throughout the task. Theoretical knowledge is apparent and applied to concrete situations as the student successfully demonstrates a comprehensive understanding of core concepts throughout the task.

Score Point 3: Proficient

The student demonstrates a broad understanding of polynomial function behavior. The student uses perceptive mathematical reasoning most of the time. He or she is able to use precise and appropriate mathematical language most of the time. Theoretical knowledge is apparent and applied to concrete situations as the student attempts to draw conclusions based on his or her investigations.

Score Point 2: Apprentice

The student demonstrates an understanding of polynomial function behavior. He or she uses mathematical reasoning at times during the task. He or she uses appropriate mathematical language some of the time. Student attempts to apply theoretical knowledge to the task but may be able to draw conclusions from his or her investigation.

Score Point 1: Novice

The student demonstrates a basic understanding of polynomial function behavior. He or she uses appropriate mathematical language some of the time. Theoretical knowledge is extremely weak and many responses are irrelevant or illogical. He or she may fail to follow directions and has great difficulty in communicating his or her responses.

Score Point 0: Unsatisfactory

Student fails to make an attempt to complete the task and his or her responses are just an attempt to fill the page or restate the problem.

Cumulative Assessment Free-Response Grids

Exercise ________________

	/	/	/	/	
·	·	·	·	·	·
	0	0	0	0	0
1	1	1	1	1	1
2	2	2	2	2	2
3	3	3	3	3	3
4	4	4	4	4	4
5	5	5	5	5	5
6	6	6	6	6	6
7	7	7	7	7	7
8	8	8	8	8	8
9	9	9	9	9	9

Exercise ________________

	/	/	/	/	
·	·	·	·	·	·
	0	0	0	0	0
1	1	1	1	1	1
2	2	2	2	2	2
3	3	3	3	3	3
4	4	4	4	4	4
5	5	5	5	5	5
6	6	6	6	6	6
7	7	7	7	7	7
8	8	8	8	8	8
9	9	9	9	9	9

Exercise ________________

	/	/	/	/	
·	·	·	·	·	·
	0	0	0	0	0
1	1	1	1	1	1
2	2	2	2	2	2
3	3	3	3	3	3
4	4	4	4	4	4
5	5	5	5	5	5
6	6	6	6	6	6
7	7	7	7	7	7
8	8	8	8	8	8
9	9	9	9	9	9

Exercise ________________

	/	/	/	/	
·	·	·	·	·	·
	0	0	0	0	0
1	1	1	1	1	1
2	2	2	2	2	2
3	3	3	3	3	3
4	4	4	4	4	4
5	5	5	5	5	5
6	6	6	6	6	6
7	7	7	7	7	7
8	8	8	8	8	8
9	9	9	9	9	9

Exercise ________________

	/	/	/	/	
·	·	·	·	·	·
	0	0	0	0	0
1	1	1	1	1	1
2	2	2	2	2	2
3	3	3	3	3	3
4	4	4	4	4	4
5	5	5	5	5	5
6	6	6	6	6	6
7	7	7	7	7	7
8	8	8	8	8	8
9	9	9	9	9	9

Exercise ________________

	/	/	/	/	
·	·	·	·	·	·
	0	0	0	0	0
1	1	1	1	1	1
2	2	2	2	2	2
3	3	3	3	3	3
4	4	4	4	4	4
5	5	5	5	5	5
6	6	6	6	6	6
7	7	7	7	7	7
8	8	8	8	8	8
9	9	9	9	9	9

Exercise ________________

	/	/	/	/	
·	·	·	·	·	·
	0	0	0	0	0
1	1	1	1	1	1
2	2	2	2	2	2
3	3	3	3	3	3
4	4	4	4	4	4
5	5	5	5	5	5
6	6	6	6	6	6
7	7	7	7	7	7
8	8	8	8	8	8
9	9	9	9	9	9

Exercise ________________

	/	/	/	/	
·	·	·	·	·	·
	0	0	0	0	0
1	1	1	1	1	1
2	2	2	2	2	2
3	3	3	3	3	3
4	4	4	4	4	4
5	5	5	5	5	5
6	6	6	6	6	6
7	7	7	7	7	7
8	8	8	8	8	8
9	9	9	9	9	9

Exercise ________________

	/	/	/	/	
·	·	·	·	·	·
	0	0	0	0	0
1	1	1	1	1	1
2	2	2	2	2	2
3	3	3	3	3	3
4	4	4	4	4	4
5	5	5	5	5	5
6	6	6	6	6	6
7	7	7	7	7	7
8	8	8	8	8	8
9	9	9	9	9	9

PORTFOLIO HOLISTIC SCORING GUIDE

An individual portfolio is likely to be characterized by some, but not all, of the descriptors for a particular level. Therefore, the overall score should be the level at which the appropriate descriptors for a portfolio are clustered.

		NOVICE	APPRENTICE	PROFICIENT	DISTINGUISHED
PROBLEM SOLVING	Understanding/Strategies	• Indicates a basic understanding of problems and uses strategies	• Indicates an understanding of problems and selects appropriate strategies	• Indicates a broad understanding of problems with alternate strategies	• Indicates a comprehensive understanding of problems with efficient, sophisticated strategies
	Execution/Extensions	• Implements strategies with minor mathematical errors in the solution without observations or extensions	• Accurately implements strategies with solutions, with limited observations or extension	• Accurately and efficiently implements and analyzes strategies with correct solutions, with extension	• Accurately and efficiently implements and evaluates sophisticated strategies with correct solutions and includes analysis, justifications, and extensions
REASONING		• Uses mathematical reasoning	• Uses appropriate mathematical reasoning	• Uses perceptive mathematical reasoning	• Uses perceptive, creative, and complex mathematical reasoning
MATHEMATICAL COMMUNICATION	Language	• Uses appropriate mathematical language some of the time	• Uses appropriate mathematical language	• Uses precise and appropriate mathematical language most of the time	• Uses sophisticated, precise, and appropriate mathematical language throughout
	Representations	• Uses few mathematical representations	• Uses a variety of mathematical representations accurately and appropriately	• Uses a wide variety of mathematical representations accurately and appropriately; uses multiple representations within some entries	• Uses a wide variety of mathematical representations accurately and appropriately; uses multiple representations within entries and states their connections
UNDERSTANDING/CONNECTING CORE CONCEPTS		• Indicates a basic understanding of core concepts	• Indicates an understanding of core concepts with limited connections	• Indicates a broad understanding of some core concepts with connections	• Indicates a comprehensive understanding of core concepts with connections throughout
TYPES AND TOOLS		• Includes few types; uses few tools	• Includes a variety of types; uses tools appropriately	• Includes a wide variety of types; uses a wide variety of tools appropriately	• Includes all types; uses a wide variety of tools appropriately and insightfully

PERFORMANCE DESCRIPTORS	WORKSPACE/ANNOTATIONS
PROBLEM SOLVING • Understanding the features of a problem (understands the question, restates the problem in own words) • Explores (draws a diagram, constructs a model and/or chart, records data, looks for patterns) • Selects an appropriate strategy (guesses and checks, makes an exhaustive list, solves a simpler but similar problem, works backward, estimates a solution) • Solves (implements a strategy with an accurate solution) • Reviews, revises, and extends (verifies, explores, analyzes, evaluates strategies/solutions; formulates a rule)	
REASONING • Observes data, records and recognizes patterns, makes mathematical conjectures (inductive reason) • Validates mathematical conjectures through logical arguments or counter-examples; constructs valid arguments (deductive reasoning)	
MATHEMATICAL COMMUNICATION • Provides quality explanations and expresses concepts, ideas, and reflections clearly • Uses appropriate mathematical notation and terminology • Provides various mathematical representations (models, graphs, charts, diagrams, words, pictures, numerals, symbols, equations)	

UNDERSTANDING/CONNECTING CORE CONCEPTS

- Demonstrates an understanding core concepts
- Recognizes, makes, or applies the connections among the mathematical core concepts to other disciplines, and to the real world

Place an X on each continuum to indicate the degree of understanding demonstrated for each core concept.

DEGREE OF UNDERSTANDING OF CORE CONCEPTS (Basic → Comprehensive with connections)

- NUMBER
- MATHEMATICAL PROCEDURES
- SPACE & DIMENSIONALITY
- MEASUREMENT
- CHANGE
- MATHEMATICAL STRUCTURE
- DATA: STATISTICS AND PROBABILITY

PORTFOLIO CONTENTS

- Table of Contents
- Letter to Reviewer
- 5–7 Best Entries

BREADTH OF ENTRIES

TYPES

- ○ INVESTIGATIONS/DISCOVERY
- ○ APPLICATIONS
- ○ NON-ROUTINE PROBLEMS
- ○ PROJECTS
- ○ INTERDISCIPLINARY
- ○ WRITING

TOOLS

- ○ CALCULATORS
- ○ COMPUTER AND OTHER TECHNOLOGY
- ○ MODELS-MANIPULATIVE
- ○ MEASUREMENT INSTRUMENTS
- ○ OTHERS

GROUP ENTRY

The Kentucky Mathematics Portfolio was developed by the Kentucky Department of Education for use by school districts throughout that state.

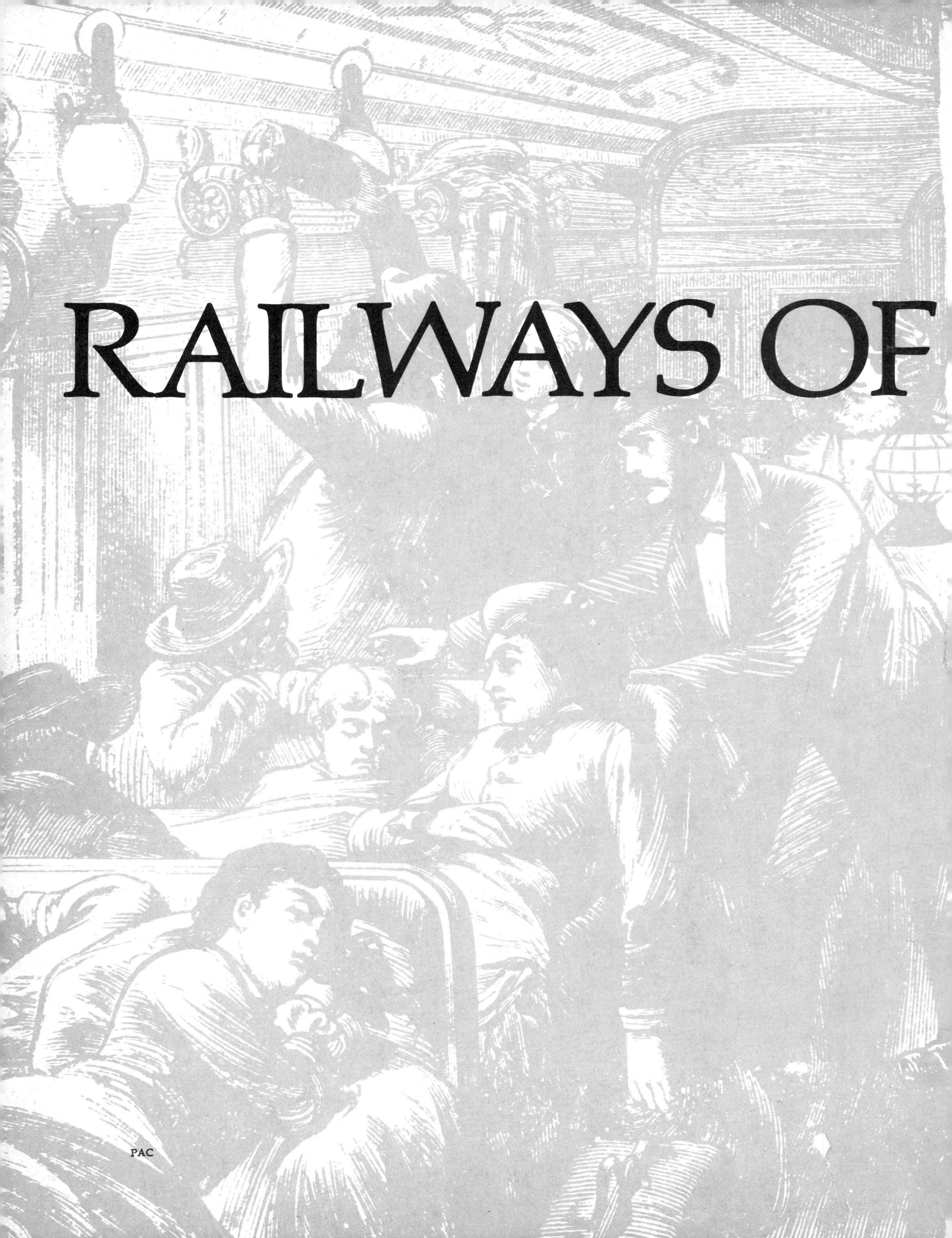

RAILWAYS OF

PAC

CANADA
A Pictorial History

Nick and Helma Mika

McGraw-Hill Ryerson Limited Toronto Montreal

RAILWAYS OF CANADA

Library of Congress Catalog Number 72-3861

ISBN 0-07-092776-6

1 2 3 4 5 6 7 8 9 EP 72 0 9 8 7 6 5 4 3 2

Printed and bound in Canada

PAC

Contents

CP

Introduction

The story of Canada's railways is many things. It is an account of the country's early growth, for railways opened up land to agriculture, tapped our wealth of resources, shifted large segments of population and brought immigrant settlers to remote areas. It is the miracle of scattered sleepy villages mushrooming into bustling cities because of freight trains moving farm produce and manufactured goods to consumer markets. It is the fate of towns and hamlets doomed to stagnation because the tracks of steel had passed them by.

Woven into the colourful tapestry of the railways' story is the history of Canada's birth as a nation, for it was the shimmering band of steel that first welded together the far-flung regions of the young dominion.

The story of Canadian railways is an epic too of man's dreams and ambitions, of daring adventure and back-breaking labour clearing the rights of way in the wilderness, bridging lakes and canyons, blasting through solid rock and filling seemingly bottomless swamps to lay the tracks. It is the story of men who built the Iron Horses which first blazed a trail through virgin forest; of engineers who braved danger and the elements guiding their trains safely through blizzard and darkness; of men who stoked the hungry fires, braked the cars in hazardous circumstances, walked and watched the tracks on duty around the clock and gave their lives riding the rails as the pioneers of progress.

And, last but not least, the story of railways tells of spectacular change and technological achievements in man's never ending quest for faster means of transportation. From the teetering little woodburners puffing through the backwoods of Canada evolved the giant steam locomotives which crossed the continent and conquered the towering mountain ranges.

Made obsolete by even more powerful and efficient diesel engines, steam locomotives no longer rule the rails. Only a few of the faithful oldtimers are still preserved — a part of our heritage and proud reminders of a glorious past — but the story of railways goes on as they serve the ever growing needs of a modern nation.

Canada's Oldest Iron Horse

On Archimedes Street in New Glasgow, Nova Scotia, next to the public library stands a modern structure of glass, steel and concrete which attracts visitors from every corner of the country. Brightly lighted at night, the building serves as the home of the *Samson*, Canada's oldest surviving Iron Horse.

Today, in the age of giant diesel engines, the *Samson* looks like a toy. In her heyday, however, she was one of the largest and most powerful locomotives in North America. Capable of pulling thirty-two cars with a weight of three tons each up a steep hill, she could easily move a load of four hundred tons on level track. A regular workhorse, she travelled at an average speed of six miles per hour and, according to her driver, as the engineer in the old days was called, she "seldom caused trouble and never left the track." She was in fact so well constructed that she hardly ever needed repairs during her forty-odd years of service. Today one of the few survivors of the pioneer steam era, she still consists almost entirely of original parts.

The *Samson*, which was the first six-coupled steam engine in British North America, can also claim the distinction of having been the first locomotive on the continent to burn coal and the first to run over all-iron rails. She is twenty feet long and fourteen feet high to the top of her smoke stack; her three driving wheels on either side measure forty-eight inches in diameter; she weighs approximately 38,000 pounds. Fuel and water used to be kept in a tender in front of the engine. The fire was made on a grate in the bottom of the flue and coal was fed directly through the stack. A back draught caused immense clouds of black smoke to emerge each time the fire door was opened. Iron baskets filled with glowing coal served as headlights at the front of the engine.

The regular driver of the *Samson* was a hardy Scotsman by the name of Donald Thompson. He did not have the comforts of a cab but sat on a small iron chair at the rear of the locomotive, fully exposed to the elements. When he had to check on the boiler during the run, he climbed over one of the cylinders and clambered along the narrow running board of the engine. He was extremely fond of his locomotive and, by his own admission, took better care of her than he did of his wife!

The *Samson* was built during the summer of 1838 in New Shildon, England, at the works of Timothy Hackworth. On completion, she and two other locomotives were shipped on a sailing vessel to Pictou, Nova Scotia, where she is said to have arrived in the fall of 1838. The exact arrival date, however, seems more likely to have been May 27,

The Samson, *first steam engine in Nova Scotia*

PAC

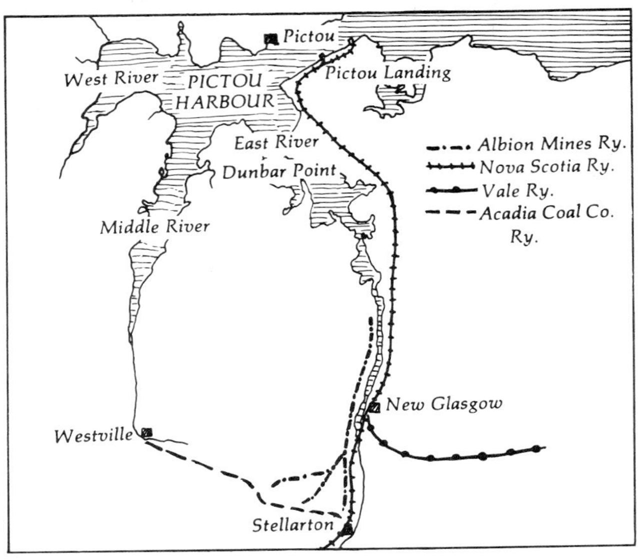

1839 as on that day the Pictou customs register lists under the heading "Ships entered" the

Brig Ythan with Captain Trotter from Newcastle-on-Tyne carrying locomotive engines for the Mining Associaton . . .

The mining company had ordered the three Hackworth engines for service at the Albion Mines near what is now the town of Stellarton. A railway track was being built at the time from the mine to the so-called loading ground on the East River, a distance of about six miles. The engines were to haul coal from the pithead to the river, where the cargo was to be transferred onto light sailing vessels bound for Pictou Harbour.

In the early days of mining operations in the area, coal was carted to the river bank. In 1829 a tramway operated by horses was opened from the pits to a wharf then located below the cemetery of Lourdes. The tramway eventually was extended downstream to New Glasgow and employed over one hundred horses along the road. The rails of this early line were of the cast-iron, fish-belly type and were made at the Albion Mines. They are considered by experts to have been the first iron rails manufactured on the North American continent. No trace of the tramway itself remains today, but several lengths of the historic rails have been preserved and are on display in museums of Nova Scotia.

Aware of the need for a more economic method of transporting their coal, the General Mining Association which ran the Albion Mines ordered the construction of a locomotive railway in the spring of 1836. It was to become Nova Scotia's first railroad, and the plans for it were drawn by Peter Crerar, a Scottish-born schoolteacher and land surveyor who had never seen a railway in his life. Experienced railway engineers, however, were so impressed with his know-how that they recommended he be employed as supervisor of construction for the project.

The line was built on a higher level and further removed from the river bank than the original tramway. Malleable iron rails of the bull-head type, set in cast-iron chairs and secured by tapered iron wedges, were used to support the weight of heavy locomotives. The total length of the track measured 10,694 yards, or slightly more than six miles, and the road ran nearly straight through uneven countryside. Bridges and culverts along the line were constructed of cut stone, and some of the latter are still in use today.

The *Samson* was chosen for the first test run over the newly built stretch of road. George Davidson, who had accompanied her from

England, served as the driver while David Floyd was the fireman on the historic run. Brakeman Patrick Kerwin, the third member of the crew, lifted his little daughter Maggie aboard the *Samson* to let her have the "first ride and make her remembered in the land." Well, Maggie has gone down into history as Nova Scotia's very first railway passenger.

Officially the railway was not opened until September 19, 1839, when thousands flocked to Albion Mines on foot, by boat, on horseback or by carriage to celebrate the big event. Bands and parades entertained the crowd, while the main attractions of the day were the company's locomotives—*Samson, Hercules* and *John Buddle*—lined up on the track outside the engine house. Not many people had ever seen a steam locomotive before, and the three Iron Horses were a sight to behold.

At the engine house a giant dinner party was given for employees and guests of the mining company. An ox had been roasted in a brick oven especially built for the occasion. The menu included, among other things, a half-ton of beef-and-mutton stew. Rum flowed like water and helped keep spirits high.

In the afternoon seven hundred happy passengers piled into thirty-five cleaned-up chaldron wagons and, hauled by the *John Buddle*, went for the first train ride in their lives. The *Hercules* ran next with an equal number of people. Each train made two round trips and a good time was had by all. The *Samson*, it appears, was not put into action.

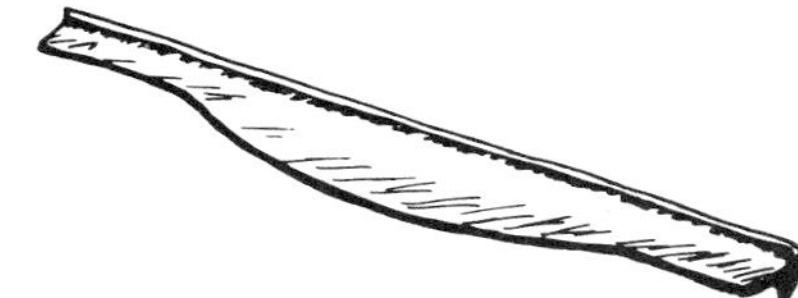
Rail of the cast-iron fish-belly type made at the Albion Mines

Some time after the railway commenced regular operations, a small wooden passenger coach was added to the rolling stock to convey company officials to and from the mines. One of the first passengers to ride in the elaborately furnished coach was the Governor-General's new bride. Thus the carriage came to be called "The Bride's Coach" and young lady passengers who rode in it without uttering a word for twenty minutes would, it was said, find a husband within the year. Long since, the original "Bride's Coach" came into the possession of the Baltimore and Ohio Railroad Company but an exact replica of it is on display, together with the *Samson*, in New Glasgow, Nova Scotia.

In regular service for nearly thirty years, the *Samson* hauled coal from the pithead of the Albion Mines to the loading pier at Dunbar Point. Semi-retired, she continued to operate in the area until 1884 when she seemed destined for the scrap yard. But nine years later, under the ownership of the Baltimore and Ohio Railroad, she made a triumphant appearance at the Chicago World's Fair. In 1927 she was once more exhibited, this time at the Fair of the Iron Horse at Halethorpe on the outskirts of Baltimore. Eventually the Baltimore and Ohio Railroad company gave the *Samson* back to Nova Scotia, and now the pioneer locomotive has a place of honour in the middle of bustling New Glasgow, not far from the area where she spent her working days.

Pioneer Railways

CNR

Dorchester, *the first railway engine in Canada: Champlain and St. Lawrence Railroad, 1836 (from a print at the Château de Ramezay)*

What might have been part of the earliest railway in Canada was discovered not long ago on Cape Breton Island. It was an ancient roadbed believed to date back to the 1720s when the French were building the fortress of Louisburg. Whether the roadbed, which extended from an abandoned gypsum mine to the bank of the Mira River, was in fact that of a mine tramway, no one has been able to prove. But it seems likely that French engineers would have copied the kind of railways operating in their homeland at that time—primitive roadways with two rails to guide the wheels of vehicles pulled by horses.

During the construction of Quebec's Citadel in the 1820s an incline railway ran up the side of the cliff to transport stone to the top. A stationary steam engine supplied the power for two cable cars which operated on double tracks, one going up loaded with building material, the other coming down empty.

At that time a Quebec merchant by the name of James George advocated the construction of a wooden tramway to accommodate carriages drawn by horses or pulled by a steam engine. His proposition was considered too outlandish to warrant public support. He eventually proceeded to build a private tramway with wooden rails for lumbering operations on his property near the present town of Richmond, Quebec.

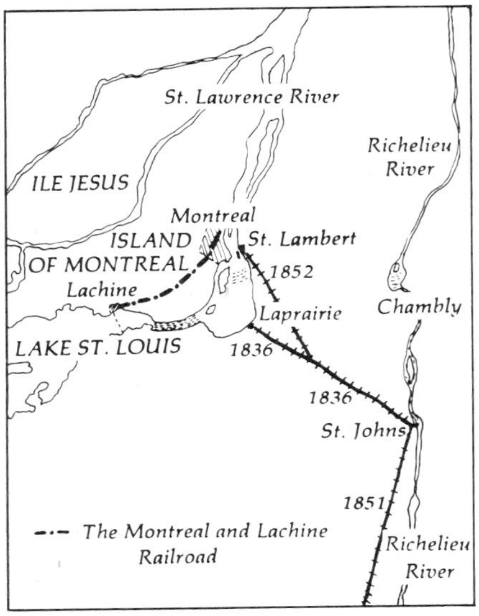

Another early tramway was in use during the construction of the Rideau Canal under Colonel By and his Royal Engineers. The canal along the Ottawa River was started in 1826 as an alternative route of communication between Upper and Lower Canada. To transport the stone needed for its weirs and locks, a five-mile wooden tramway ran from a quarry at Hog's Back to the construction site. Operated by horses, it served until the completion of the canal in 1832 when it was abandoned.

In those days rivers, canals and lakes were still the main arteries of transportation. Passenger boats regularly plied the waterways. Few good highways existed. Most of them were impassible in the winter, muddy with axle-deep ruts in the spring, and choked with dust during the rest of the year. Overland travel by stage coach was time-consuming and anything but comfortable by our standards. As for motive power on land, it was the horse which ruled the road until the year 1836, when a new chapter began in Canada's history. That year the first Iron Horse, puffing and snorting along a wooden track, made its appearance in the country and started the wheels of progress rolling.

Champlain and St. Lawrence Railroad

CHAMPLAIN AND ST. LAWRENCE RAILROAD

Canada's first pioneer locomotive belonged to the Champlain and St. Lawrence Railroad Company and was affectionately known as the *Kitten*, because of her short wheel base and her occasionally unpredictable ways. She had a definite tendency to teeter. Her official name was *Dorchester*, after the town of Dorchester. Now known as St. Johns, Quebec, the town at the time was the terminus of the first passenger railway line in Canada.

All that is left today of the famous *Dorchester* is her brass name plate which was found by a farmer while ploughing his field. He gave it to the College Museum of Joliette, Quebec. No authentic contemporary pictures of the locomotive have ever been found, other than an engraving on the gold watch of George W. Pangborn, the *Dorchester's* first regular engineer. But from descriptions and specifications listed by her builders, experts are reasonably sure that she looked like the full-scale model which can be seen among the exhibits in the Château de Ramezay in Montreal, together with a small portion of an old wooden railway track like the one on which she used to run.

Canada's first railway was designed as an overland link in the water route from Montreal to New York. It covered the stretch between

Laprairie on the St. Lawrence River and St. Johns on the Richelieu. Before the line was built, travellers, having been ferried across the St. Lawrence to Laprairie, were transported by stage to the boat connection at St. Johns over a sixteen-and-a-half-mile portage road, rough at the best of times and a real nightmare in a rainstorm when it turned into a veritable quagmire.

In the early days of settlement a journey from Montreal to New York was a tiresome, vexing experience that took two weeks in the summer, involving boats, overland stages and sailing clippers. During the winter frozen lakes and rivers made excellent highways for travelling by fast sleighs, but bitter cold and blizzards often hampered the journey. By 1810 steamboats began to ply the waters of Lake Champlain and the Hudson River making it possible to reach New York from Montreal in a matter of days.

Concerned with replacing the portage road between Laprairie and St. Johns, which remained one of the most troublesome portions in the overall travel route, a group of Montreal businessmen began advocating the construction of a railway in the 1830s. Public opinion was not in favour of the idea. Railways were still largely a mystery, and a canal was suggested as a better solution. Most likely an artificial waterway would have been built instead of a railway, had costs not made the project prohibitive. On February 25, 1832 the "Company of Proprietors of the Champlain and St. Lawrence Railroad" obtained a charter from the legislature of Lower Canada authorizing it "to construct a railway to facilitate and dispatch the carriage and conveyance of goods, passengers, etc. between the navigable waters of Lake Champlain and the River St. Lawrence opposite the city of Montreal." Seventy-four names were listed as the original proprietors of the proposed railroad.

Passenger cars (first- and second-class) of the Champlain and St. Lawrence Railroad

Soon newspaper advertisements offered railway shares to the public, and company headquarters were set up in the Exchange Coffee House at Montreal. But the outbreak of a cholera epidemic which caused two thousand deaths in the city brought matters to a temporary halt. Political strife, election riots and uncertain financial policies resulting in a general depression further delayed the start of the project.

Not until 1835 did the railway finally get under way, thanks to the relentless efforts of St. Johns merchant Jason C. Pierce, and the financial support of Montreal brewer and steamboat operator John Molson.

At the first company stockholders' meeting, which took place on January 12th of that year, the exact location of the proposed road was discussed and two American experts, William R. Casey and Robert F. Livingston, were appointed as chief engineer and assistant respectively. Tenders were called almost immediately for lumber, iron and fencing material, and construction started in the spring. Although severely hampered by bad weather, Mr. Casey reported by the end of the year that fencing, masonry and bridge work, a wharf at Laprairie, and parts of two station houses were completed.

Meanwhile, the company ordered a steam locomotive from the firm of Robert Stephenson of Newcastle-on-Tyne in England at a cost of fifteen hundred pounds sterling. Four passenger cars were expected to come from the United States, while baggage wagons and other initial rolling stock for the railway were being manufactured in Montreal.

As soon as the snow had melted in the spring of 1836, the laying of the rails began simultaneously on different sections of the 14½-mile stretch of the line. The single wooden track was 4'8½" wide, a width which has long since become the standard railway gauge. The rails consisted of six-inch pine squares joined by iron splice plates and bolts,

CNR

Lamp and torch used by the Champlain and St. Lawrence, Canada's first railway

and were anchored firmly to the ties by wooden blocks. Iron straps, half an inch thick, were spiked to the upper surface of the rails for protection. The counter-sunk spikes, however, had a tendency to pull out or snap off under stress causing the thin iron straps to curl upward; hence they came to be known as "snakehead rails." Although potentially dangerous, the rails caused only one minor accident on the Champlain and St. Lawrence line during the years they were in use. In the 1850s they were replaced by fifty-six-pound iron T-rails.

The exact arrival date of the railway's first locomotive is not known. In all likelihood the engine, dismantled and crated, entered Canada under a bill of lading marked "boiler and machinery." From the port of entry, probably Quebec City, the crates were brought by barge to Laprairie where the locomotive was reassembled.

The *Dorchester*'s initial trial runs were apparently staged in the moonlight, in secrecy, so as not to frighten the public. After all, hardly anyone had ever seen one of these fire-spewing iron monsters in action. Another reason for the secrecy may have been the desire to avoid embarrassment should anything go wrong. Actually a mishap did occur during one of these trial runs, just before the official opening of the line. The fireman had let the water out of the boiler and kept the fire going until the flues were burned. The damage was repaired and the *Dorchester* was ready when the day arrived to celebrate the opening of Canada's first railway.

It was Thursday, July 21, 1836. The new, powerful steamboat, *Princess Victoria,* built to serve in conjunction with the railway line, was on her way from Montreal to Laprairie carrying three hundred ladies and gentlemen, among them the Earl of Gosford (the Governor-General of Canada), Sir George and Lady Gipps, Sir Charles Grey, the Honourable Peter McGill (the president of the railroad company), members of the provincial legislature and officers of the garrison. The weather was excellent and the band of the 32nd Regiment entertained the passengers during the fifty-minute trip. The main events of the memorable day are best described by excerpts from the contemporary report of the Montreal *Morning Courier*:

> *After landing at the Railroad Wharf, which runs out into the river a considerable way, the company proceeded to the cars which were waiting at the termination of the Railway to convey them to St. Johns. Before starting, the locomotive engine made two short trial trips with its tender, and as the accident which occurred lately to it had not been thoroughly repaired, it was deemed advisable to attach it to only two of the passenger cars, all of which were very comfortably fitted up and elegantly painted outside; while the other cars with the rest of the company, were drawn each by two horses. The locomotive with its complement soon shot far ahead of the other cars, which passed along the road, just as fast as the nags, which were none of the fleetest, could drag them. The motion was easy, and elicited from many, comparisons far from favorable to the usual comforts of travelling by the stage road. In less than two hours from starting all the company had arrived at St. Johns in good time, and in excellent mood for a cold collation in the Railway Station House, which was pleasantly cool and decorated with green branches. The repast, with its accompaniments of sparkling champagne and Madeira, was not more enjoyed, than it was universally admitted to be in itself, suitable and excellent.*

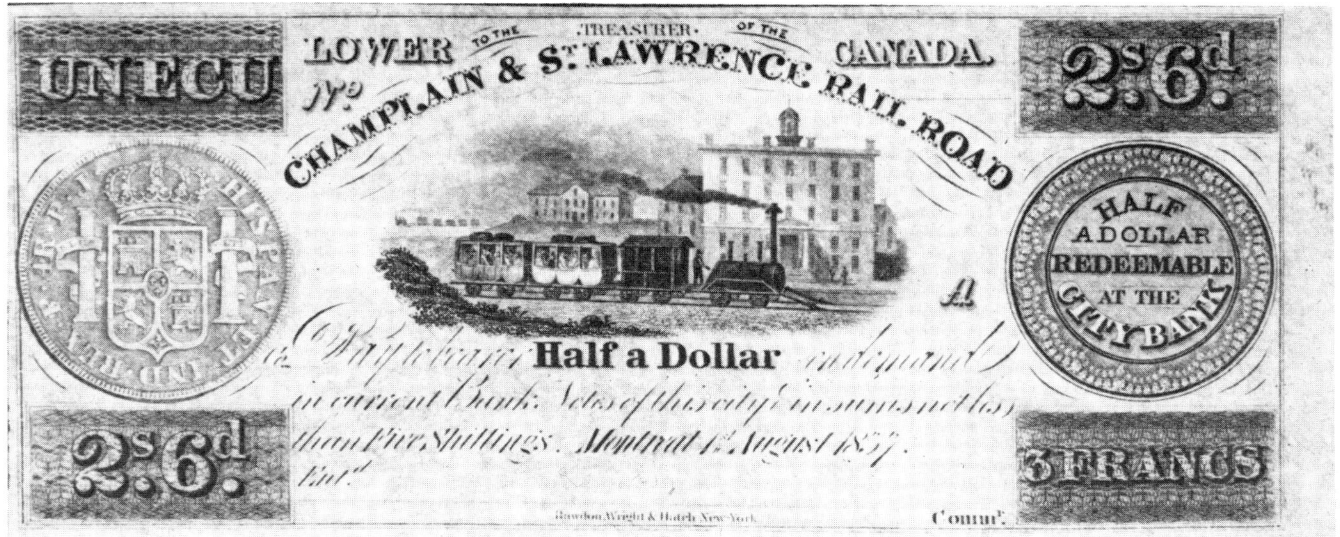

CNR

Currency issued in 1837 by the Champlain and St. Lawrence Railroad. It consisted of three bills with face values of 12½, 25 and 50 cents

> *After the company had appeased both their hunger and thirst, Mr. McGill proposed in succession three toasts—"The King," "The President of the United States" and "His Excellency the Earl of Gosford" and the "ladies who honored the day's proceedings with their presence."*

Two hours later the locomotive returned to Laprairie. This time she pulled four cars while the other twelve were drawn by horses. According to the paper:

> *There would have been almost a surfeit of enjoyment, had nothing occurred to break in upon the pleasures of the day. It was pretty far advanced in the afternoon, before the company got re-embarked on board the "Princess Victoria" for Montreal, and it unfortunately happened that, in consequence of a strong easterly wind, and the depth of the boat in the water, she grounded on leaving the wharf. When at length she was got clear and had proceeded a little way on her voyage, she was again detained by being compelled to lie-to, till a man who had fallen overboard was picked up. By this time it was so dark that it was considered dangerous to continue the voyage and the boat returned to Laprairie.*

On July 23, 1836 the *Montreal Gazette* carried Canada's first railroad timetable. It announced that, in connection with the steamer *Princess Victoria,* the Champlain and St. Lawrence Railroad Company was prepared to convey passengers between Montreal and St. Johns starting on Monday, the 25th instant. Departure times of the steamer and the locomotive were given for both directions but significantly no mention was made of the time of arrival!

During most of the first year of operation the company employed horses rather than the locomotive to haul passenger cars. The line had yet to be properly ballasted and the engine was apt to jump the track. When work on the roadbed was completed, the Iron Horse was put into regular service.

The *Dorchester* was about thirteen feet long and weighed 11,275 pounds. Originally she had four forty-eight-inch wooden wheels, but when her short wheel base and high centre of gravity were found to account for much of her eccentric behaviour on the track, the company's directors decided to have her rebuilt, changing to a more flexible 4-2-0 wheel arrangement.

CNR

THE FIRST RAILWAY TRAIN TO COME INTO MONTREAL, MONTREAL & LACHINE RAILROAD, NOV. 19, 1847.

FROM AN ENGRAVING BY J. WALKER, MADE AT THE TIME. FROM THE COLLECTION OF THE LATE ALD. DOUGAL MAC DONALD, MONTREAL.

Montreal and Lachine Railroad timetable printed in 1852 in American Railway Guide

MONTREAL & LACHINE RAILROAD.

W. R. Coffin, Pres., Montreal, Canada. John Farrow, Supt., Montreal.

Miles.	Fares.	MONTREAL To LACHINE.	1st Trn	2d Tr'n	3d Tr'n	4th Trn	5th Trn	6th Trn
		TRAINS LEAVE	AM.	AM.	M.	PM.	PM.	PM.
		Montreal*	8 00	10 00	12 00	4 00	5 30	7 00
		Tanneries						
8	37	Arr at Lachine.	8 20	10 20	12 20	4 20	5 50	7 20

* Connects at this point with St. Lawrence & Atlantic R.R. see below. Also, St. Lawrence & Champlain R.R., see below.

Miles.	Fares.	LACHINE To MONTREAL.	1st Trn	2d Tr'n	3d Tr'n	4th Trn	5th Trn	6th Trn
		TRAINS LEAVE	AM.	AM.	PM.	PM.	PM.	PM
		Lachine	8 30	10 30	12 30	4 30	6 00	7 30
		Tanneries						
8	37	Arr Montreal* ..	8 50	10 50	12 50	4 50	6 20	7 50

ST. LAWRENCE & ATLANTIC RAILROAD.

A. T. Galt, Pres., Sherbrook Canada. A. C. Webster, Sec'y, Montreal.

Miles.	Fares.	MONTREAL St HYO'THE.	1st Trn		
		TRAINS LEAVE	PM.		
		Montreal*	3 00		
3		Longueuil	3 30		
7		Charons			
13		Montarville ...			
17	50	Beloeil & St. Hil			
33	1 00	St. Hyacinthe.	4 55		
		Arr Richmond	7 00		

Miles.	Fares.	St HYO'THE MONTREAL	1st Tr'n		
		TRAINS LEAVE	AM.		
		Richmond ...	7 00		
		St. Hyacinthe.	9 05		
		Beloeil & St. Hil			
		Montarville ...			
		Charon			
		Longueuil			
		Ar. Montreal*	10 30		

* Connects at this point with Montreal & Lachine R.R., see above. Also, Champlain & St. Lawrence R.R., see below.

CHAMPLAIN & ST. LAWRENCE RAILROAD.

John Malsov, Pres., Montreal, L. C. A. H Brainard, Supt., Montreal, Ca.

Miles.	Fares.	ROUSE'S PT MONTR'AL.	1st Tr'n	2d Tr'n	3d Trn
		TRAINS LEAVE	AM.	M.	PM
		Rouse's P'nt*	7 15	12 00	3 00
		Lacolle			
		Stott's			
		Grand Ligne .			
22	1 00	St. John's ...	8 40	1 20	5 00
37	1 50	Laprairie	9 30	2 10	6 15
46	1 50	Ar Montreal	10 00	3 10	7 30

Miles.	Fares.	MONTRAL ROUSE'S P.	1st Tr'n	2d Tr'n	3d Trn
		TRAINS LEAVE	AM.	AM.	PM.
		Montreal† ..	6 00	11 00	4 00
9		Laprairie ...	7 30	12 15	5 15
24	1 00	St John's ...	9 00	1 10	6 15
		Grand Ligne			
		Stott's			
		Lacolle			
46	1 50	Rouse's P't	10 30	2 40	8 00

* Connects at this point with Vermont Central R.R., see page 82; Also Northern (N.Y.) R.R., see page 66.

† Connects at this point with Montreal & Lachine R.R., see above. Also St. Lawrence & Atlantic R.R., see above.

CNR

On her passenger run the wood-burning engine hauled a small tender with cordwood and a puncheon of water. She was capable of reaching the precarious speed of thirty miles an hour, puffing thick clouds of smoke and showering the neighbourhood with fiery sparks. In the open coaches behind her men hung on to their hats when the train rounded a curve. Ladies put up their umbrellas to guard against flying cinders, but many an umbrella had to be hastily discarded when it caught fire. Despite the obvious hazards, the novelty of a train ride attracted so many passengers that at times the railway could barely handle the crowd. Many Montrealers took the train for the sheer thrill of having a picnic in the middle of the railroad tracks while the engine crew was replenishing fuel supplies at wooding-up stations along the way. No one was in a rush, and the Iron Horse simply had to wait. Smoking was strictly forbidden aboard the train and regulations stated "no climbing on top of the engine or the railway cars." Offenders could be fined as much as ten shillings.

The price for a ticket from Montreal to St. Johns was five shillings for first class and two shillings and sixpence for second class. Children travelled at half the fare. Freight was conveyed at "two shillings a barrel for ashes, one shilling for beef and pork, six pence for flour and meal and five shillings per thousand feet of boards and planks."

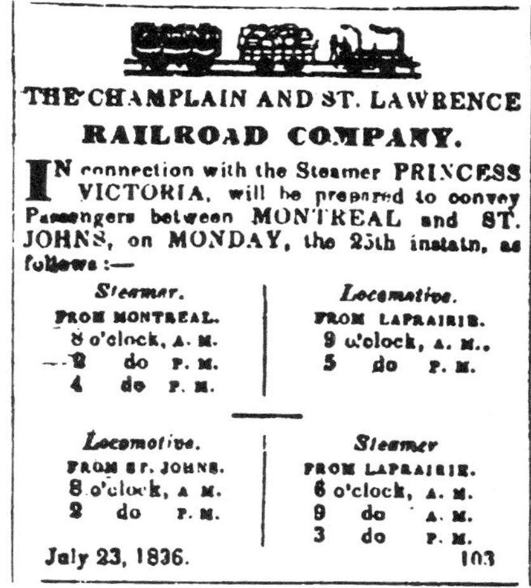

THE CHAMPLAIN AND ST. LAWRENCE RAILROAD COMPANY.

IN connection with the Steamer PRINCESS VICTORIA, will be prepared to convey Passengers between MONTREAL and ST. JOHNS, on MONDAY, the 25th instatn, as follows:—

Steamer. FROM MONTREAL.	*Locomotive.* FROM LAPRAIRIE.
8 o'clock, A. M.	9 o'clock, A. M.
2 do P. M.	5 do P. M.
4 do P. M.	

Locomotive. FROM ST. JOHNS.	*Steamer* FROM LAPRAIRIE.
8 o'clock, A. M.	6 o'clock, A. M.
2 do P. M.	9 do A. M.
	3 do P. M.

July 23, 1836. 103

In 1837 the Champlain and St. Lawrence Railway Company decided to issue its own money. It consisted of three notes with face values of 12½, 25 and 50 cents printed on each in Halifax (Canadian), American and French currency. The railway money was to combat an extreme shortage of small change which developed after banks had suddenly suspended specie (coin) payments.

As required by its company charter, the railway's level crossings were protected with double swing gates placed across the track and opened only when a train was passing. The first level crossing accident on the line, and for that matter anywhere in Canada, occurred a year after the railway was opened at Côte St. Raphael. A team of oxen collided with the train throwing it off the track. Swing gates were removed from the tracks during the 1840s and a watchman was put on duty.

During the winter months traffic on the line frequently came to a halt. Section men kept tracks clear of snow near the station houses, but on the road the only snow-fighting equipment available to the engine driver were two large birch brooms fastened to the buffer beam of the locomotive. Needless to say, they were useless in a blizzard.

In time the Champlain and St. Lawrence Railway acquired two additional engines, the *Laprairie* and the *Jason C. Pierce,* both 4-2-0 wheelers (see note on page —) built in Philadelphia by the Norris Company. They were eventually joined by the *Montreal,* a powerful Baldwin locomotive and the first eight-wheeler seen in Canada. The *Montreal* weighed about ten tons and was eighteen feet long. Her driving wheels had a diameter of 54 inches, her truck wheels 30 inches.

In 1849, thirteen years after her first trip, the *Dorchester* was sold to the Lanoraie and Industry Railroad, then under construction. Under her new owners she continued to operate until 1864 when a boiler explosion and consequent derailment near the village of St. Thomas, Quebec, smashed her up so badly as to make repairs impossible. Thus, rather suddenly and sadly, ended the life of the *Kitten,* Canada's first locomotive.

The Champlain and St. Lawrence Railway which had started out as a primitive portage line was extended in 1851 to Rouses Point, New York, and a year later the northern terminal was moved to St. Lambert,

Quebec. Growing competition from other lines and financial difficulties eventually forced an amalgamation with the Montreal and New York Railroad; the company operated after that under the name of Montreal and Champlain Railroad, until it was finally absorbed by the Grand Trunk Railway of Canada.

Token of the Montreal and Lachine Railroad CNR

THE MONTREAL AND LACHINE RAILROAD

By 1847 Montreal had grown into a bustling city of fifty-seven thousand. Early in November of that year residents watched a steam locomotive drawn by eighteen horses move along Montreal's St. Antoine Street. It was the *Lachine,* a 4-4-0 type engine which had arrived from Philadelphia to be placed in service on the newly completed Montreal and Lachine Railroad. On Friday, November 19th, a train headed by the *Lachine* pulled out of Bonaventure Street Station, then little more than a spacious open shed known as the Griffintown Terminal. Throngs of people were present to witness the first train to leave the city of Montreal. Eight passenger cars attached to the engine carried company officials and dignitaries, among them Lord Elgin, the Governor-General of Canada. The trip to the Lachine terminal, situated on a wharf, took twenty minutes and was enthusiastically described by those aboard as having been extremely smooth.

First-class cars, made in Montreal, were elaborately furnished with soft cushions, luxurious satin hangings and silk blinds. Second-class coaches had comfortable leather seats and, in contrast to third-class cars, were fully enclosed to protect passengers from the elements. Doors were securely locked before the train left the station, much to the dismay of some passengers who were afraid of getting trapped in case of accident. Ladies vehemently objected to being locked into compartments with total strangers.

Built to bypass the Lachine Rapids, the Montreal and Lachine Railroad was a mere seven and a half miles long. But it took nearly two years to construct the line through terrain made up largely of morass and swamp. Thousands of loads of rock and earth were required to gain a foothold, with nearly all of the fill coming from the Lachine Canal which was being enlarged at the time.

Alexander Millar, an engineer and former locomotive superintendent of a Scottish railway, supervised the construction of the line. He had come from Scotland at the request of wealthy Montreal merchant James G. Ferrier, himself a Scot by birth, who had arrived in Montreal literally penniless some twenty years earlier. It was Mr. Ferrier who had first promoted the idea of a railway from Montreal to Lachine. Prominent in public life, he eventually became the president of the Montreal and Lachine Railroad Company.

After the inauguration of the line late in 1847 the railway closed down for the winter months. The following spring the company's rolling stock was bolstered by two additional engines arriving from Scotland. Named *James G. Ferrier* and *Montreal,* both were 2-2-2-type locomotives (later changed to 4-4-0 arrangement), claimed by their manufacturers as being extremely fast.

One of the new engines was to be tested when traffic on the line resumed. Company officials and three American engineers, anxious to observe the locomotive's performance, boarded a three-car special train. Alexander Millar took the throttle. Determined to show what a Scottish-built engine could do, he drove her at breakneck speed and before his passengers could catch their breath they had arrived at Lachine. The time: all of eleven minutes. Sandy Millar's pride in his engine was equalled only by the fury of the officials. One of them had literally "gone through the roof" during the trip and only a miracle had saved him from being injured. The others were badly shaken from continuously bobbing up and down in their seats. Horrified at the thought of possible derailment, they threatened to hire post chaises for the return trip to Montreal, unless the engineer promised to take it easy. Well, Sandy promised. But bent on bettering his record, he tried for a speed of nearly a mile a minute and arrived in Montreal in nine minutes' time. President Ferrier was furious and swore to fire the engineer first thing in the morning. But the next day the episode was forgotten and soon afterwards Alexander Millar was promoted to general manager of the line.

Among the engine drivers employed by the company was one Patrick Kelly whose reputation for recklessness frightened many a potential passenger into taking the stage coach instead of the train. Those who dared travel with him usually got their money's worth of excitement. At a certain level crossing he would stop his train, climb down from his engine and pull the chair from under watchman O'Reilly who usually sat sound asleep near the swing gates failing to do his duty. The crossing located on the Upper Lachine Road at the western end of Ville St. Pierre soon became known as "O'Reilly's Crossing." Patrick Kelly was killed some years later when his engine exploded.

A unique feature of the Montreal and Lachine Railroad was the use of metallic tokens instead of paper tickets for third-class passengers. The size of fifty-cent pieces, the tokens had a hole in the centre so the conductor could string them up on wire. One side of the token bore the inscription of the company's name along with a picture of a four-wheeled locomotive, while the other side portrayed a beaver. Extremely rare, these tokens are now collectors' items.

In August of 1850 the Montreal and Lachine merged with the Lake St. Louis and Province Railway. The latter had been chartered to build from Caughnawaga, opposite Lachine, to the international boundary near the village of Hemmingford. The new company took the name of Montreal and New York Railroad. At the same time that the Caughnawaga division was under construction, a connecting line was built by an American company from Plattsburg to the border. Regular through service from Montreal to Plattsburg was inaugurated on September 20, 1852. Passengers at first were conveyed by boats across the St. Lawrence River, but a few months later the car ferry steam boat *Iroquois* commenced running between Lachine and Caughnawaga. Capable of carrying a locomotive and three cars, it was the first car ferry to operate in Canada.

In 1857 the Montreal and New York Railroad lost its identity. Amalgamated with the Champlain and St. Lawrence, it emerged as the Montreal and Champlain Railroad. Six years later this new company, unable to cope with financial difficulties, leased its properties to the Grand Trunk Railway, and in time Canada's first two pioneer lines became part of the vast system of the Canadian National Railways.

Link with the Sea

If all the railways which have been incorporated in Canada over the years were running today, our transportation system would resemble a giant spider's web. Many of the early charters however were never followed up, and nothing more remains of them than a record in the archives. By 1850, only sixty-six miles of railroad were actually operating in British North America; but ten years later the picture had drastically changed. During the railway boom of the fifties Canada's railroad mileage climbed to more than two thousand. Politicians, aware of the importance of communications to the growth of the country, started to make railways their business. In 1849 an Act was passed by the Province of Canada (in 1841 the provinces of Upper and Lower Canada had been united into the Province of Canada) guaranteeing as a loan the interest on any sum required for the construction of a railroad seventy-five miles or more in length, provided half of the proposed line was completed. Two years later this guarantee was extended to the principal as well as the interest, but now government assistance was restricted to the construction of a main trunk line.

Early railroads invariably were short and served primarily as connecting links between two waterways. Not until the middle of the century were railways planned as roads of communication between Canada East and Canada West. Open all year round, they were destined to eliminate the isolation of scattered villages during the long winter months. Through-connections being established between major centres of British North America and the northern United States were to promote closer trade relations between the two countries.

Locomotive No. 6, Atlantic & St. Lawrence Railway, built in 1856 in Portland, Maine

The world's first international railway was inaugurated on July 18, 1853, when trains began operating on a continuous line from Longueuil, opposite Montreal, to the American city of Portland, Maine, on the Atlantic seaboard. The event climaxed ten years of planning, promotional activities and fierce rivalries.

In February of 1843, a group of citizens from the eastern townships of Quebec first met at Sherbrooke to discuss the feasibility of an international railway connecting their isolated area with Montreal and to the south with the Atlantic seaboard. During subsequent discussions Boston emerged as the likely choice for the coastal terminus and promoters from that city did their best to reach an agreement with the Canadians. In 1845 a delegation went to Montreal to convince the Board of Trade of Boston's advantages as the terminus of the proposed railway.

They had almost succeeded and were about to sign an agreement, when a tall, handsome gentleman entered the board room and introduced himself as John A. Poor from Portland.

Mr. Poor had travelled five days through a raging blizzard with howling winds and giant snow drifts that likely would have killed a less determined man. But Mr. Poor had a mission. For years he had been advocating a railroad from Portland to Montreal in the hope of developing the wilderness area between the two cities and making Portland an important harbour. Visiting logging camps, villages, hamlets and towns along the three-hundred-mile stretch between Portland and Montreal, he had aroused an enthusiastic response to his proposals and spread the railroad fever throughout the countryside. When he heard of the railway promoters' meeting in Montreal, he left Portland at once. Despite unfavourable weather conditions, he arrived just in time to stop his Boston rivals from getting the nod.

While Montrealers listened to Mr. Poor's forceful address and pondered his arguments in favour of Portland as the terminus, a second visitor arrived at the meeting. Judge Preble of Portland had followed Poor's trail through the backwoods to announce that a charter had been granted by the legislature of Maine incorporating the Atlantic and St. Lawrence Railway Company, for the purpose of constructing a line from Portland to the international border, over a route which had already been surveyed by Mr. Poor. Although two further charters would have to be obtained later in order to cross the states of New Hampshire and Vermont, the Canadians were impressed.

As a final test, so the story goes, a mail race was suggested between Boston and Portland to determine which was the better route to Montreal. The outcome may not have been the deciding factor in choosing Portland as the terminus but, as it was, that city won easily. The mail from Portland arrived in Montreal fifteen hours ahead of the Boston dispatch. Knowing the route at first-hand, Mr. Poor is said to have seen to it that fresh horses were stationed at intervals and the road was kept open.

The Canadian portion of the line was incorporated a month later. Its name, the St. Lawrence and Atlantic, was the reverse of its American counterpart.

Construction of both the Montreal and the Portland section began simultaneously, in the summer of 1846. A gauge of 5′ 6″, today known as the broad gauge, was adopted for the rail. This gauge subsequently became the official standard gauge in the Maritimes and in the Canadas until the early 1870s, when the decision was repealed and today's standard gauge of 4′ 8½″ was instated.

From the beginning the St. Lawrence and Atlantic was plagued by financial troubles. Money was so scarce that some Quebec subscribers were allowed to pay for their railway shares by supplying butter, milk and eggs to construction gangs. Subscriptions raised in England fell far short of expectations. More than once the project seemed doomed to fail for lack of funds, and it took considerable business ability and persuasive talent on the part of an energetic management to carry on.

The first stretch from Longueuil to the Richelieu River was completed in November 1847. By 1852 the line reached Sherbrooke and residents there celebrated the arrival of the first train. A year later, at Island Pond, the rails of both the Canadian and the American sections were joined. From Longueuil a ferry boat provided connection with Montreal, linking the city by rail to the sea. The first international railway had become a reality. Portland with its ice-free harbour began to prosper, becoming a trans-shipping centre for Canadian goods on the way to European markets.

The first steam engine bought by the St. Lawrence and Atlantic Railway was the *Longueuil*. Little more is known about her, other than the fact that she probably was a second-hand American engine. Most later locomotives serving on the line were of the 4-4-0 type, built in Portland, the first of these being the *A. N. Morin*, named after one of the company's directors.

Immediately after it had been opened the American portion of the line was leased to the Grand Trunk Railway for a period of 999 years. The Canadian counterpart was sold to the Grand Trunk Company the following year. Thus the Montreal-Portland rail connection formed the beginning of the "Old Grand Trunk," Canada's greatest pioneer railway.

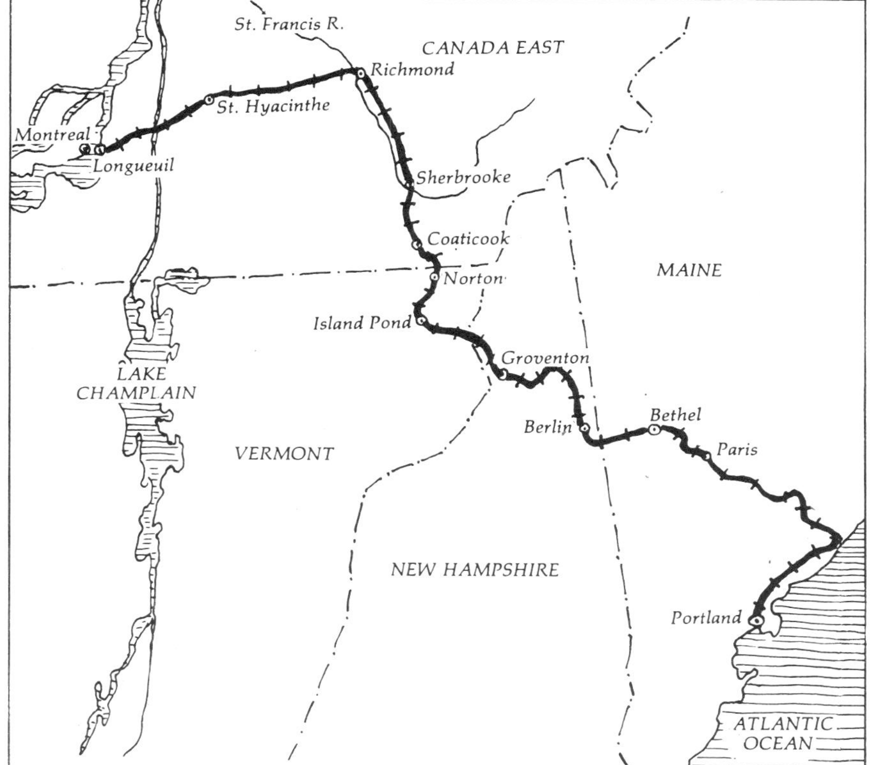

MONTREAL TO PORTLAND.

THE ST. LAWRENCE & ATLANTIC and the ATLANTIC & ST. LAWRENCE RAILWAYS being now completed and connected together at Island Pond, these Sections of the Grand Trunk Railway of Canada will be open for public traffic on and after

Monday, the 18th July, instant.

A DAILY MAIL TRAIN WILL

Leave Montreal at 3, p.m.; arrive at Sherbrooke at 7:30, p.m.; leave Sherbrooke at 6:30, a.m.; arrive at Portland at 3, p.m.

Leave Portland at 1:15, p.m.; arrive at Sherbrooke at 8:30, p.m.; leave Sherbrooke at 6:30, a.m.; arrive at Montreal at 11, a.m.

A DAILY EXPRESS TRAIN WILL

Leave Montreal, at 7, a.m.; arrive at Island Pond at 12:30 Noon; arrive at Portland at 6:30, p.m.

Leave Portland at 7, a.m.; arrive at Island Pond at 1, p.m arrive at Montreal at 6:30, p.m.

Fare Between Montreal and Portland,

SIX DOLLARS.

Passengers by the Express Train from Montreal will reach Portland in time for the Steamer, landing at Boston early the following morning—Steamboat Fare, Portland to Boston 3s. 9d.

Passengers by the Mail Train from Montreal will reach Portland in time to take the Boston Train at 3:30 p.m., arriving at Boston at 8, p.m.

Through Fare by Rail, Montreal to Boston,

SEVEN DOLLARS.

The Steamer *L'Aigle* plies to and from the Jacques Cartier Basin, in connection with all Passenger Trains.

. C. WEBSTER,
Superintendent.

Montreal, July 853. 1047-tf

Advertisement from Montreal newspaper, 1853

Railways of Upper Canada

By the middle of the last century railways were the topic of the day. Speculation and enthusiasm ran wild. Dormant charters were revived, and promoters of shares painted glowing pictures of immense profits. Some schemes collapsed before they got off the ground, while others emerged as far-sighted ventures. Survey parties pushed their way through dense forests, construction sites turned into beehives of activity. Railroad building in Canada had begun to take on a definite pattern. As one newspaper of that period put it:

> *The days of stage coaches have come to an end, and everywhere is to be heard the snorting of the iron horse, and the shrill blast of the steam whistle warning the thoughtless of the danger if they have ventured upon its path. Distance must be abridged, no matter what the labour or expense.*

Among the earliest railways chartered in Upper Canada was the Erie and Ontario. Incorporated in 1835, it became the first railroad company to operate in what is now the Province of Ontario. The short road of iron-strapped timber rails running from Queenston to Chippewa was opened in 1839. Trotting horses were employed to draw passenger coaches, draught horses to haul freight cars. Locomotives available at the time would not have been powerful enough to climb the long steep grade of Queenston Heights which was part of the route. Consequently, steam was not introduced until 1854 after the line had been reconstructed with an easier grade. By then an extension had also been built to Niagara-on-the-Lake. It was opened for traffic on July 3rd of that year. The only steam locomotive of the line came from the Amoskeag Company of Manchester, New Hampshire. She was a 4-4-0 type named *Clifton.*

In 1862 the line was purchased by William A. Thomson who in turn sold it to the Fort Erie Railway Company which had received a charter to construct a line from Fort Erie to Chippewa. When the two companies merged they became known as the Erie and Niagara Railway. In the course of time Upper Canada's oldest railway completely lost its identity after being absorbed by the Canada Southern, which in turn was leased to the Michigan Central and eventually by the latter to the New York Central. Of the old horse-operated line only traces remained.

ONTARIO, SIMCOE AND HURON RAILWAY

Residents of Upper Canada, or Canada West as it was then called, did not see their first steam locomotive until October 1852.

To Contractors.

SEALED TENDERS, will be received at the Office of the undersigned Contractors, in the City of Toronto, until the 21st day of March, inst., for the construction of the line of the Ontario, Simcoe and Huron Union Railroad, embraced between the Town of Barrie, and the terminus on Lake Huron, a distance of about 35 miles.

The work to be contracted for, will include the grading, bridging, and masonry, the furnishing of timber for superstructure, and the fencing required on the line,

Plans, profiles, and specifications of the work, will be exhibited at the Office of the Engineers, in the City of Toronto, on and after the 15th inst.

M. C. STORY & CO.

Toronto, March 3, 1853. 69.

A cheering crowd welcomed her at the wharf in Toronto when she arrived by schooner. Built in Portland, Maine, she had coupled driving wheels, five feet in diameter, and featured inside cylinders and the "hook" motion type of action. Her owners named her *Lady Elgin* after the wife of the Governor-General of Canada. The sound of her steam whistle echoing across Toronto Bay heralded the beginning of a new era for Upper Canada.

The Toronto *Patriot* described the event as follows:

HURRAH FOR THE NORTH. — We have the pleasure to announce the arrival in this City, of the first locomotive for the great Northern Railroad [chartered as the "Ontario, Simcoe and Huron Railway," the name was later officially changed to "Northern Railway of Canada"]; which was conveyed yesterday afternoon from Oswego, by the steamer "Forwarder" of St. Johns, Canada East. The Locomotive is called the "Lady Elgin". In a few days therefore, we may expect to offer our congratulations to our northern friends, on seeing the "Lady Elgin" polish-

Toronto, *the first locomotive manufactured in Ontario. On April 16, 1853 she was moved out of the shop on temporary wooden rails to her permanent track on Front Street, near the foot of Bay Street*

CNR

CNR

Lady Elgin, built in Portland, Maine, named after the wife of the Governor-General of Canada. First locomotive of the Ontario, Simcoe & Huron Railway

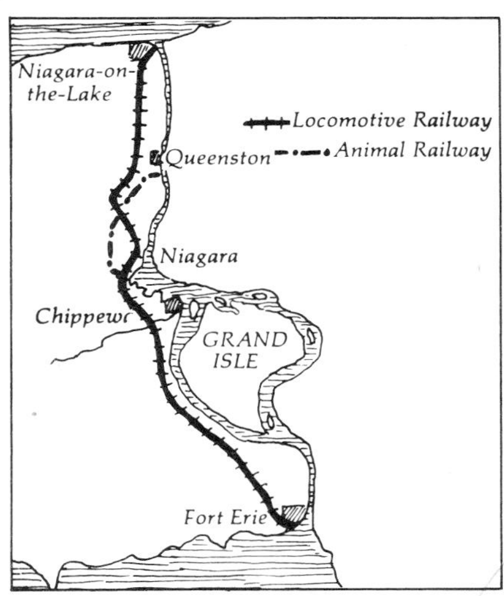

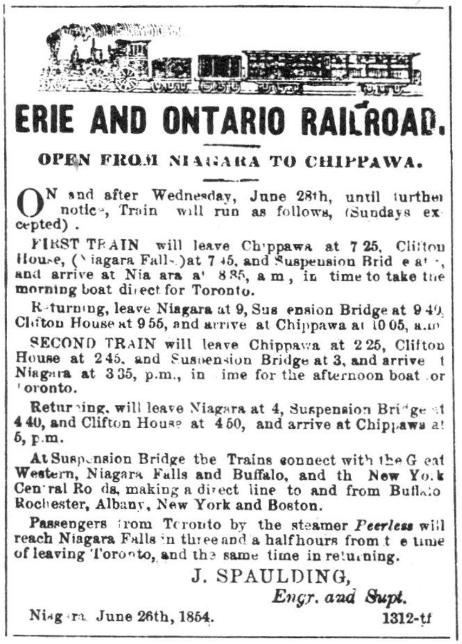

ERIE AND ONTARIO RAILROAD.

OPEN FROM NIAGARA TO CHIPPAWA.

ON and after Wednesday, June 28th, until further notice, Train will run as follows, (Sundays excepted).

FIRST TRAIN will leave Chippawa at 7 25, Clifton House, (Niagara Falls) at 7 45, and Suspension Bridge at 8, and arrive at Niagara at 8 35, a m, in time to take the morning boat direct for Toronto.

Returning, leave Niagara at 9, Suspension Bridge at 9 40, Clifton House at 9 55, and arrive at Chippawa at 10 05, a.m.

SECOND TRAIN will leave Chippawa at 2 25, Clifton House at 2 45, and Suspension Bridge at 3, and arrive at Niagara at 3 35, p.m., in time for the afternoon boat for Toronto.

Returning, will leave Niagara at 4, Suspension Bridge at 4 40, and Clifton House at 4 50, and arrive at Chippawa at 5, p.m.

At Suspension Bridge the Trains connect with the Great Western, Niagara Falls and Buffalo, and the New York Central Roads, making a direct line to and from Buffalo Rochester, Albany, New York and Boston.

Passengers from Toronto by the steamer *Peerless* will reach Niagara Falls in three and a half hours from the time of leaving Toronto, and the same time in returning.

J. SPAULDING,
Engr. and Supt.

Niagara, June 26th, 1854. 1312-tf

Erie and Ontario Railroad announcement in the Toronto Globe, *1854 ing the iron bands, which now unite Lake Simcoe to Lake Ontario, — on hearing her shrill whistle rouse the dormant energies of the backwoodsman to action, and proclaiming to the owls, and the bats, the wolves and the bears, that the bushes, and the forests, in which they have hitherto found shelter in Simcoe, can no longer continue to be their abode. "Lady Elgin's" shrill whistle will announce to the forests, that from henceforth she will call daily, to convey the rich treasures of the maiden soil which they cover, to the marts of traffic, commerce and wealth.*

In connection with this subject, it would be ungrateful in a guardian of the public interests, did we omit to mention, that to the energy and enterprise of Mr. Capreol, are the public mainly, if not solely indebted, for this great public work, which may justly be termed the Pioneer Railroad of Western Canada. And now that the great enterprise, which his mind conceived, and his energy accomplished, has been all but completed, that public would indeed be an ungrateful one, which would hesitate to make honorable mention of his name in connection with it.

Soon after her arrival the *Lady Elgin* was placed on construction work a few miles outside of Toronto, where the Ontario, Simcoe and Huron Railway had begun to thrust its rails northward to Collingwood on Georgian Bay. Officially named after the three lakes which the railway was designed to link as a portage line, it was more commonly known as the "Oats, Straw and Hay Railway."

Plans for this railway had been discussed for the first time back in 1834, when a group of interested citizens gathered at the British Coffee House on York Street in Toronto. Nothing came of the meeting other than a company charter and eleven years of "talking, surveying, scheming, sleeping and issuing prospectuses," to use the words of one of the Toronto papers of the day. The man who finally got things rolling was Frederick Chase Capreol from Toronto, "Mad Capreol," they called him, because he talked of nothing but railways. To raise some of the capital required, he planned to hold a giant lottery, offering 100,000 raffle tickets at the price of twenty dollars each. The proceeds of two million dollars were to be converted into railway shares and given as prizes to the winners. Capreol went to England to obtain royal assent for a charter granted by the Canadian legislature in 1849, which permitted him to raise the funds either by subscription or by lottery. Torontonians, however, voted down the gambling project. Consequently, the lottery idea had to be abandoned, and Capreol was forced to concentrate on more conventional methods of raising the necessary funds. Eventually he succeeded in qualifying the company for government assistance under the so-called Guarantee Act, which had recently been passed in the legislature. By then he had spent more than twelve thousand pounds of his own money to promote the railway, but in the end he received little thanks for his unremitting efforts. The day before the first sod was to be turned on the line he was dismissed as office manager of the company.

To celebrate the breaking of the ground for the railway, the city fathers of Toronto declared a public holiday on October 15, 1851. The Governor-General's wife, Lady Elgin, turned the first sod with a silver spade on a spot located opposite the Parliament Buildings, which then stood on Front Street in Toronto. The mayor of the city, wearing knee breeches, silk stockings, buckled shoes, a cocked hat and a sword, assisted in wheeling the sod away in a handsomely carved wooden barrow. More than twenty thousand people witnessed the ceremony.

ONTARIO, SIMCOE AND HURON RAILROAD.

In order to connect with the Boat on Lake Simcoe, the hours for dispatching Trains have been changed.

Until further notice, a Train will leave Toronto daily (Sunday's excepted) at 10, A.M.

Returning, will leave Machell's Corners at 4 h. 30 min., P.M.

Stages will be in readiness to convey Passengers to and from the Boat.

Fare to Machell's............ 3s. 1½d.
Fare to Bradford............ 5s. 0d.

For terms of Freight apply at the Office,

A. BRUNELL,
Superintendent.

(Papers copying former advertisement will change to this.)

Toronto, May 16, 1853 90.

ONTARIO, SIMCOE AND HURON RAILROAD.

COMMENCING on Monday the 20th instant, the Passenger Train will leave the *Foot of Bay Street*, at 7 A. M., and 2 P. M., for Bradford, connecting with the Steamer *Morning*, on Lake Simcoe, Returning will leave Bradford at 9 45 A. M.,

To the Subscribers to the Capital Stock

OF THE

ONTARIO, SIMCOE, & HURON

Railroad Union Company.

NOTICE is hereby given that the Instalments heretofore called in by advertisement up to and inclusive of the Instalments due on the 15th of April, 1854, are required to be paid forthwith.

Future Instalments will be at the rate of 5s per share and are hereby required to be paid as follows:—

On the 20th July, and 21th October, 1854.

20th January, 20th April, 20th July and 20th Oct., 1855. 20th January, 20 April, 20th July and 20th October, 1856, and 20th January 1857.

By order of the Board of Directors.

W. SLADDEN,
Secretary & Treasurer.

Toronto, May 20, 1854. 1280-6t

Colonist and *Leader* to copy for one week.

CNR

Locomotive Josephine, *the fastest engine of her day in Ontario. Her driver was Cyrus Huckett. Built in 1853 in Paterson, New Jersey, she operated until 1880*

While the railway was under construction, a steam engine was being built at James Good's foundry in Toronto. It was to be the first locomotive manufactured in what is now Ontario. On April 16, 1853 she was moved out of the shop on sections of temporary wooden rails, part of which had to be lifted up behind her and placed again in front of her, as she slowly rolled along Queen and York Streets to her permanent track on Front Street, near the foot of Bay Street. Incredible as it may sound, the moving procedure took all of five days, and everyone had ample opportunity to admire the locomotive which was named *Toronto* after the city of her origin.

A month later, on May 16th, the *Toronto* hauled the first regular steam train ever to run in Ontario from Toronto to Machell's Corners (now Aurora), inaugurating a service between the two locations. A pilot train had been sent over the track the day before the event. North of Davenport Road it lost a car. The only passenger aboard escaped injury but was very upset over losing his spectacles when tumbling down the embankment.

The Toronto station where the inaugural train started was a small wooden shed situated at Front and Bay Streets. William Huckett, master mechanic of the road, drove the engine which hauled two box cars, one combined passenger and baggage car, and one passenger coach. A three-foot cast figure of a Highlander holding up the Union Jack was mounted behind the cowcatcher and served as the engine's mascot. Conductor John Harvey sold the first ticket and the man who purchased it is said to have been a shoemaker named Maher; the fare was one dollar. The historic trip to Machell's Corners took two hours. Today a plaque outside Toronto's Union Station commemorates this event.

One month after opening the train service was extended to Bradford and by October the railway had reached Allandale. The company referred to the latter as "Barrie Station," perhaps in an effort to appease the citizens of Barrie who were angry because the station had been built a mile outside their town limits. According to some accounts the reason for this was Barrie's refusal to grant a bonus to the railway, whereupon the company's chief engineer reputedly vowed to "make grass grow in Barrie streets and pave Allandale streets with gold." A more likely explanation, however, appears to be that the hills around Barrie presented an engineering problem.

Northern Railroad!

COLLINGWOOD HARBOR.

NUMEROUS APPLICATIONS having been made for

BUILDING LOTS

AT THE

"HEN-AND-CHICKENS,"

the subscriber takes this method of informing the applicants and the public that as the survey is being made, and plans prepared, the Lots will shortly be open

For Sale by Auction in Toronto,

of which further notice will be given.

The terms will be one-half down, and the balance in two equal annual instalments with interest, secured by mortgage or otherwise, at the option of the owner. A liberal discount will be made to those who prefer paying in full.

B. W. SMITH.

Barrie, May 16, 1853. 91-w:s.

To the Subscribers to the Capital Stock

OF THE

ONTARIO, SIMCOE & HURON

Railroad Union Company.

TAKE NOTICE,

THAT in pursuance of a resolution of the Board of Directors passed at a meeting held on the 11th day of March, 1854, the following calls upon every £5 share are made payable to me at the Offices of the said Company, at the times under mentioned, viz :

Ten Shillings per £5 Share,	on the	15th day of	March inst.
Ten Shillings	" "	15th "	April, 1854.
Ten Shillings	" "	15th "	May, 1854.
Ten Shillings	" "	15th "	June, 1854.
Ten Shillings	" "	15th "	July, 1854.
Ten Shillings	" "	15th "	Aug. 1854.
Five Shillings	" "	15th "	Sept. 1854.

Notice is also hereby given that all subscriptions in arrear, are required to be paid to me forthwith. By order of the Board.

W. SLADDEN,
Secretary and Treasurer,

Toronto, March 13th, 1854. 1222-td

RAILROAD SPIKES!

FOR SALE BY

THOMAS PECK & Co.,

MANUFACTURERS,

No. 155½, St. Paul Street

Montreal, Nov. 24, 1853. 1131-tf

Toronto Public Libraries

Davenport Station on the Northern Railway, four miles from Toronto

Early in 1855 the entire stretch to Collingwood, including a branch line to the shore of Lake Simcoe, was completed and the first train made the ninety-four-mile run connecting Lake Ontario with Huron waters, replacing the long roundabout water route that once had been the only link. Prior to the railway, schooners sailing from Toronto depended on favourable weather to reach Georgian Bay within three weeks. After crossing Lake Ontario, they had to pass through the locks of the Welland Canal, and be towed by teams of oxen or horses. The rest of the first week was spent on Lake Erie, while the second week was taken up by the passage through the Straits of Detroit. During the third week schooners sailed up Lake Huron, around the Bruce Peninsula and into Georgian Bay.

Before being chosen as the northern terminus of the railway line, Collingwood was known as "Hen and Chickens Harbour" on account of one large and four small off-shore islands, which have since become part of the mainland. A group of engineers, who surveyed the site, decided a more suitable name was desirable for a railway terminus and so right then and there they opened a bottle of wine and christened the place "Collingwood."

The first railway station in Collingwood was a picturesque building of Spanish-style architecture. It is believed to have been designed by Frederick W. Cumberland, at that time chief engineer of the railway, and also an outstanding architect responsible for many of Toronto's impressive public buildings, including St. James' Cathedral and Osgoode Hall.

With the railroad came prosperity to Collingwood. The place soon turned into a busy trans-shipping centre for grain, which arrived in steam barges from the midwestern United States to be forwarded by rail to Toronto. Lumber rafts from the northern forests began to fill Collingwood's harbour awaiting transport to the south. Simcoe County with its rich timber stands became the chief lumber producer of Ontario, and the railway carried millions of feet of timber and lumber to Toronto. Some days as many as fifteen trainloads passed over the line. Incidentally, the first freight shipment carried by the Ontario, Simcoe and Huron Railway after the line was opened consisted of a chest of tea, one dozen brooms and a barrel of salt consigned from Toronto to Bradford.

The railway, more than any other factor, also was responsible for the beginning of Toronto's phenomenal growth as a centre of trade and commerce and a prospering shipping port.

L. NIPISSING
North Bay
Huntsville
GEORGIAN BAY
Bracebridge
Gravenhurst
Meaford
Penetang
Washago
Orillia
Collingwood
Barrie
L. SIMCOE
Angus
Bradford
Newmarket
Aurora
Toronto
LAKE ONTARIO

Northern Railway

Toronto Public Libraries

Allandale, Ontario: the Northern Railway station, not far from Barrie

Among the pioneer locomotives of the Ontario, Simcoe and Huron Railway was a high-speed passenger engine called *Josephine*. No one seems to know the exact speed she was capable of, but contemporaries agree that she was the fastest engine of her day in Canada West. Her driver was William Huckett's brother Cyrus. A dashing figure, Cyrus was just as popular as his engine. As soon as his train pulled into the station, the womenfolk crowded around to get a glimpse of "Dandy Cy of the Josephine," which was the title of a poem written about him:

I dressed myself from top to toe
and out from Toronto I did go.
My hair all combed so slick and fine
I looked as prim as the Josephine . . .

The *Josephine*, like most early engines, was coated with tallow and polished to a high gloss at the end of each trip. Built in 1853 in Paterson, New Jersey, she operated until 1880 when she had become obsolete and was broken up. Two other well-known locomotives of the Ontario, Simcoe and Huron Railway were the *Samson* and her sister engine the *Hercules*.

Busy as the railway was, financially it was anything but a success. Shareholders grumbled because the company continuously failed to pay dividends. Workmen received no pay. The track lacked ballasting and the line had never been fenced. Angry farmers along the route complained about the loss of livestock which had wandered onto the unprotected track. The railway's liabilities by far exceeded its profits. On the verge of bankruptcy, the company was reorganized in 1860. The name was changed to Northern Railway of Canada, recognizing the important role the line had played in opening up the fertile district north of Toronto and tapping the rich timber resources of northern Ontario. A period of expansion followed and the length of the line grew to nearly five hundred miles. Eventually the Northern was amalgamated with the Hamilton and North Western Railway, which had built a line from Lake Erie via Hamilton to Collingwood and Barrie. Today Northern Ontario's first railway is part of the Canadian National system.

The Great Western Railway

The Great Western Railway had its origin in a charter granted in 1834 by the legislature of Upper Canada. Under the name of London and Gore Railway, the company was authorized to

> *construct a single or double track, wooden or iron railroad from London to Burlington Bay; and also to the navigable waters of the Thames and Lake Huron; and to employ thereon the force of steam or the power of animals, or any mechanical or other power.*

Nothing happened for many years thereafter. The charter, about to lapse, was renewed and amended in 1845 to allow the company to construct a railway from any point at the Niagara River to the Detroit River. At the time, the corporate name was changed to Great Western Railway Company.

The proposed line was designed to shorten by approximately 125 miles the cumbersome water route from Lake Ontario to Lake Huron by way of the Welland Canal, Lake Erie and the Detroit River. The railway was to run from the Suspension Bridge at Niagara Falls via Hamilton, Woodstock, London and Chatham to Windsor, opposite Detroit. Original surveys planned that the line, from Hamilton westward, should pass through the town of Brantford, but when short-sighted civic officials of Brantford declined to grant a bonus to the railway company, the route was diverted.

The ground-breaking ceremony for the Great Western Railway took place at London, Ontario, on October 23, 1849. Visitors began pouring into town from the early morning hours. At noon stores and shops closed and a procession began to form on the Courthouse Square. Headed by the Rifle Company, the Artillery and a band, it moved along Dundas to Richmond Street and, after turning north, passed the barracks and crossed the bridge at Lake Horn. About a mile from the courthouse it came to a halt on a spot where the forest had been cleared and stands for spectators had been erected. The assembled crowd numbered between four and five thousand. Among the dignitaries present were Sir Allan MacNab, president of the company and at the time leader of the Conservative opposition in the legislative assembly, and Colonel Thomas Talbot, famed colonizer of southwestern Ontario and father of the Talbot settlement, the capital of which — St. Thomas — commemorates his name. Colonel Talbot was given the honour of turning the first spade-full of sod. In his speech he recalled that fifty-five years earlier he had slept near this spot in the forest, with a porcupine as his friend. "We were often

Toronto Public Libraries

hungry in those days," he said, "but never so hungry as on the night when we ate the porcupine." Well, times had changed! After the ceremony a lavish banquet was held at London's Western Hotel and the menu listed no less than seventeen varieties of meat.

The Great Western Railway Station, Toronto, Ontario

The construction of the railway did not start for some time. In fact, it was not until November 1853 that part of the line was finally opened for traffic, with the first train running from Hamilton to Niagara Falls.

On December 15th of the same year Londoners were able to celebrate the opening of the central section with the arrival of the first train from Hamilton. It consisted of the locomotive and two cars carrying, among other passengers, leading company officials. The eighty-mile trip from Hamilton, with several stops along the way, had occupied the greater part of the afternoon. By stage coach the same journey would have taken anywhere from twelve to twenty hours depending on the condition of the road. Nevertheless, at least one passenger on the first train was not impressed. He found the rocking of the coaches over the rough roadbed frightening, and was appalled that no platforms had as yet been erected at the stations. Apparently the mud was so deep that he got stuck as he left the train.

With the railroad came wealth and prosperity to the little town of London. Two years after the arrival of the first train London was incorporated as a city, and from then on it developed rapidly into the manufacturing and distributing centre of western Ontario.

On Tuesday, January 17, 1854 the Great Western Railway was officially opened for traffic between Niagara Falls and Windsor. A steam passenger ferry from Windsor to Detroit provided connection between the Great Western terminus and that of the Michigan Central across the Detroit River on the American side. Celebrations at all stations along the route marked the occasion as an event of international significance. With the completion of the Great Western continuous railway communication was now available from the Atlantic Ocean to the Mississippi River. The City of Detroit declared a holiday and, according to the *Toronto Globe* of January 23, 1854:

> *. . . cards were despatched far and wide by the City Corporation, inviting persons from other States and Cities to join them in their festivities. Invitations from the Great Western Company accompanied the cards of the Corporation, enabling the parties to pass to and fro over the Railroad.*

The "rendezvous" for the guests from the east was the City of Hamilton, and all were invited to be ready to start at six o'clock on Tuesday morning. On Monday afternoon and evening, the "ambitious little city" gave ample proof that the summons would be largely responded to. Every stage, steamboat and Railroad train arrived loaded with guests; the hotels were crowded to excess, and the merry groupes [sic] on the streets showed that the whole population was in a high state of excitement about the approaching celebration.

The fated morning came in clear and pleasant as the most sanguine could wish; cabs, carriages, and omnibuses were in full requisition long before day-break, and by six o'clock the station-houses and platforms were thronged with eager travellers. And a strange motley groupe they certainly presented! From Boston, New York, Philadelphia, Albany, Rochester, Buffalo, and many other cities and towns in the United States, were to be found deputations more or less numerous; mayors, judges, fire-engine men, contractors, politicians, clergymen, newspaper-editors, the southern slave-holder, and the genuine Yankee — all were to be found in the "eastern deputation." And the collection from our own Province was not less diversified. From Toronto there were some forty persons, and from every town and hamlet between it and Hamilton one or more individuals might be found. From the Niagara District, and in fact from every county around Hamilton came many persons to show their interest in the event of the day.

Timetable of the Great Western Railway, published in Hamilton newspapers in 1853 and 1854

GREAT WESTERN RAILWAY!

Open from London to Niagara Falls.

ON and after WEDNESDAY, the 4th JANUARY, 1854, Trains will run as follows:—

GOING EAST.

Leave London at - - - - 8 00, A. M.
" Hamilton at - - - - 12 25, P. M.
Arrive at the Falls at - - - 2 45, P. M.

GOING WEST.

Leave the Falls at - - - - 11 45, A. M.
" Hamilton at - - - - 2 15, P. M.
Arrive at London at - - - 6 20, P. M.

The above Trains connect with Trains to and from Buffalo, New York, Boston, Albany, Schenectady, Utica, Syracuse, Rochester and intermediate places.

Omnibusses will be in readiness to convey Passengers across the Suspension Bridge.

Passengers can now purchase Through Tickets at Hamilton or London, for New York, and the principal Stations on the New York Central Line.

FARE FROM HAMILTON TO NEW YORK $9.50
" " LONDON " " $11 75.

Passengers going East will arrive in New York at 9, a.m., next morning.

The Buffalo Trains will in future arrive at, and start from the Suspension Bridge and Niagara Falls.

C. J. BRYDGES,
Managing Director.

Hamilton, 31st Dec., 1853. 1161-tf

About half-past 6, the first train of six large cars moved off from the Station, amid loud cheers from the spectators, heartily returned by the passengers on the train; immediately afterwards a second train followed with the same number of cars. There were probably between 600 and 700 in the two trains, which number gradually swelled as the several stations contributed their quota in passing, until the whole number at Chatham amounted to upwards of one thousand persons. Every seat was occupied, and a great many had to stand; but the utmost good humour and enjoyment seemed to prevail throughout the crowded apartments.

The arrangement for the day, was that the locomotive should arrive at Detroit by half-past one; and had the contractors been enabled to complete their work as early as was anticipated, this would have been easily accomplished. But unfortunately, with all the energy that could be applied, the last rails were only laid on a portion of the route, by lamplight, during the night preceding the departure of the train. No trial trip had been made, and the directors with most commendable prudence resolved that every precaution should be taken against accident. The cars therefore arrived at Dundas at 7.15; at St. George at 8; Woodstock, 9.45; London, 11.30; Mosa, 2.10; Thamesville 2.45; Chatham, 3.30; and it reached Detroit at half-past four. At each of these places crowds of persons were assembled to welcome the visitors, and considerable stoppages were made at them all. . . .

While the cars were proceeding towards the shore of Michigan, the citizens of Detroit were all alive preparing to receive their guests. The Detroit "Advertiser" says, that the streets were early thronged. Thousands of strangers had come in from all parts of the great west, to participate in the joyful festivities, and more were constantly arriving each moment, adding to the

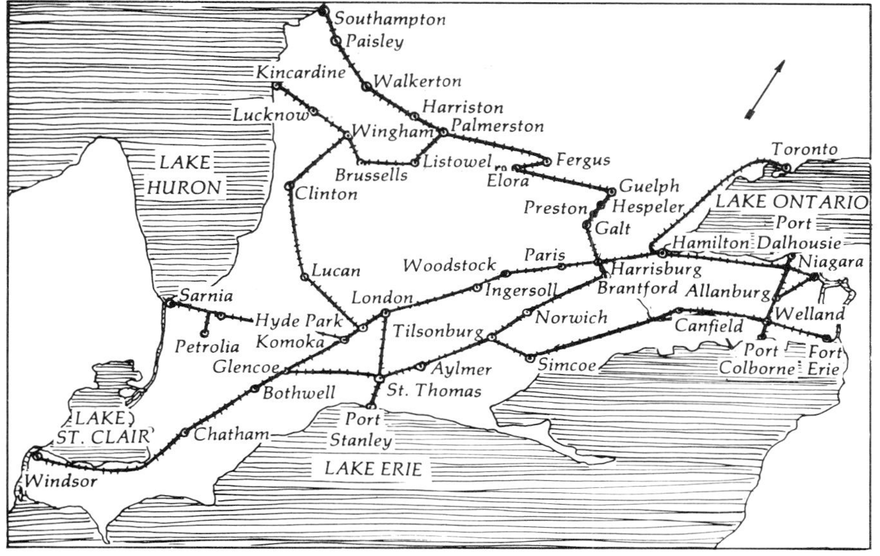

gaiety of the scene. Flags were floating from prominent places, mottoes stretched across the streets, and in front of the City Market the stars and stripes and the flag of Great Britain were waving side by side. Early in the afternoon the Fire Companies commenced forming on Jefferson avenue, and a splendid show they made with their uniforms, engines, bells and banners. Crowds of gaily dressed spectators thronged the side-walk, windows and house-tops. At 2 o'clock, P.M., the citizen soldiery in their brilliant uniforms were seen hastening from every portion of the city to their respective rendezvous. And very soon thereafter the military of the city and the several fire companies had taken up the positions assigned them on Woodward avenue stretching from the dock to Campius Martius in one continuous line. The whole of that broad and spacious street was filled with a dense crowd of citizens and strangers representing every county and town in Michigan, Indiana, Illinois, and Wisconsin, with the cities of Chicago, Milwaukee and Michigan.

The steamer "Dart" and the ferry boats were constantly plying between the city and Windsor, carrying over to the Canada shore the thousands who were desirous of greeting the arrival of the train over the Great Western Railway.

The train came in about half-past 4 p.m., and most gracefully did it sweep alone [sic] the Canadian shore, where the track lies below the bank and in full view of Detroit, for nearly a mile. The citizens, who had gone over, united with their Canada friends in giving the hero of the day, the iron steed, a welcome worthy of his grand achievement. At 5 o'clock, the booming of cannon from both sides of the river, announced the arrival of the second train. Shouts, mingled with the roar of cannon, welcomed them, and clouds of steam and smoke went up, amid the waving banners of the boats which dotted the river. . . .

GREAT WESTERN RAILWAY!

Open from Windsor to Niagara Falls.

ON and after MONDAY, the 30th JANUARY, 1854, Trains will run as follows :—

GOING EAST.

EXPRESS TRAIN.

Leave Windsor at	10 00, A. M.
" London at	2 30, P. M.
" Hamilton at	6 10, P. M.
Arrive at the Falls	8 00, P. M.

ACCOMMODATION TRAIN.

Leave London at	7 00 A. M.
" Hamilton at	10 50 A. M.
Arrive at the Falls	1 30 P. M.

GOING WEST.

EXPRESS TRAIN.

Leave the Falls at	10 30, A. M.
" Hamilton at	12 30, P. M.
" London at	4 05, P. M.
Arrive at Windsor	8 40, P. M.

ACCOMMODATION TRAIN.

Leave the Falls at	12 15 P. M.
" Hamilton at	2 45 P. M.
Arrive at London	6 40 P. M.

The above trains run in direct connection with Trains on the New York Central and Michigan Central Railroads.
Through Tickets may be obtained at Hamilton, London, Windsor, or Detroit, for New York, Boston, Albany, Buffalo, and Chicago.

C. J. BRYDGES,
Managing Director

Hamilton, Jan., 1854. 1186-tf

As the western terminus of the railway, Windsor soon began to flourish. So did other hamlets and towns along the line. New settlements were springing up overnight. New industries were born. The value of land within several miles of the railroad's right of way doubled and tripled. What was once referred to as "waste land." became a rich farming district.

A few years after completion of the line, the powerful paddle-wheeler *Union* was put into service between Windsor and Detroit. Designed to withstand heavy ice in winter, the *Union* could haul a trainload of passengers on a single trip. Eventually a car ferry with two tracks was employed to transport an entire train across the river and in 1870 the Great Western and the Michigan Central obtained power to construct a railway tunnel under the Detroit River.

Meanwhile, at the eastern end of the Great Western a railway suspension bridge across the Niagara River was opened for traffic in 1855. A 368-ton freight train, spanning the bridge from end to end, had been made up to test the structure. The bridge is said to have settled ten inches under the enormous weight but returned to normal once the train had passed. Fifteen thousand miles of wire were absorbed within the structure. Its four towers, two on either end, rose eighty feet above the ground, the span between them being eight hundred feet. At the time it was the only bridge in the world carrying a main railway line in addition to a highway. Tracks ran on the upper deck, 215 feet above the water, and an enclosed lower deck served pedestrians and horse-drawn vehicles. The suspension bridge was in use until 1897 at which time the Whirlpool Rapids Bridge took its place.

Like all pioneer lines, the Great Western Railway was built almost entirely with picks, shovels, wheelbarrows and back-breaking labour. Grading was done by hand, as was the laying of the track. A locomotive ordered to help in the construction had to be hauled by six teams of oxen from Port Stanley to the railhead near London. In the vicinity of Dundas it was necessary to divert a canal in order to lay foundations for the abutments of a bridge. In the same area work gangs were kept busy for two years trying to fill a seemingly bottomless marsh. Labour unrest flared up occasionally as wages were governed entirely by supply and demand. Local workers were scarce, since the population was still very small. The majority of the thousands of navvies employed on the Great Western were transients.

Despite the many problems and engineering difficulties, construction proceeded rapidly, but not fast enough to suit the public who demanded the opening of the line before it was ballasted. Reluctantly, management agreed, a decision which was to be a costly one. Not only did operations prove extremely hazardous, but the uneven track was hard on the rolling stock. Consequently a fifth of the available passenger cars was usually under repair. Cuttings, not properly sloped yet, turned into mud several feet deep, forcing the engineer to brake his train and causing delays. Signal lights, if they had been installed at all, had not been tested and did not work. Claims mounted from farmers whose livestock was killed on the right of way which still lacked fencing.

Partly as a result of the premature opening of the line, such a high number of accidents plagued the Great Western in the beginning that some people began to wonder if progress was worth the risk of travelling by train. Of course, some of the accidents were caused by inexperienced crews, others could be blamed on sheer carelessness. Fifteen accidents occurred during the first year of operation. Fifty-two people were killed and almost as many injured at St. Baptiste where a mail train, running seven hours late, collided with the gravel train engine *St. Lawrence.*

A mechanical defect caused one of the Great Western's worst disasters. It happened on a wooden swing bridge over the Desjardin Canal near Hamilton in the late afternoon of March 12, 1857. As the regular passenger train from Toronto approached the bridge, the leading axle of its engine truck broke close to the right wheel. The locomotive

CNR

Great Western Railway *Scotia, the first engine in Canada with a steel boiler. Built by the Great Western Railway in 1861*

Oxford toppled, and, the train plunged sixty feet down into the icy water. Fifty-nine people either drowned or were burned to death as they fell against the heating stoves in the two passenger cars. For some time after the bridge was repaired and reopened for traffic, passengers were allowed to detrain and cross on foot if they so desired.

The first passenger locomotives of the Great Western came from different builders in the United States. Among the earliest ones were the *Canada,* the *Niagara,* the *London* and the *Hamilton,* all 4-4-0 type locomotives with inside cylinders and coupled wheels. Eight ballast or shunting engines were purchased for construction and general work from the Globe Works in Boston. From England came twenty freight engines of the 0-6-0 type with coupled wheels, no trucks and inside cylinders connected to the middle axle. Among them were the *Elephant,* the *Buffalo* and the *Tiger.* Twelve engines arrived from Manchester. The *Spitfire* and *Firefly* belonged to this group of 2-4-0 type locomotives. By the end of 1857 the Great Western already owned more than seventy engines.

As time went on the Great Western Railway gained a reputation for progressiveness. After the Westinghouse air brake had become available, the company at once proceeded to equip its passenger engines and rolling stock on the main line with the new invention. In 1854, to expedite postal service, it inaugurated the sorting of mail on trains, a first for Canada.

In its Hamilton shops the company turned out the first sleeping car in the country. It was designed by car superintendent Samuel Sharp in 1857 and featured two rows of berths, three tiers high. One of the company's latest models was described by the press as follows:

> *The G. W. R. has just turned out another one of those travelling luxuries which are now so prominent a feature in railway travel. . . .*
>
> *The car which the G. W. R. has just converted into a sleeping car is that which was placed at the disposal of the Prince of Wales during his sojourn in these parts of Canada. It has accommodation for forty-four passengers and may be used either as a day or night car. Partitions of solid walnut, beautifully polished, divide the berths into eleven compartments, four berths each. The beds are spring stuffed. They are enclosed by curtains which secure complete privacy.*

In 1860 the Great Western began to build its own locomotives. The first to leave its Hamilton shops was the *George Stephenson*, built by Richard Eaton, locomotive superintendent of the company. The latter was a genius in his own right and his experiments brought about many innovations in the Great Western's motive power. He designed the first coal-burning locomotive and experimented with oil as a locomotive fuel. The *Scotia*, built at the Hamilton shops, was the first engine in Canada to have a steel boiler. In 1864 the company set up a rolling mill in Hamilton for the purpose of re-rolling worn iron rails; but before long the Great Western began to introduce superior quality steel rails on its line.

Government regulations at the time of construction had forced the Great Western Railway, however reluctantly, to adopt the broad gauge of 5′ 6″, a fact which soon proved a serious handicap in the interchange of traffic with American lines operating on 4′ 8½″ tracks. To speed up freight handling by accommodating the American cars, the company eventually laid down a third rail. Great Western engines, rebuilt to run on the narrower gauge, were identified by a large plate on the front of the buffer beam reading "NG." The outside rail was later taken up and by 1873 the main line and branches had a uniform gauge of 4′ 8½″.

By then the Great Western had grown into a system serving a large territory throughout southwestern Ontario. It now either worked or owned branch lines from Hamilton to Toronto, from Harrisburg to Guelph, from Fort Erie to Niagara, from Komoka, west of London, to Sarnia, and from Wyoming to the oil region of Petrolia.

With the opening of the Sarnia branch residents of that city were able to enjoy some hitherto scarce commodities from eastern manufacturers. The Sarnia newspaper of the time happily reported that a large quantity of freight from London was being shipped to Sarnia, part of which consisted of "thirty barrels of beer and fifty dozen of ale from the celebrated brewery of Mr. Labatt." In 1860 the Great Western had built a grain elevator at the Sarnia waterfront to capture the grain trade.

The Galt and Guelph branch, running northward from Harrisburg, started out as an independent company but was soon absorbed by the Great Western. Construction on the Hamilton and Toronto Railway began in 1854. Leased to the Great Western during the same year, it was eventually amalgamated with the latter. Among the Great Western affiliates were also the London, Huron and Bruce Railway from London to Wingham, and the Wellington, Grey and Bruce, a line running from Guelph to Palmerston and Southampton with an extension from Palmerston to Listowel, and thence northward to Kincardine.

The Great Western authorized the construction of the Canada Air Line Railway, a loop line from Glencoe to Fort Erie, Ontario, opposite the city of Buffalo, to meet the growing competition of the Canada Southern Railway as a connecting link between Canada and the United States. The Canada Southern, later leased to the Michigan Central, was chartered in the late 1860s under the name of Erie and Niagara Extension Railway. Backed by American interests, this formidable competitor of the Great Western ran from Fort Erie via St. Thomas to Windsor, Ontario.

Added trackage, however, did not always mean increased profits. More often the acquisitions turned out to be financial burdens. And at the same time a powerful giant, the Grand Trunk Railway of Canada, was spreading its empire from east to west, gradually swallowing its competitors along the way. Eventually, the Great Western too was absorbed by this great rival. Amalgamation with the Grand Trunk Railway became effective on August 12, 1882.

GREAT WESTERN RAILWAY

1876. TIME TABLE 1876.

FROM MONDAY, 3rd JULY, 1876, UNTIL FURTHER NOTICE.

GOING WEST.

Distance from Clifton	STATIONS.	* Mixed.	Morning Express.	Kincardine Accom.	Pacific Express.	Windsor Accom.	Accom.	St. Boat Express.	Chicago Express.
	NEW YORK .... dep	....	11.50	....	8.30	....	....	11.00	10.30
	BOSTON .... "	....	8.30	....	6.00	....	....	9.00	8.30
	ROCHESTER .... "	....	4.50	....	10.25	....	P.M.	5.35	10.30
	BUFFALO .... "	....	7.00	....	11.50	....	2.45	8.30	12.05
	SUS. BRIDGE (N.Y.C.) arr	....	8.05	....	12.55	....	3.25	9.05	1.10
	CLIFTON (G.W.R.) dep	....	8.30	....	1.35	....	3.45	9.50	1.35
9	Merritton	....	8.50	....	1.54	....	4.05	10.14	....
11½	St. Catharines	....	8.56	....	2.00	....	4.11	10.20	2.00
17	Jordan	....	9.06	....	....	....	4.23	....	....
22¼	Beamsville	....	9.18	....	....	....	4.35	....	....
26¾	Grimsby	....	9.27	....	....	....	4.45	....	....
31¾	Winona	....	9.38	....	....	....	4.56	....	....
43½	HAMILTON, arr	....	10.00	P.M.	2.55	....	5.20	11.25	3.00
	HAMILTON, dep	....	10.10	3.35	3.05	....	5.35	11.35	3.10
49¾	Dundas	....	10.27	3.55	....	....	5.54	11.55	....
54¾	Copetown	....	10.39	4.09	....	....	6.08	....	....
59	Lynden	....	10.48	4.20	....	....	6.17	A.M.	....
62¼	Harrisburg	....	11.02	4.30	3.50	....	6.27	12.28	....
64¾	St. George	....	11.08	P.M.	....	....	6.32	....	....
72	PARIS, arr	....	11.25	....	4.12	....	6.47	12.50	4.10
	PARIS, dep	....	11.30	....	4.17	....	6.52	12.55	4.15
79	Princeton	....	11.45		....	....	7.08	....	....
81¾	Gobles	....	11.51		....	....	7.14	....	....
86½	Eastwood	....	12.02		....	....	7.28	....	....
90¾	Woodstock	....	12.16		4.56	....	7.41	1.40	4.52
95¾	Beachville	....	12.27		....	....	7.51	....	....
100	Ingersoll	....	12.41		5.17	....	8.04	2.02	5.12
109¾	Dorchester	....	1.02		....	....	8.24	....	....
119½	LONDON, arr	....	1.25		5.55	P.M.	8.45	2.45	5.45
	LONDON, dep	6.05	1.40		6.10	6.30	P.M.	2.55	5.55
129¾	Komoka	6.45	2.04		....	7.00	....	....	....
134	Mount Brydges	7.05	2.14		....	7.13	....	....	....
140	Longwood	7.43	2.27		....	7.30	....	....	....
145	Appin	8.10	2.40		....	7.50	....	....	....
149¾	Glencoe	8.35	2.51		....	8.02	....	....	....
155¾	Newbury	9.30	3.06		....	8.22	....	....	7.03
161	Bothwell	10.00	3.19	....	....	8.37	....	4.20	7.14
168½	Thamesville	10.35	3.33	....	....	8.59	....	....	....
174¾	Lewisville	11.05	....	....	....	....	....	....	....
183½	CHATHAM, arr	11.50	4.00	....	8.00	9.40	....	5.05	7.50
	CHATHAM, dep	P.M. 12.35	4.05	....	8.10	A.M. 7.00	....	5.10	8.10
197¼	Jennette's Creek	1.35	4.32	....	....	7.35	....	....	....
203	Stoney Point	2.00	4.44	....	....	7.53	....	....	....
212	Belle River	2.45	5.03	....	P.M.	8.25	....	A.M.	....
229	WINDSOR .... arr	4.00	5.35	....	9.30	9.20	....	6.40	9.35
	DETROIT (M.C.R.R. Dep.) arr	P.M.	5.55	....	10.00	A.M.	....	7.00	9.55
	CHICAGO .... arr	....	A.M. 6.30	....	A.M. 8.00	....	....	P.M. 7.30	P.M. 8.00

This Train Runs to Kincardine and Southampton on the W. G. & B. Railway.

TORONTO BRANCH.

Connecting with Main Line Trains at Hamilton.

From Hamilton	TO TORONTO, &c.	Mixed.	Express.	Accom.	Accom.	Exp.	Mail.	Mixed.
	HAMILTON .... dep	7.00	9.00	10.05	11.35	3.17	5.25	9.25
4	Waterdown	7.18	....	10.14	11.50	....	5.35	9.42
7	Wellington Sq.	7.28	....	10.19	11.56	....	5.42	9.55
13½	Bronte	7.45	....	10.31	12.1	....	5.56	10.13
17½	Oakville	8.03	....	10.40	12.20	....	6.06	10.28
25¼	Port Credit	8.27	....	10.58	12.38	....	6.24	10.53
32	Mimico	8.44	....	11.12	12.53	....	....	....
34½	Sunnyside	....	....	11.19	....	....	....	....
	North R'y Junct.	....	....	....	....	4.22	....	....
39½	TORONTO (Union) arr	9.10	10.10	11.35	1.10	4.25	6.55	11.45
	TORONTO (Yonge St.)	9.15	10.15	11.40	1.15	4.30	7.00	11.50

From Toronto	TO HAMILTON, &c.	Accom.	Exp.	Accom.	Mail.	Exp.	Mixed.
	TORONTO (Yonge St.) dep	7.10	9.50	12.55	3.20	6.35	11.20
	TORONTO (Union)	7.15	9.55	1.00	3.25	6.40	11.25
	Northern R'y Junction	....	10.03	....	....	....	....
4½	Sunnyside	....	....	....	3.37	....	....
7¼	Mimico	7.31	....	....	3.45	....	11.53
14	Port Credit	7.45	....	1.32	3.59	....	12.13
22	Oakville	8.03	....	1.50	4.18	7.25	....
26½	Bronte	8.13	....	2.00	4.28	....	....
32½	Wellington Square	8.27	....	2.14	4.42	....	....
35½	Waterdown	8.35	....	2.20	4.50	....	....
39½	HAMILTON .... arr	8.45	11.10	2.30	5.00	7.55	1.40

WELLINGTON, GREY & BRUCE BRANCH.

Connecting with Main Line Trains at Harrisburg.

From Harrisburg	GOING NORTH.	[illegible]	[illegible]	[illegible]	[illegible]	[illegible]
	Harrisburg .... dep	8.55	11.05	4.40	....	6.35
6¼	Branchton	9.17	11.20	4.58	....	6.53
11¼	Galt	10.05	11.34	5.15	....	7.15
16	Preston	10.20	11.43	5.26	....	7.30
19½	Hespeler	10.34	11.52	5.36	....	7.42
27½	Guelph, arr	11.00	12.08	5.55	....	8.05
	Guelph, dep	11.25	12.22	6.10	....	8.15
40¼	Elora	12.15	12.53	6.48	....	9.03
43	Fergus	12.25	1.00	6.56	....	9.30
49¼	Alma	....	1.16	7.14	....	....
54¼	Goldstone	....	1.28	7.27	....	....
18¼	Drayton	....	1.36	7.37	....	....
62½	Moorefield	....	1.45	7.48	....	A.M.
69¼	Palmerston	....	2.15	8.20	11.10	7.50
75	Harriston	....	2.27	8.35	P.M.	8.35
81¾	Clifford	....	2.42	8.53	....	9.20
90½	Mildmay	....	3.02	9.15	....	10.05
96½	Walkerton	....	3.17	9.30	....	11.05
101	Dunkeld	....	3.29	9.42	....	11.30
105¼	Pinkerton	....	3 40	9.54	....	12.00
111½	Paisley	....	4.00	10.10	....	12.25
118¾	Turners	....	4.24	10.28	....	P.M.
124¼	Port Elgin	....	4.42	10.40	....	....
129	Southampton .... arr	....	4.55	10.50	....	....

From Southampton	GOING SOUTH.	Accom.	Accom.†	Mixed.	Accom.	Mixed.
	Southampton .... dep	....	A.M. 5.05	....	P.M. 1.35	....
4¾	Port Elgin	....	5.14	....	1.50	....
10¼	Turners	....	5.27	....	2.08	P.M.
17¾	Paisley	....	5.46	....	2.30	3.00
23¾	Pinkerton	....	6.00	....	2.49	3.40
28	Dunkeld	....	6.12	....	3.02	4.05
32½	Walkerton	....	6.25	....	3.17	4.55
38½	Mildmay	....	6.40	....	3.34	5.35
47¼	Clifford	....	7.00	....	3.56	6.25
54	Harriston	....	7.16	....	4.15	7.25
59¾	Palmerston	A.M. 4.00	7.45	A.M. 10.30	4.40	7.45
66¾	Moorefield	....	8.02	11.05	5.01	....
70¾	Drayton	....	8.12	11.30	5.14	....
74¾	Goldstone	....	8.20	11.47	5.25	....
79¾	Alma	....	8.32	12.15	5.40	P.M.
86	Fergus	6.00	8.48	1.20	6.02	1.00
88½	Elora	6.07	8.54	1.45	6.10	1.10
101⅞	Guelph, arr	....	9.25	2.40	6.45	1.55
	Guelph, dep	6.45	9.30	3.35	6.55	2.00
109¾	Hespeler	7.04	9.47	4.10	7.18	2.30
113	Preston	7.14	9.56	4.30	7.30	2.44
117¾	Galt	7.27	10.05	5.15	7.44	3.00
122¾	Branchton	7.44	10.20	5.45	8.02	3.23
129	Harrisburg .... arr	8.00	10.35	6.10	8.20	3.45

† This train connects at Toronto Junction with train for Toronto and intermediate Stations; also with Atlantic Express for all points East.

W. G. & B. SOUTH EXTENSION.

STATIONS.	Going North: Mixed.	Accom.	Accom.	STATIONS.	Going South: Accom.	Accom.	Mixed.
	A.M.	P.M.	P.M.		A.M.	P.M.	P.M.
Palmerston	6.20	2.15	8.20	Kincardine	4.40	1.10	2.20
Gowansto'n	6.40	2.32	8.35	Ripley	4.58	1.30	2.55
Listowel	7.30	2.44	8.45	Lucknow	5.18	1.55	3.35
Newry	7.55	3.00	9.00	Whitechur'h	5.32	2.13	4.20
Henfryn	8.20	3.10	9.10	Wingham	5.48	2.35	5.05
Ethel	8.40	3.25	9.20	Blue Vale	6.00	2.49	5.30
Brussels	9.25	3.40	9.35	Brussels	6.17	3.08	6.15
Blue Vale	10.05	3.54	9.53	Ethel	6.32	3.25	6.45
Wingham	10.50	4.06	10.07	Henfryn	6.40	3.36	7.10
Whitechur'h	11.10	4.20	10.22	Newry	6.50	3.47	7.40
Lucknow	11.50	4.35	10.38	Listowel	7.06	4.05	8.45
Ripley	12.25	4.55	11.00	Gowanstown	7.16	4.15	9.00
Kincardine	12.55	5.15	11.20	Palmerston	7.30	4.30	9.20

GOING EAST.

Distance from Windsor	STATIONS.	Accom.	Atlantic Express (Sundays included)	* Mixed.	Kincardine Accom.	Day Express.	Buffalo Express.	London Express.	New York Express.
	CHICAGO .... dep	....	P.M. 5.15	....	....	P.M. 9.00	....		A.M. 9.00
	DETROIT (M.C.R.R. Dept.) dep	....	A.M. 3.50	A.M.	....	A.M. 8.40	P.M. 12.30		P.M. 7.05
	WINDSOR .... dep	....	5.00	7.45	....	9.30	1.10		7.55
17	Belle River	....	....	9.00	....	10.05	....		....
26	Stoney Point	....	....	9.40	....	10.26	....		....
31¾	Jennette's Creek	....	....	10.00	....	....	....		....
45½	CHATHAM, arr	....	6.20	10 50	....	11.05	2.22		9.10
	CHATHAM, dep	....	6.25	11.50	....	11.10	2.26		9.15
54½	Lewisville	....	....	P.M. 12.35	....	....	....		....
60½	Thamesville	....	....	1.05	....	11.40	....		....
68	Bothwell	....	....	1.45	....	11.56	....		9.54
73¼	Newbury	....	7.12	2.15	....	P.M. 12.08	....		....
79¼	Glencoe	....	7.25	2.51	....	12.23	3.20		....
84	Appin	....	....	3.10	....	12.34			....
89	Longwood	....	....	3.35	....	12.48			....
95	Mount Brydges	....	....	4.00	....	1.01			....
99¾	Komoka	....	....	4.20	....	1 13			....
109¾	LONDON, arr	A.M.	8.20	5.00	....	1.40		P.M.	11.10
	LONDON, dep	6.00	8.40	P.M.	....	2.05		6.10	11.20
119¼	Dorchester	6.20	....	....	....	2.28		6.33	....
129	Ingersoll	6.39	9.19	....	....	2.55		6.54	11.56
133¼	Beachville	6.47	....	....	....	3.08		7.04	A.M.
138¼	Woodstock	6.58	9.38	....	....	3.23		7.16	12.13
142¼	Eastwood	7.06	....	....	....	3.35		7.28	....
147¼	Gobles	7.17	....	....	....	....		7.39	....
150	Princeton	7.24	....	....	....	3.55		7.47	....
157	PARIS, arr	7.37	10.12	....	....	4.10		8.00	12.45
	PARIS, dep	7.40	10.20	....	....	4.15		8.05	12.50
164⅞	St. George	7.56	....	....	A.M.	....		8.22	....
166¼	Harrisburg	8.05	10.40	....	11.02	4.35		8.35	....
170	Lynden	8.14	....	....	11.10	4.41		8.42	....
174¼	Copetown	8.25	....	....	11.19	....		8.52	....
179¼	Dundas	8.37	11.07	....	11.30	4.59		9.04	....
185½	HAMILTON, arr	8.55	11.20	....	11.45	5.15		9.20	1.50
	HAMILTON, dep	9.10	11.30	....	....	5.25		9.30	2.00
197¼	Winona	9.38	....	....	....	5.48		9.54	....
202¼	Grimsby	9.48	....	....	....	5.59		10.06	....
206¼	Beamsville	9.57	....	....	....	6.08		10.16	....
212	Jordan	10.08	P.M.	....	....	6.20		10.32	....
217¾	St. Catharines	10.21	12.27	....	....	6.34		10.45	3.00
219¾	Merritton	10.27	....	....	....	6.40		10.51	....
229	CLIFTON .... arr	10.50	12.55	....	....	7.05		11.15	3.30
	SUS. BRIDGE (N.Y.C.) arr	10.55	1.15	....	....	7.15		11.25	3.40
	BUFFALO	P.M. 12.50	P.M. 3.20	....	....	P.M. 9.25	P.M. 8.40	P.M.	A.M. 6.00
	ROCHESTER .... arr	....	4.15	....	....	11.03	11.03	....	7.05
	BOSTON .... arr	....	A.M. 10.00	....	....	P.M. 3.25	P.M. 3.25	....	P.M. 11.00
	NEW YORK .... arr	....	A.M. 6.30	....	....	A.M. 10.30	A.M. 10.30	....	P.M. 7.00

Connects via St. Thomas with Train leaving Detroit at 12.30 p.m. — VIA LOOP LINE.

LOOP LINE.

From Ft. Erie, miles	Going West.	FROM BUFFALO: Express.	Accom.	* Mixed.	Going East.	TO BUFFALO: Accom.	Buffalo Express.	* Mixed.
	Buffalo, (Exchange St. Depot.)	A.M. 8.00	P.M. 4.20		Glencoe .... dep	....	P.M. 3.20	A.M. 8.10
	" (Black Rock)	8 30	4.50		Ekfrid	....	....	8.35
	Fort Erie	8.45	5.05		Middlemiss	....	....	8.55
7	Stevensville	‡8 57	5.20		Lawrence	....	....	9.13
13	Humberstone	....	5.34	A.M.	Bairds	....	....	9.25
16½	Welland Junction	9.14	5.43	8.05	Paynes	....	....	9.43
23½	Marshville	9.31	6.04	8.35	St. Thomas, arr	A.M.	4.10	10.00
31¾	Moulton	9.45	6.19	9.00	St. Thomas, dep	7.30	4.15	11.00
33⅞	Diltz	‡9.49	6.25	9.12	New Sarum	7.43	....	11.30
40	Darling Road	‡10 00	6.44	9.35	Aylmer	7.55	....	12.13
48	Cayuga	10.20	7.15	10.30	Corinth	8.11	....	P.M. 12.55
53½	Nelles' Corners	10.30	7.27	11.10	Tilsonburg	8.27	....	1.30
61½	Jarvis	10.44	7.47	11.55	Courtland	8.34	....	1.55
67½	Renton	....	8.01	12.20	Delhi	8.50	....	2.30
72	Simcoe	11.05	8.14	12.40	Nicksville	9.00	....	2.55
76½	Nicksville	11.13	8.24	P.M. 1.00	Simcoe	9.12	5.34	3.30
81	Delhi	11.24	8.37	1.25	Renton	9.22	....	3.52
88¾	Courtland	11.37	8.54	1.55	Jarvis	9.35	....	4.25
92	Tilsonburg	11.46	9.05	2.25	Nelles' Corners	9.55	....	5.05
99	Corinth	11.59	9.25	3.00	Cayuga	10.15	6.30	5.55
107½	Aylmer	12.13	9.41	3.45	Darling Road	10.33	....	6.44
112	New Sarum	....	9.52	4.25	Diltz	10.45	....	7.04
117½	St. Thomas, arr	12.35	10.10	4.45	Moulton	10.51	....	7.12
	St. Thomas, dep	P.M.	P.M.	5.20	Marshville	11.05	....	7.40
122¼	Paynes	....	....	5.38	Welland Junction	11.24	....	8.10
126¼	Bairds	....	....	6.08	Humberstone	11.32	....	
129¾	Lawrence	....	....	6.20	Stevensville	11.45	....	
134	Middlemiss	....	....	6.40	Fort Erie	12.00	7.55	
139½	Ekfrid	....	....	7.00	Buffalo, Black Rock	12.15	8.10	
145	Glencoe .... arr	....	....	P.M. 7.20	" (Exchange St. Arrive.)	P.M. 12.45	P.M. 8.40	

BRANTFORD BRANCH.

Connecting at Harrisburg with Main Line Trains and Trains on the W., G. & B. Branch

TO BRANTFORD					STATIONS.	TO HARRISBURG				
P.M. 9.10	P.M. 7.05	P.M. 5.10	A.M. 11.30	A.M. 8.50	arr ... Brantford ... dep	A.M. 7.30	A.M. 9.50	A.M. 10.30	P.M. 3.10	P.M. 5.45
P.M. 8.40	P.M. 6.35	P.M. 4.40	A.M. 11.05	A.M. 8.20	dep ... Harrisburg ... arr	A.M. 8.00	A.M. 10.20	A.M. 10.55	P.M. 3.40	P.M. 6.15

PETROLIA BRANCH

Miles	TO PETROLIA.	A.M.	A.M.	P.M.	P.M.	FROM PETROLIA.	A.M.	A.M.	P.M.	P.M.
0	Wyoming .. dep	9.05	11.10	3.50	8.40	Petrolia .... dep	6.00	10.10	1.15	7.50
5½	Petrolia .... arr	9.35	11.45	4.20	9.10	Wyoming .. arr	6.30	10.45	1.45	8.20

LONDON, HURON & BRUCE RAILWAY.

Miles	GOING NORTH: STATIONS.	Mixed.	Mail.	Miles	GOING SOUTH: STATIONS.	Mail.	Mixed.
	LONDON .... dep	A.M. 7.30	P.M. 5.00		WINGHAM, ... dep	A.M. 7.30	A.M. 11.00
4½	Hyde Park Junction	7.50	5.10	6¾	Belgrave	7.53	11.30
11½	Ilderton	8.20	5.25	13¼	Blyth	8.15	12.00
15¾	Brecon	8.45	5.40	17¼	Londesborough	8.30	12.25
20½	Clandeboye	9.15	5.55	24	Clinton	8.55	1.15
26½	Centralia	9.35	6.20	30¾	Brucefield	9.15	1.40
31½	Exeter	10.50	6.30	34¾	Kippen	9.30	1.55
37	Hensall	11.15	6.50	37	Hensall	9.40	2.05
39½	Kippen	11.35	7.00	42¾	Exeter	10.00	2.45
43½	Brucefield	11.55	7.15	47½	Centralia	10.13	3.05
50	Clinton	12.50	7.40	53½	Clandeboye	10.33	3.25
56¾	Londesborough	1.15	8.00	58¼	Brecon	10.45	3.45
60¾	Blyth	1.35	8.15	62¾	Ilderton	10.57	4.05
67¼	Belgrave	2.00	8.35	69¾	Hyde Park Junction	11.10	4.35
74	WINGHAM, ... arr	P.M. 2.30	P.M. 9.00	74	LONDON .... arr	A.M. 11.20	P.M. 4.50

LONDON & PORT STANLEY BRANCH.

From London, miles	GOING SOUTH: STATIONS	Accom.	Mail.	Accom.	GOING NORTH: STATIONS	Accom.	Mail.	Accom.	Exp.
	LONDON .... dep	A.M. 9.00	P.M. 2.30	P.M. 6.35	Pt. STANLEY. dep	A.M. 7.00	P.M. 12.05	P.M. 4.20	....
5¼	Westminster	9.20	2.50	6.55	White's	7.12	12.17	4.32	....
9¼	Glanworth	9.35	3.00	7.10	ST. THOMAS, arr	7.25	12.30	4.45	P.M.
13¼	Yarmouth	9.45	3.10	7.22	ST. THOMAS, dep	7.35	12.40	5.00	4.35
15¼	St. THOMAS, arr	9.50	3.15	7.30	Yarmouth	7.40	12.45	5.06	....
	St. THOMAS, dep	10.00	3.20	7.35	Glanworth	7.55	12.55	5.20	....
19¼	White's	10.10	3.30	7.45	Westminster	8.10	1.05	5.35	....
23¾	Pt. STANLEY .. arr	10.20	3.40	7.55	LONDON, ... arr	8.30	1.25	5.55	5.20

SARNIA BRANCH.

GOING EAST: Accom.	Mixed.	Accom.	Accom.	From Sarnia, miles	STATIONS.	From London, miles	GOING WEST: Accom.	Mixed.	Accom.	Accom.
P.M. 10.55	P.M. 5.30	P.M. 1.05	A.M. 8.30	61	arr ..... London ..... dep		A.M. 6.45	A.M. 7.10	P.M. 1.45	P.M. 6.15
10.20	4.45	12.35	8.11	51	Komoka	10	7.15	8.05	2.12	6.46
9.50	4.10	12.08	7.43	40¾	Strathroy	20¼	7.43	8.55	2.37	7.15
9.30	3.40	11.50	7.27	34¾	Kerwood	26¼	8.00	9.20	2.52	7.35
9.10	3.10	11.30	7.10	27½	Watford	33½	8.20	9.50	3.10	8.00
8.48	2.25	11.08	6.51	19¾	Wanstead	41¼	8.45	10.25	3.30	8.23
8.35	2.05	10.55	6.40	15½	Wyoming	45½	8.58	11.00	3.43	8.35
8.18	1.20	10.35	6.22	9¾	Mandaumin	51¼	9.15	11.23	4.00	8.49
P.M. 7.55	P.M. 12.40	A.M. 10.10	A.M. 6.00		dep .... Sarnia .... arr	61	A.M. 9.40	A.M. 12.00	P.M. 4.25	P.M. 9.10

* MIXED TRAINS ARE AT ALL TIMES SUBJECT TO BE CANCELLED. ‡ FLAG STATIONS, TRAINS STOP ONLY WHEN PASSENGERS AT OR FOR.

PALACE SLEEPING CARS ON ALL NIGHT TRAINS

Parlor Cars on Day Express Trains. Dining Cars on Atlantic Express going East and Pacific Express going West.

ALSO, PARLOR CARS ON THE ACCOMMODATION TRAIN BETWEEN TORONTO, HAMILTON AND BUFFALO.

TRAINS RUN BY HAMILTON TIME.

TIME CHART.

Montreal—Faster than Hamilton Time	26	minutes.
Albany " "	25	"
New York " "	24	"
Buffalo " "	4	"
Toronto " "	3	"
Detroit—Slower "	12	"
Chicago " "	31	"

CHAS. STIFF,
SUPERINTENDENT, HAMILTON.

F. BROUGHTON,
GENERAL MANAGER, HAMILTON.

Great Western Railway Offices, Hamilton, June, 1876.

FREE PRESS PRINTING CO., LONDON.

PAC

Of Pioneer Railroad Days

Small though they were, the wood-burning locomotives of early railroad days had a tremendous appetite. Since their tenders had a limited capacity, wooding-up stations were required at regular intervals along the right of way to replenish fuel. Many a settler earned his first cash income in Canada by supplying the railway with cordwood, which was recut at the station, usually with a circular saw powered by a horse on a treadmill.

Hundreds of cords of eighteen-inch wood were stacked in racks, each one holding one cord. The man in charge at the station tallied the consumption of each locomotive. Engineers were awarded bonuses by most employers for efficiency in operating their engines, a practice which caused considerable rivalry between the different crews. Farmers frequently discovered that their woodpile, stacked and ready for sale near the railway tracks, had been raided by an engine crew trying to "save" on fuel. If the locomotive happened to run out of fuel between stations, it was customary for the conductor to hand out axes to male passengers and ask them to help cut down a few trees.

Wood served not only as locomotive fuel, but quite often the engineer or fireman would have to grab a heavy chunk and throw it at an obstinate cow which had wandered onto the track and could not otherwise be induced to move. The large cowcatcher of early American and Canadian locomotives was designed to remove obstacles from the track, but an encounter with a fair-sized animal would not only kill the unfortunate beast but occasionally derail the engine, if not part of the train.

Frequent water stops were a must for the early Iron Horse. If need be, water for the engine had to be scooped from a ditch along the right of way. Wooden water-storage tanks in the railroad yards became landmarks of the steam era. They now have all but disappeared. Before the advent of electricity, steam pumps or windmills used to pump the water into the tanks from where spouts conveyed it to the engine.

As the pioneer locomotive puffed and snorted along the road belching thick clouds of smoke and hot cinders, the air was filled with the scent of burning maple, elm, tamarack or beech. Characteristic of the early woodburners were their immense smoke stacks, fitted with netting screens and hoppers for cinder removal. The fireman would strike the stack with a pole at intervals to loosen the cinders which had accumulated in the wire screen. In dry weather wayside fires kindled by the flying sparks of a passing locomotive were a constant menace, and section men often had to walk for miles along the track beating out the smouldering flames with their shovels. Years went by before engine designers succeeded in creating a chimney that would not allow sparks and cinders to fly. As engines grew bigger, the smoke stack became shorter and shorter until it almost disappeared, swallowed up by a large boiler.

The railroader of yesteryear had to be a tough and daring man. Engine, train and line crews worked hard, during long hours, sometimes around the clock, under primitive conditions. Driver and fireman in the beginning occupied an open deck, exposed to the elements, their vision blurred by smoke. Even the subsequent cab with flapping canvas awnings, and the closed-in vestibule that followed were still a far cry from today's all-weather steel cab with its sophisticated equipment.

Trains were braked by hand. When the engineer whistled "on brakes," the men hurried over the tops of moving cars to twist and turn the awkward "Armstrong" brakes on the platforms. In the winter, when sleet and ice made the car tops slippery, the job, risky at the best of times, was outright dangerous. Rules and regulations of the Great Western Railway specified that "brakemen must examine their brakes before starting, to see that they are in proper working order, and ride outside the cars so as to be in a position to apply their brakes immediately upon the proper signals for putting on all the brakes on the train."

There were few brakemen in the old days who could not tell some never-to-be-forgotten incident connected with their life on the rail. One such story we find in a newspaper of 1877:

> *Three miles from the station, the whistle blew for brakes, and in a mighty short time we had the train stopped. Going back about three hundred feet, we found that one of the rails had got loose and was out of place, but as we had been going slow we had run over the spot safely. Our conductor looked up, and seeing me, said: "Jim, get back and signal the passenger train. She will be along in a short time now; and take this," he said, handing me a red-light lantern; "we'll go on; you can come on with the other train."*
>
> *With that, all hands got on board, and soon there was nothing but myself and the lantern left.*
>
> *A cold gust brought me to myself with a quick turn, and then I remembered what I had to do. Holding the lantern up I saw*

the light was flickering, and shaking it found it almost empty. Then I began to feel the responsibility of my position. A lamp with no oil in it, the train due in ten minutes, with the chances of it being thrown off the track, and no telling how many people killed or wounded! In a case of this kind, sir, every brakeman will do his best to save human life, although he sometimes loses his own in the attempt, and all he gets for it is having his name in the paper and being a brave fellow.

Quicker than I tell it, I made up my mind that the train must be signaled, lamp or no lamp. But how to do it was the question. If I ran ahead without a light, the engineer might think I wanted to stop the train for robbery—for such things have often been done, you know—and would not only dash on faster than ever, but maybe try to scald me as the locomotive rushed by. I tell you I felt like praying just then; but brakemen are not selected for their religious feelings—so I didn't pray much, but looked around and saw a light shining in a window some distance off. I laid down my lantern carefully on the track, made a bee-line for the house, and soon my knock brought a woman to the door, who looked more frightened than I was at my excited appearance. It was useless to ask for sperm oil—the only sort we use—so I cried out:

'For God's sake give me some straw!'

She seemed to realize the position, and quickly brought a bundle. Feeling in my pocket, I found three matches, and, grabbing the straw, I made my way back to the track.

Laying the straw between the rails, I struck a match and shoved it into the bundle. It flickered an instant and then went out. I felt and found the straw damp.

Just then a dull, faint, rumbling sound came down on the wind, and I knew she was coming—the train would soon be there.

I struck the second match, and it touched off the straw. A blaze, a little smoke, and it was dark again, and, raising my eyes, I saw the head-light of the approaching train away in the distance. But trains don't crawl, and the buzzing along the rail told me to be lively. The red light was burning but faintly; five minutes more and it would go out. For an instant I stood paralyzed, when a shrill scream from the engine brought me to my senses, and I saw that inside of two minutes she would be there.

Seizing the lantern with one hand, I struck the last match, and bending down laid it carefully inside the straw, and then dashed forward, waving the red light. The glare from the head-light shone down the track, and the engineer saw me, but did not notice the red light—the sudden waving had put it out—only screeching he came straight on. When the train was almost on me, I jumped to one side, and slinging the lantern over my head, dashed it into the cab. The engineer saw the lamp as it broke on the floor, and seeing the red glass and battered lantern, whistled the danger signal and tried to check up.

Looking down the track, I almost screamed from excitement. The last match had found a dry spot, and the straw was blazing up bright. The train came to a stand-still. She was saved, that's all I remember.

One of the brakeman's jobs was the coupling and uncoupling of cars. Many a man was crushed between the wooden buffers or lost some of his fingers in the process of coupling the cars together with the old-fashioned chains and steel hooks. Sometimes the locomotive would jerk the hooks, splitting the train apart and leaving part of it standing on the line. Not until the invention of the Westinghouse air brake and the automatic coupler did railway operations become safer and more efficient.

In the 1860s and 1870s the standard colour of passenger coaches on Canadian railways, as a rule, was yellow. The Great Western's cars back then were painted a bright canary shade. Engines were decked out in different colours: black, red, green, even striped in some instances. Engine crews took pride in their clean wood-burning locomotives and intensely disliked the use of dusty coal when it was first introduced. The highly polished Iron Horses of yesteryear, with their shining brass trimmings and bells, never failed to attract an admiring crowd when they pulled into a station.

Each engineer used to look after his own locomotive and keep her in good repair. At the station stop he would get down from his cab to adjust a bearing, tighten a nut or test the big drivers with his lead hammer. He knew his engine inside out; to him she was a living thing, not just a piece of machinery.

Not everyone took kindly to the railways as they spread their net of tracks across the country. One old lady disliked trains so much that she took the trouble of soaping the rails on a steep hill near her house every night, just before the regular freight was due to pass. Of course, the heavy train would slide back downhill faster than it could climb up. A week went by before the engineer realized what was the cause of his continuous trouble.

Winter with its ice and snow was the worst enemy of Canada's pioneer railways. Lacking snow-fighting equipment, most lines had to suspend services for the winter months. Even the wedge snowplough of later years was of little use in a blizzard. Snow drifts and sub-zero temperatures could stop a train for hours, sometimes for days. But a writer of the 1870s seemed to think that being stuck on a train in the middle of nowhere during a snowstorm was not all that bad:

> *March 7, 1874:*
> *The man who fumes and rages at the delay of eight or ten hours on a railway would accept with comparative composure nearly a week's interruption to his journey if his progress depended on the stage coach or other old-fashioned conveyance. The secret of this unreasonable temper may be found in the fact that, when a railway train is stuck in a snow drift, the passengers as a rule are made exceedingly comfortable. The stoves burn brightly, if night, the lamps are kept well trimmed, a good supply of fresh water is generally on hand, and the hope or certainty of deliverance so near, that even the dread of suffering from hunger finds no room in the impatient passengers' hearts. When the snow plow ceases to be able to perform its work because of the immense accumulation of driven snow, it is no more than two to three hours of delay to await the arrival of an engine from the nearest station. Passengers grumble as the conductor politely presents every able-bodied man with the loan of a spade or snow shovel and*

invites him to assist in digging out the train. But this is really not any worse than taking the rail from the nearest fence and assisting in prying the stage coach out of a mud hole.

Not until the introduction of the rotary snowplough was it possible for trains to run on schedule in the wintertime. The brainchild of a Toronto dentist, who described it as a "revolving snow shovel," it had been patented in the 1860s. At the time he failed to interest anyone in the manufacture of his device, and nearly two decades passed before others developed and perfected his idea.

Travelling over the early tracks, which lacked the rigidity of today's rails, was rough. Wheels clattering over the joints produced a continuous roar. Broken rails were a constant threat, particularly in winter, and track walkers had to keep watchful eyes on the rails at all times. Railway accidents occurred with such frequency that people came to accept them almost as a way of life. Papers of bygone days regularly gave accounts of such disasters, many of them caused by fire. One such gruesome story reads as follows:

The scene was on the line of the Great Western Railway, midway between London and Komoka. On Saturday evening the Sarnia Express left the former place at twenty minutes past six, with several petroleum and baggage cars and one coach crowded with passengers. About midway between London and Komoka an oil lamp in the closet fell from where it was suspended to the floor. In a moment the oil ignited and the whole interior of the closet was on fire. Panic at once seized the passengers. The great speed at which the train was going, reckoned at over thirty miles an hour, fanned the fire to such a degree that no hope was left but an immediate stoppage of the train. But there being no bell rope attached, no communication could be passed, until the conductor under much personal risk ran forward and gave the word. By this time the fire had gained full sway and the frightened passengers were throwing themselves headlong from the platform and out of the windows. In a few minutes the car was consumed and those who could not escape were burned to a crisp.

Safety regulations of early railways make interesting reading. Engineers were warned to hold their speed at five miles an hour when crossing bridges. If a train became irregular, it was to run with caution and send a flag or lamp ahead, around curves and where the view was obstructed. In case of accident, men had to be dispatched each way to warn approaching trains. Should two trains happen to meet between stations, on a single-track line, the Northern Railway required the train nearest to the station to back up and let the other one pass. A rule of 1863 specified:

When engines or trains upon the Northern and Great Western, or Northern and Grand Trunk lines approach their respective crossings at the same time, the Northern engine or train will have the right of track in all cases and will proceed.

Most of the railroad crossings had no signs of any kind. When the Intercolonial Railway introduced such a sign on its line in the 1870s, one newspaper complained:

One can hardly read it, it appears to us that putting up crossing signs is a waste of money. In a dark night we cannot see them soon enough to benefit by them.

The sign was red, with the words RAILWAY CROSSING written in black.

Trains used to run on "local time." Between Halifax and Toronto existed five different standards of time. Clocks varied in nearly every community across the country depending on the passage of the sun across a meridian of longitude. Montreal time, for instance, was twenty-three minutes faster than Toronto time. To the railway traveller time-tables were confusing, especially if he wished to make connections. One Canadian, however, who missed a train once because of this bewildering system, decided to do something about it. He proposed a zone system of standard time not only for North America, but for the world. In 1884 standard time was adopted at a world conference held in Washington, D.C. The originator, later knighted in recognition of his great contribution, was Sandford Fleming, one of Canada's greatest railway engineers, of whom we shall hear more.

Accident on the Great Western Railway between London and Komoka, Ontario

Toronto Public Libraries

The Grand Trunk Railway of Canada

CNR

Bonaventure Station, Montreal, Grand Trunk Railway, c. 1870

When Canadian National Railways' new turbo train was put into trial service between Montreal and Toronto in 1969, it was hailed as the fastest intercity train in North America. Scheduled to cover the 334-mile distance between the two cities in three hours and fifty-nine minutes, the train's gas turbo engine would have been capable of operating at the speed of 120 miles per hour, had it not been for other traffic on the line. A far cry from the first through train inaugurating regular service between Montreal and Toronto on October 27, 1856, over the Grand Trunk Railway which was being constructed at the time!

That day, at 7:30 in the morning, a train made up of a baggage car and three first- and second-class coaches each painted a lively yellow and headed by a GTR woodburner, left Montreal for Toronto. Crowded with passengers, the train also carried Her Majesty's mail, and copies of four different Montreal newspapers. A similar train, bound for Montreal, had started out from Toronto station half an hour earlier. The two trains passed each other at Kingston about two o'clock in the afternoon. After a thirty-minute stop for lunch, each continued on its journey over the single-track line.

Along the route people from every hamlet and town gathered to cheer the first through-train, which stopped at sixty-four locations before reaching its ultimate destination after more than fifteen hours. Station agents had their hands full restraining enthusiastic crowds from blocking the track. Shopkeepers in the larger towns who had planned to close their stores for the day so they could see the train pass, found a large influx of country people kept them so busy that they missed the event. At all major centres, mayors and local dignitaries greeted the first train with welcoming speeches, and in Toronto the city fathers sat down to a lavish banquet with Grand Trunk officials to mark the historic occasion.

Bonaventure Station, which had seen the departure of Montreal's first pioneer train in 1847, served as the Montreal terminus of the Grand Trunk. The original wooden structure stood until the 1880s when it was replaced by a more impressive building. In Toronto, Grand Trunk trains arrived at a little brick station house at the foot of York Street before Union Station was built in 1873.

The Grand Trunk Railway had been planned to provide a continuous line from the Atlantic seaboard through the Canadas to Sarnia and thence to the fertile regions of the midwestern United States.

CNR

The Grand Trunk Railway Bonaventure Station, Montreal, during spring floods, 1880

Several company charters were utilized to carry out the immense project, started at a time when Canada's population had barely reached the two-million mark. As the correspondent of the *Coburg Star*, September 10, 1856 expressed it:

> *We cannot but feel a lively interest in the rapid approach to completion of the Grand Trunk Railway, a line which for magnitude and importance has not its equal in the world . . . A gigantic undertaking truly for the most established country, but how formidable for a young one, even when backed by capitalists of old England . . .*

A pioneer railroad in the true sense of the word, construction of the Grand Trunk created intense public excitement throughout the Canadas. It was the first comprehensive transportation system, significant enough in scope to influence profoundly the country's future and become the basis for its prosperity. The project had not been conceived by its promoters merely as a speculative venture, but as a deliberate means of developing the vast resources of the Canadas, and providing the stimulus needed for the industrial growth of what is now Quebec and Ontario by linking the two provinces and branching out across the borders into the United States.

Sections of the line were opened to traffic as soon as they were completed, and local railway celebrations usually accompanied the occasion. The first part of the railway to be operational was the line from Montreal to Portland, which opened in 1853. A year later a reporter from Portland stated:

> *The railway to Montreal has turned the forests along its line into gold . . . The value of the railway to the landowner and the lumberman has far surpassed any previous estimate or conception of it. The whole region has been touched with new life, realizing for those owning lands or water-power, fortunes of which they little dreamed.*

By then the line had also been opened between Richmond and Quebec City.

When the Toronto-Oshawa section was officially put into service on August 25, 1856, the Toronto *Globe* reported the events of the day, which included a dinner party for invited guests at Whitby:

> *After the last toast had been drunk, the last speech delivered, and the last farewell exchanged with the hospitable people of Whitby, the company left to resume their seats in the cars. They*

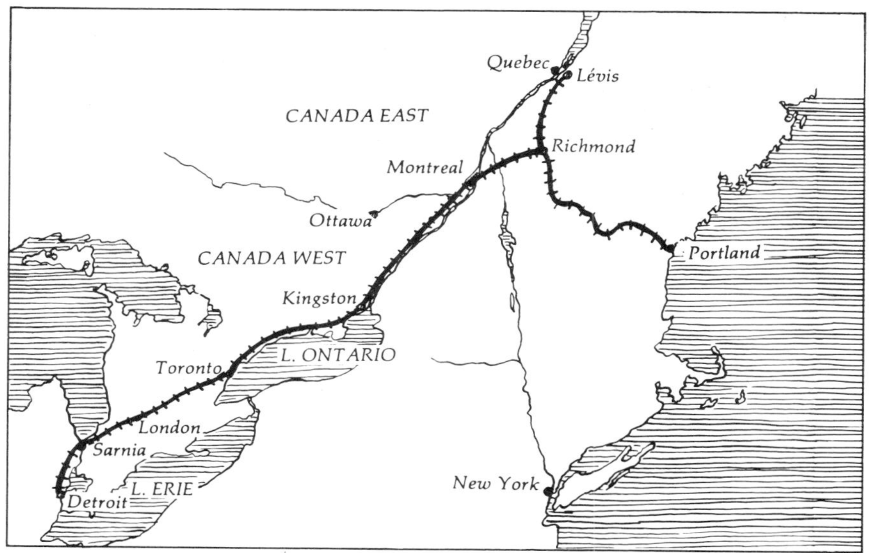

found, however, that most of the space had been occupied by a motley crew of persons belonging to places along the line who willy nilly were determined to make their way in comfort to their various destinations, without regard to the rights of previous incumbents.

Strenuous efforts were made for their dislodgment by the officials, but with little success, and the train proceeded with over 500 persons on board. It proceeded very slowly, however, the engine being a miserably bad one, and the upgrades being not only very heavy but very numerous even in places where a little cutting would have removed them. Slower and slower became the progress of the train until at length it stopped, and began to recede gently down the hill. After a little rest, steam was got up in sufficient quantity and the grade was surmounted in safety.

By and by another engine came up behind to assist the first, but the speed was but little greater, and it was not till the train began to approach the city, [Toronto] and to run down hill that speed was obtained. When the train reached the station it was half past seven o'clock, two hours and a half behind time, and the company separated, well wearied of the first class English road and the wonderful English punctuality.

By the time the line reached Stratford towards the end of 1856, the Grand Trunk had become the longest railway in the world. To mark the achievement a "Railway Celebration" was to be staged in Montreal on November 12th and 13th of that year, the like of which the country had never seen. The Grand Trunk sent out invitations throughout Canada and the New England States accompanied by certificates for free rail transportation to and from Montreal, not only via the Grand Trunk Railway but also over several American lines.

The night before festivities were due to start, special trains began arriving from Toronto and Quebec as well as Portland and Boston carrying thousands of guests. Large passenger boats tied up in the harbour not only had brought additional visitors, but were utilized as temporary hotels.

After a giant trade procession on the morning of the 12th, some 4,600 guests gathered for a seven-course dinner at the Grand Trunk's Point St. Charles machine shops. A number of speeches and toasts were followed by a colourful military review and a visit to the Grand Trunk's

Hastings Chronicle.

Belleville, Thursday, May 5, 1853.

Railroad Intelligence.

By the kindness of John Bell Esq., Solicitor for the Grand Trunk Railway, we have received a copy of the Prospectus of the "Grand Trunk Railway Company of Canada," which amalgamates the Quebec and Richmond Railway Company, the St. Lawrence and Atlantic Railway Company, the Grand Junction Railway Company, and Toronto and Guelph Railway Company,—forming 1112 miles of Railroad, including a Bridge over the St. Lawrence at Montreal,—with a capital of £9,500,000, and to be constructed on a uniform gauge of 5 feet 6 inches. The conditions of these contracts contemplate the construction of a first-class single track Railway, with the foundations of all the large structures sufficient for a double line, equal in permanence and stability to any Railway in England,—including stations, sidings, work-shops, ample rolling stock, and every requisite essential to its perfect completion; in short, the arrangements made with the contractors ensure its completion to the satisfaction of the Government. The Eastern section of this line will commence at Montreal, and proceed westward through the following towns and villages:—Lachine, St. Clair, St. Anne, New Longueil, Lancaster, Charlottenburgh, Cornwall, Osnabruck, Williamsburgh, Matilda, Edwardsburg, Augusta, Elizabethtown, Yonge, Lansdowne, Leeds, Pittsburgh to Kingston, passing through Ernestown, Napanee, Shannonville, Belleville, Trenton, Brighton, Colborne, Grafton, Cobourg, Port Hope, Bond Head, Bowmanville, Whitby, Pickering, Scarboro', to the city of Toronto, where it will connect with the Great Western, and thence to Port Sarnia, on Lake Erie.

This amalgamation scheme is certainly a noble one, and one which the Hon. John Ross has, with untiring perseverance, taken the lead in bringing to a completion. It will ere long confer incalculable advantages on this Province,—opening up our western country by a chain of Railways, through which the natural products of the Province will find at all seasons of the year a ready and speedy transit to the seaports.

We have not time at present to enter upon the subject matter of this Prospectus as fully as we would wish, feeling as we do that it is one of those extensive undertakings which will ere long place Canada in such a position that she will be, in a commercial and agricultural point of view, second to no other country.

The Hastings Chronicle *of May 5, 1853, describes the proposed Grand Trunk Railway system*

Victoria Bridge which was still under construction at the time. The following night a display of fireworks and a grand ball brought celebrations to a close.

In 1859 the Grand Trunk main line was opened to Sarnia with a branch extending to London, Ontario. In the east, Rivière du Loup was reached the following year. The Grand Trunk now operated a total of 872 miles of railway or nearly half the mileage in service in all of Canada. It also leased a line between Huron, opposite Sarnia, Ontario, and Detroit.

Prior to the great railway construction boom of the 1850s, Canadian governments spent millions of dollars improving lake and river navigation and building canals, a fact which made Canada's inland waterways the finest in the world. Steamboats navigating from Lake Superior to the Atlantic Ocean helped develop a flourishing trade; but during the winter months all navigation ceased. Considering the superior speed and advantage of year-round communication possible by rail, it soon became evident that the old ways of transportation were outdated. Furthermore, railways spreading rapidly throughout the adjacent United States

Admission card to the ball celebrating the completion of the Grand Trunk Raiway between Montreal and Toronto, November 1856

CNR

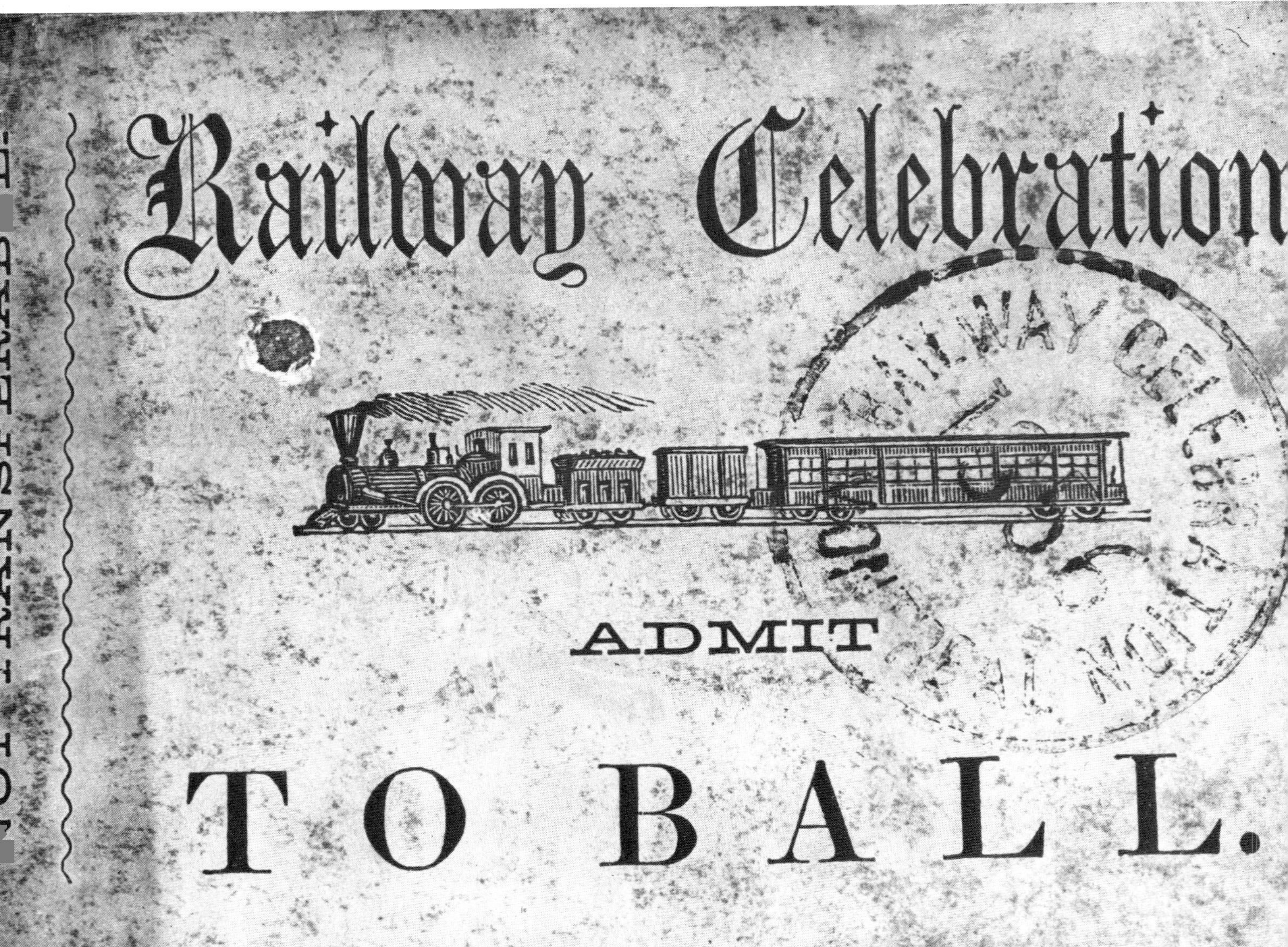

threatened to divert from Canadian channels major portions of western traffic to Atlantic ports. Consequently, a Canadian system of railways had to be devised that would allow Canada to compete with her neighbour to the south. Combined with a splendid inland navigation, such a railway would be able to provide rapid communication among large centres of population, commerce and industry, and ensure the continued growth of the Province of Canada.

Sponsored by Francis Hincks, Inspector General (the equivalent of today's Minister of Finance) under the Baldwin-Lafontaine government, an Act was passed by the Canadian parliament in 1849 guaranteeing, under certain conditions, financial assistance to new railway enterprises. One condition required the proposed road to have a minimum length of seventy-five miles, half of which had to be completed.

Francis Hincks continued to play an important role in the development of Canada's first key railway line and became known as the "Father of the Grand Trunk." In recognition of his services, Queen Victoria bestowed upon him the honour of knighthood. His political foes, however, accused him later of having made a personal fortune.

CNR

NO. (NOT TRANSFERABLE.)

CITY OF MONTREAL.

TORONTO. MONTREAL.

Montreal, 20th October, 1856.

Sir,

It being the intention of the Citizens of Montreal to celebrate the Completion of the Grand Trunk Railroad, connecting this City with the City of Toronto, the honour of your company, and that of the Ladies of your family, is requested at the Commemorative Festivities, to be held in Montreal, on the 12th and 13th days of November next.

DAVID KINNEAR, *Chairman.*
HENRY STARNES, *Mayor of Montreal.*
L. H. HOLTON, M.P.P. *President Board of Trade.*
CHARLES GARTH, *President Mechanics' Institute.*
A. A. DORION, M.P.P.
HENRY LYMAN, *City Councillor.*
HENRY BULMER, *City Councillor.*
W. WORKMAN.
JOHN LEEMING.
AUGUSTUS HEWARD.
THOMAS S. BROWN.
THOMAS CRAMP.
WALTER JONES, M.D.
C. J. COURSOL.
BROWN CHAMBERLIN.
THOMAS WILY.
ALFRED PERRY.
W. RODDEN.
J. G. DINNING.
THOMAS MORLAND.

Members of the Executive and Invitation Committees.

Secretary.

RAILWAY CELEBRATION MONTREAL 1856

To Hon [illegible]

QUEBEC & RICHMOND RAILWAY COMPANY.

TO SUB-CONTRACTORS.

MESSRS. JACKSON, BRASSEY, PETO, and BETTS, Contractors for the Works on the Line of Railway from Quebec to Richmond, are prepared to receive proposals for CHOPPING GRUBBING, EXCAVATING, MASONRY, and various descriptions of Work connected with Railway construction. Payment will be made in Cash every fortnight.

Mr. RECKIE (Resident Agent) will be in attendance at the Railway Company's Office, Quebec, after the 15th September, to receive proposals.

August 28, 1852. 16-tf.

WANTED,

TWO Hundred Men to work on the G. T. Railroad and Bridges in Belleville, Canada West. Work to be commenced about the 20th of July next.

C. R. MILKS & Co.

Belleville, June 24, 1854. 44

Interior view of the old Union Station, Toronto

Toronto Public Libraries

In 1851 parliament passed an Act concerning the construction of a main trunk line "throughout the entire length of the Province of Canada, and from the eastern frontier thereof, through the Provinces of New Brunswick and Nova Scotia to the city and port of Halifax."

The Honourable Joseph Howe of Nova Scotia who had gone to England to try to convince the imperial government of the necessity for such a trunk line in the colonies, returned home with the promise of a £7,000,000 guarantee to construct the proposed railroad in the provinces.

When it later became evident that the imperial promise would not be fulfilled, Nova Scotia and New Brunswick proceeded to build their own railways, each province within its own boundaries. Francis Hincks, during a visit to London, negotiated with the noted English railway builders Brassey, Peto, Betts and Jackson, who agreed to promote the Grand Trunk venture if they were assured of a sizeable construction contract on the line.

Meanwhile, in 1851, the Montreal and Kingston Railway and the Kingston and Toronto Railway had been empowered to construct a line from Montreal to Kingston and from Kingston to Toronto respectively. But as the Grand Trunk scheme began to take shape, the Acts pertaining to both these projects were repealed, and instead the Grand Trunk Railway of Canada was incorporated in 1852 to build a railway from Toronto through Port Hope, Cobourg, Belleville, Kingston, Brockville and Prescott to Montreal. Shareholders of the Montreal and Kingston Railway and the Kingston and Toronto Railway were to be reimbursed by the Grand Trunk for organizational expenses and surveys already carried out, and the government promised a guarantee of £3,000 per mile towards the construction of the line. The legislature also authorized the incorporation of the Grand Trunk Railway of Canada East for the purpose of building a line from opposite Quebec City to Trois Pistoles, Quebec.

Since it was impossible to raise sufficient capital for a grandiose railway scheme such as the Grand Trunk within the colony, the company decided to appeal to the British public. Accordingly, a prospectus of the Grand Trunk Railway, issued in London, England, in April of 1853, proposed to raise £9,500,000 sterling "to complete and construct 964 miles of track in Canada, extending through the whole province, and to annex the Atlantic and St. Lawrence Railway, a line of 148 miles from Island Pond to Portland, Maine, making a total length of 1,112 miles."

Companies by then in the process of being amalgamated under the name of "Grand Trunk Railway of Canada" included the Grand

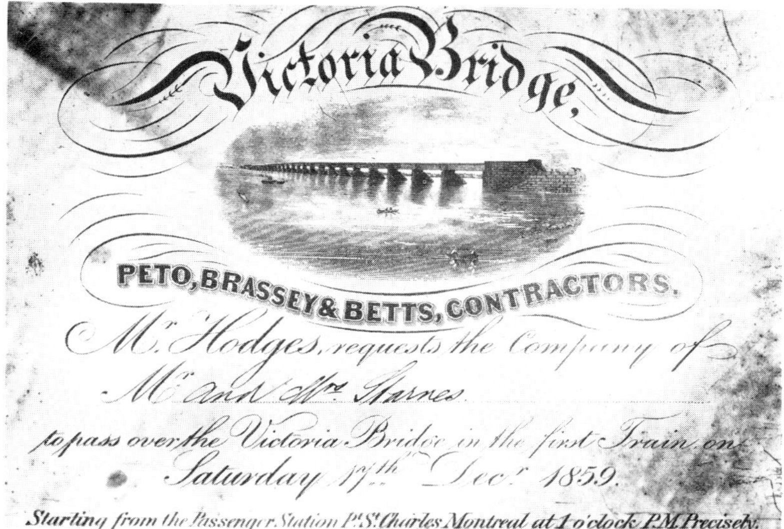

☞ The Grand Trunk Company offer to every mason, labourer, blacksmith, and carpenter who will engage to come to Canada to work on the Railway, an advance of £2 10s., and £1 for every member of his family, should he have any, on condition of his furnishing security to the local agents for the re-payment of such advance at the rate of 1s. per week. The wages offered are—for masons, 7s. 6d. to 8s. 9d per day; labourers, 5s.; blacksmiths and carpenters, 7s. 6d.

Grand Trunk Railway overseas advertising, 1853

Trunk Railway of Canada East, the St. Lawrence and Atlantic Railway, the Quebec and Richmond Railway, the Grand Junction Railway and the Toronto and Guelph Railway, the latter having power to extend to Sarnia.

The St. Lawrence and Atlantic, in operation since 1853 together with its American counterpart, the Atlantic and St. Lawrence, from Montreal to Portland, Maine, thus formed the beginning of the Grand Trunk system.

The Grand Junction Railway, chartered to build from Belleville to Peterborough, Ontario, did not materialize until years later when the Grand Trunk surrendered the charter and the original company was revived. Not until after the Grand Junction had been consolidated with the Midland Railway system did it eventually come into the possession of the Grand Trunk.

As for the Quebec and Richmond Railway, it had been incorporated in 1850 to build from Point Lévis, opposite the city of Quebec, to Richmond on the St. Lawrence and Atlantic Railway. The line, already well advanced at the time of takeover, was completed by the Grand Trunk in 1854.

The Brassey firm, contractor for the Quebec and Richmond Railway, was engaged to build the Montreal-Toronto section as well as an extension of the Grand Trunk eastward to Rivière du Loup. The contract for construction of the line from Toronto westward to Sarnia went to the Canadian firm of Gzowski & Company. Both builders agreed to construct their portion of the road in a manner "superior to any of the American or Canadian railways then in existence." The English contractor received £9,000 per mile for the Montreal-Toronto section, while Gzowski & Co. were paid £8,000 per mile for the western division. The latter company, more familiar with Canadian conditions, appeared to have made a fortune on the job, while the English firm claimed to have lost a considerable amount.

Work on the Grand Trunk proceeded vigorously and the railway moved swiftly from city to city. Navvies brought over from England swelled the labour force. During the height of the activities fourteen thousand men and two thousand horses were employed on the line in Canada West alone.

The first stone of the Victoria Tubular Bridge across the St. Lawrence River at Montreal was laid on July 20, 1854. Designed jointly by Robert Stephenson, son of the inventor of the steam locomotive, and Alexander M. Ross, the Grand Trunk's chief engineer, the bridge ranked among the finest specimens of engineering skill in the world. Its super-

RAILROAD CELEBRATION.

SECOND DAY,
THURSDAY, 13TH NOVEMBER, 1856.

PROGRAMME

OF EXCURSION TO

VICTORIA BRIDGE,

AND

WHEEL HOUSE OF THE MONTREAL WATER WORKS.

Steamers will leave the Island Wharf

AT NINE O'CLOCK PRECISELY.

Those desirous of examining the Works, Work-shops and Buildings at Point St. Charles, for which every facility will be offered by the Officers of the Grand Trunk Company, and the Contractors, are vised to go at this hour.

BANDS OF MUSIC

WILL ACCOMPANY.

STEAMERS will again Leave the Same Place

AT HALF-PAST NINE O'CLOCK,

To convey his **Worship the Mayor** and **Corporation.**

Those who prefer, may wait for this Trip. TICKETS to be retained, but SHOWN when going on board.

On arrival of the Mayor and Corporation, at Point St. Charles a

TRAIN OF CARS

Will be in readiness, to convey them and those who wish to join, to

THE WHEEL HOUSE OF THE WATER WORKS.

The Party will remain there for THREE QUARTERS OF AN HOUR, and return on the Cars, to the Steamers, so as to arrive back in the City before NOON.

☞ *PERSONS without Tickets are requested not to Enter the Cars, nor ATTEMPT going on Board the Steamers.*

A PROCESSION,

Headed by the MAYOR & CORPORATION, will proceed from the Landing Wharf to Commissioner Square where the

OPENING of the MONTREAL WATER WORKS

will be Celebrated, at NOON.

☞ Strangers and Citizens along the Line, are requested to "fall in" with the Procession.

T. S. BROWN,
Chairman of Excursion Committee.

Victoria Bridge in the making: announcement of an excursion on November 13, 1856

Toronto Public Libraries

Snowplough stranded in a drift on the Grand Trunk Railway near Stratford, Ontario

structure consisted of a rectangular iron tube, 6,592 feet long, resting on two abutments and twenty-four piers founded on solid rock. The tube, which was just wide enough to permit trains to pass through the inside, weighed 9,044 tons. One hundred thousand cubic yards of stone carried to the site by twenty-five barges were used in the masonry construction.

The interior of the tube was dark and dirty, although an opening, two feet wide, ran along the centre line of the top to allow locomotive smoke to escape. Covered by a roof, it was of little use. On hot summer days temperatures of 125° inside the iron tube were by no means uncommon. One of the Grand Trunk's wealthy English shareholders found out first-hand just how unpleasant it could get. On a visit to Canada he had made it his business to inspect the line thoroughly and insisted on walking across the bridge along the track. With him was an entourage of elegant ladies in hoop skirts, assorted relatives, secretaries and valets. Although orders had been given to keep the bridge clear while the party was crossing, someone blundered, and a freight train loaded with several hundred pigs entered the tube. The terrified walkers jumped aside in the nick of time and, their faces pressed against the soot-covered rusty wall of the tube, prayed for their lives. When an upward grade caused the train to slow down to a crawl, the stench of the pigs, magnified by the intense heat in the tube, became unbearable. Many of the ladies fainted, while some of the men succumbed to violent attacks of nausea. A gang of section men finally came to the party's rescue.

Victoria Bridge had been opened for public traffic on December 17, 1859, after a test train safely crossed the structure. In his report to the Board of Railway Commissioners the Inspector of Railways described the testing procedure:

> *The test applied to the tube of the Victoria Bridge consisted of a train of 18 platform cars, loaded with stones as heavily as they would bear, and drawn by two locomotive engines coupled. This train was long enough to reach over two spans at one time, and weighed . . . about one ton to the lineal foot. In passing this train over the bridge a load of 242 tons was laid on each of the side spans, and 330 tons upon the central span. . .*

Two locomotives, however, were not sufficient to move the heavy train and a third engine was required to haul it across the bridge. The deflection under the extreme weight, double that of an average heavy freight train, was found to be a little over one inch. Each section of the tube, however, immediately returned to its former position as soon as the load was removed.

Toronto Public Libraries

Pullman Palace car, Grand Trunk Railway, Montreal

On August 25, 1860, the nineteen-year-old Prince of Wales, later to become King Edward VII, drove the last rivet into place and thus formally completed the magnificent structure of the Victoria Tubular Bridge. Considered the eighth wonder of the world in its day, the bridge served until 1898. It was replaced by the double-track Victoria Jubilee Bridge which had been built around the old tube while traffic continued uninterrupted, except for twenty hours, during the year of construction.

One of the worst accidents in Canadian railway history occurred on the Grand Trunk line near St. Hilaire, Quebec. On the morning of June 28, 1864 a train from Quebec carrying 354 German emigrants plunged off the Beloeil Bridge into the waters of the Richelieu River. The first despatch in the Toronto *Globe* gave the following account of the mishap:

> *The train stopped at St. Hilaire, about one mile from the bridge across the Richelieu River, where there is a swing bridge. The swing bridge was opened about a quarter past one, to let a number of barges, in tow of a steamer, pass. The proper signals were turned before the bridge was opened, and the red light was burning. When the man in charge of the bridge heard the whistle, he waved his red hand lamp. The standing orders are that all trains must come to a full stop before reaching the bridge. This was disregarded, and the train ran into the open draw . . .*

A later despatch describes the scene as it was found by rescue workers:

> *The cars lie mostly a pile of fragments, crushed together, resting on a barge which was fortunately passing through at the moment of the accident. Had the cars fallen into the open water, to the number of deaths caused by being crushed in the smashing of the cars must have been added a vast number of drowned. All the barges in tow of the steamer had passed except two. Those on the barge actually passing through the bridge saw the train coming, knew what must happen, and jumping on the barge behind saved their lives. The locomotive lies submerged in the water out of sight. The appearance presented by the wreck it is impossible to describe. The train consisted of two or three second class cars, and the remainder box cars fitted up with benches for the emigrants . . .*

Ninety men, women and children lost their lives in the wreck. The driver of the locomotive miraculously escaped with only slight injuries.

Another accident involving the Grand Trunk made headlines around the world. It happened in the Grand Trunk Railway Yards at St. Thomas, Ontario, on the evening of September 15, 1885, when Jumbo, the most famous circus elephant in the world owned by the equally famous P. T. Barnum, was killed by a fast freight train.

The three-toed African elephant, measuring twelve feet in height and weighing seven tons, is believed to have been one of the largest ever held in captivity. Known to be stubborn and obstinate, he actually charged the on-coming locomotive trying to knock it off the track. The impact drove one of his tusks back into his brain, but he had succeeded in damaging one of the cylinders and the smoke stack of the engine. Jumbo's hide was stuffed and, mounted on a platform wagon, he continued to travel with the circus for a couple of years as the show's major attraction. His mate, Alice, and fifteen other elephants used to follow his wagon, carrying black-bordered sheets in their trunks and wiping their eyes, as they made their entry into the arena.

Toronto Public Libraries

Trevithick, *built by the Grand Trunk Railway in 1859*

CNR

No. 40, used on the Canadian National Railway's "Pioneer Train" during the 1950s

According to information gleaned from contemporary newspapers, the Grand Trunk's rolling stock taken over from amalgamated lines at the time of the company's formation consisted of 34 engines, 33 first and second class cars, 15 baggage and 459 freight cars. By the late 1880s the Grand Trunk's motive power had grown to over 700 locomotives, and the rolling stock included 578 first- and second-class cars, some 60 post office cars, 131 baggage cars, 18,000 freight cars and 49 snowploughs.

Number One on the Grand Trunk's locomotive roster was the *A. N. Morin*, the Portland-built engine inherited with the takeover of the St. Lawrence and Atlantic Railway. The 4-4-0 type locomotive, with small headlights, a great balloon stack and plenty of ornamental brass, was followed by what became probably the most colourful assortment of steam locomotives ever possessed by any railway in the world. In addition to its own locomotives, the Grand Trunk, by absorbing a vast number of different lines, acquired steam engines of every make and model built in England, Canada and the United States. Locomotives owned by the Grand Trunk during its seventy-year history ranged from the quaint woodburners of the pioneer days to the powerful Moguls and Pacifics of later vintage.

Among the earliest locomotives of the Grand Trunk were the so-called "Birkenheads," built by Brassey, Peto, Betts and Jackson at their Canada Works in Birkenhead, England. Fifty of these engines were delivered between 1854 and 1858, painted either red, green or black. On the road they were easily distinguished from other locomotives because of their peculiar frontal aspect. Clumsy in appearance, they had a reputation for being powerful, reliable and almost indestructible, even in a wreck.

A blend of English and American characteristics, the most visible being a funnel smoke stack, a cowcatcher and a driver's cab, the Birkenheads originally featured a 2-4-0 wheel arrangement, which means a pair of leading wheels and two pairs of coupled drivers. This arrangement, no doubt, would have been suitable for smoothly graded English roads, but the humps of the rougher Grand Trunk track in Canada were another matter. Consequently, the locomotives had to be altered to the American standard 4-4-0 type.

Some drivers and firemen found the Birkenheads awkward to handle because they were left-handed engines, with the throttle and the reverse on the left side of the cab and the fire-box door opening to the

CNR

Grand Trunk Railway locomotive
Earthquake

left. Governed by English tradition, the old Grand Trunk ran its trains on the left whenever there was a double track until the early 1900s.

Among the most picturesque locomotives of bygone days were the Grand Trunk's "Kinmond" engines. With their gaudy red wheels and their brass trimmings polished to a gleam, they were a sight to behold. The Kinmond Brothers, originally of Dundee, Scotland, had established a locomotive works in Montreal in 1852, and the first steam engine to be built in that city was delivered by the Company to the Grand Trunk the following year. A total of ten Kinmonds appeared on the Grand Trunk's roster, the last one, GTR's No. 87, having been acquired in November of 1856.

One of the more famous Grand Trunk locomotives was No. 209, better known as the *Trevithick*. Named after the company's first mechanical superintendent, F. H. Trevithick, the engine was the first of many to be built in the Grand Trunk's own shops at Point St. Charles, a section of Montreal. She was completed in May of 1859, and the following year was chosen to head the special train carrying the Prince of Wales over the Grand Trunk line during his visit to Canada.

During 1868 the Grand Trunk purchased twenty-five 4-4-0 type locomotives from Neilson & Co. of Glasgow, Scotland. Their peculiar smoke stacks, six feet in diameter at the top, were dubbed "champagne cups."

The *Canadian Illustrated News* back in 1870 tells of the first Pullman Palace Car being placed in service on the Grand Trunk Railway:

> *Elegant carriages for railway travel are now being introduced on the Grand Trunk Railway. The first car, the "Montreal" was put upon the track on Monday, Aug. 22, 1870 and made the trip to Toronto that evening. It was an object of much interest and admiration. In the Pullman Palace Cars there are three state rooms and 2 drawing rooms, and one is at a loss which most to admire, the elegant black walnut cabinet work, the splendid mirrors, the warm crimson velvet upholstery, or the snug convenient tables, lit as they are, so as to take from night travelling all its gloom, and instead of a tedious night's work between Toronto and Montreal, it is an agreeable evening in a very handsome first-class drawing room.*

Sleeping cars were introduced reluctantly. Experience on American railways had shown that such accommodation frequently attracted questionable characters, and Grand Trunk officials were having none of this. The earliest version of a GTR sleeper consisted of bunks or benches along the entire length of the car without divisions or curtains. Each passenger was given a pillow and a blanket. At night, the car resembled a hodge-podge of dark objects with the occasional arm or leg sticking out and, as one observer put it, looked as if the passengers had been thrown into it in one big heap.

Following the lead of the Montreal-Portland railroad the Grand Trunk adopted a gauge of 5′ 6″, which subsequently became the standard for all new Canadian lines. As the American railway standard was 4′ 8½″, interchange of traffic at border points soon presented serious problems. Changeable devices for car wheels not only were costly but unsafe and caused many a train wreck. A third rail, later laid inside the broad gauge to accommodate American equipment over certain sections of the line, also failed to prove satisfactory. The steadily increasing volume of traffic eventually forced the Grand Trunk to change its gauge to the width of 4′ 8½″, which has remained the generally accepted standard railway gauge until today.

The task of changing over the main line between Stratford and Montreal was accomplished in twenty hours on October 3 and 5, 1873, with only sixteen hours of interruption in traffic. East of Montreal the gauge was changed with equal efficiency the following year. With the standardization of the road, the rolling stock, of course, had to be converted and locomotive power was urgently needed. One hundred and fifty new engines were ordered during this period, and many of the old locomotives were rebuilt for the new gauge.

Grand Trunk Railway

OF CANADA.

Engines for Sale.

In consequence of the change of gauge upon this Railway in the month of October next, the Grand Trunk Railway Company will have about

100 Engines to Dispose of.

They are of various sizes and all 5 feet 6 inches gauge.

Many of them are suitable for Mill and other purposes.

Full particulars will be given on application to the undersigned, or to Mr. H. WALLIS, Mechanical Superintendent, Montreal.

C. J. BRYDGES,
Managing Director.

August 14. s 9 3t

The first new standard-gauge engine, produced for the Grand Trunk in Portland, Maine, was delivered in 1872. Today this old wood-burner, known as No. 40, remains the only survivor of her class.

Sixty-one engines in all were purchased from the Manchester Works in England immediately after the change in the gauge. Named "Blood Engines" after Aretus Blood, agent of the Manchester Works, they were among the finest locomotives ever to serve on Canadian roads. Among them was GTR No. 416, known as the *Earthquake*. She was the pride and joy of engineers among the fast passenger engines on the Montreal-Toronto run. The engine was supposed to have been given to the Grand Trunk Railway by the Manchester builders as an example of their high-quality workmanship in the hope of securing orders.

As the Grand Trunk modernized its motive power, some of the old Birkenheads and Portlands were sold to smaller lines which operated them up until the early 1900s in the timber regions north of the Ottawa River, in New Brunswick and in eastern Quebec.

While the early years of the Grand Trunk had been marked by feverish construction activities, the 1860s brought a slowdown in expansion due to financial difficulties which plagued the company on and off throughout its colourful career. Following the completion of initial projects, the Grand Trunk embarked on a policy of leasing and acquiring existing lines rather than building new ones of its own. By the time the great pioneer railway observed its sixty-sixth birthday it had swallowed no fewer than one hundred and twenty-five companies, all of which had once possessed a separate legal identity.

In 1869 the Grand Trunk leased the Buffalo and Lake Huron Railway, a line extending from Fort Erie, opposite Buffalo, to the Lake Huron port of Goderich. Soon after, the road was fully integrated with the GTR.

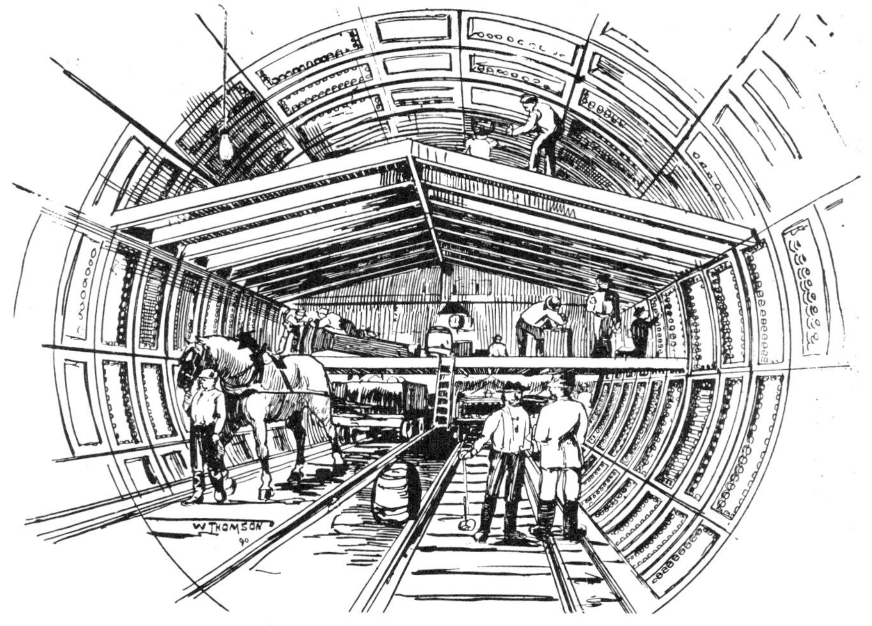

Late in the nineteenth century the St. Clair River tunnel was completed. It was the first international submarine tunnel in the world. This sketch shows the tramway used to haul away clay

The 1870s saw the construction of the International Bridge across the Niagara River to replace the ferry service hitherto in use to provide direct rail connection with Buffalo. The contractors for the bridge, Gzowski & Company, encountered considerable difficulty, due to the swiftness of the current, a treacherous river bottom and rapid fluctuations in the rise and fall of the water. Two swing bridges had to be incorporated in the structure, one over the Erie Canal, and the other over the river channel. On October 27, 1873, the first train passed over the bridge, and it was officially opened for traffic on November 3rd of that year.

By 1880 a line between Port Huron and Chicago had been added to the system. Before that time the Grand Trunk entered Chicago via the Michigan Central from Detroit. The new through line consisted of several existing railways acquired by the Grand Trunk, joined by connecting links built by the latter, and consolidated under the corporate name of the Chicago and Grand Trunk Railway. Later this company was reorganized and became known as the Grand Trunk Western.

It was on this line that Thomas A. Edison travelled back and forth as a boy selling newspapers. In his spare time he used to conduct some experiments in the rear of the baggage car, until one day he set the car on fire. The conductor was furious, but the baggage car was saved. Years later, during ceremonies marking the ninety-third birthday of the great inventor, the car travelled once more behind the tender of an ancient woodburner over the rails of the old Grand Trunk Western.

In 1881 the Grand Trunk acquired control of the Georgian Bay and Lake Erie group of railways, comprised of 171 miles of track extending from Port Dover via Stratford to Wiarton on Georgian Bay.

One of the Grand Trunk's most formidable competitors in southwestern Ontario was eliminated when the entire system of the Great Western Railway was taken over in 1882. The Great Western, which ran from Toronto via Hamilton to Windsor, had been in operation prior to the Grand Trunk and, with its numerous branch lines built in the ensuing years, had developed into a serious rival as far as traffic to and from the United States was concerned. With the acquisition of the Great Western the Grand Trunk added at once over 900 miles to its steadily growing railway empire.

Another 472 miles of track came under control of the Grand Trunk when the Midland Railway was leased in 1884. The Midland system, eventually consolidated with the GTR, consisted of a line from Port Hope, Ontario, to Midland on Georgian Bay, with a branch to Peterborough and a number of other adjacent branch lines and connections.

According to the papers of 1890, the railway tunnel being constructed under the St. Clair River at Sarnia attracted much interest among the great trade carriers between the west and the east.

> *The work has already made such progress that its completion seems assured. It is only a matter of a short time when the heavy freight will be making rapid transit under the river instead of by the old and unsatisfactory makeshift of steamboat transfers.*

Digging the tunnel was accomplished by means of shields, penetrating through the clay stratum between the river bottom and underlying bedrock from either side of the river. Surveys had been made and tests carried out since the early 1880s, but the actual tunnel work did not get into full gear until the spring of 1890, when the job was being pushed ahead day and night with three gangs of diggers, each working an eight-hour shift. The shields moved at an average rate of seven to nine feet per day.

The tunnel proper is 6,026 feet in length, 2,310 feet of which are located under the river, the balance under dry ground. The depth of the lowest part from the mean level of the river is 81 feet and the clear diameter of the tunnel's interior measures 21 feet.

In the fall of 1891 the St. Clair tunnel was officially opened for traffic and provided the Grand Trunk with an efficient underground rail connection between Canada and the United States.

During the 1890s the Grand Trunk began double tracking its main line. By the end of the century only the forty-six mile section from Port Union to Port Hope remained single track on the Montreal-Toronto stretch. With the exception of a few miles in the United States and the St. Clair Tunnel, by 1917, continuous double track was completed from Ste. Rosalie, Quebec, to Chicago, Illinois.

After gaining control of the Central Vermont Railway which served parts of the Province of Quebec and the States of Vermont, New York and Massachusetts, the Grand Trunk also acquired the Canada Atlantic, thus connecting its main line with the capital of Canada and once again adding considerable mileage to its system.

In the meantime the Grand Trunk had begun to eye with envy the success of Canada's first transcontinental railway, the Canadian Pacific, and was more than anxious to tap the growing traffic of the prairies and the west. Under its vigorous president, Charles M. Hays, it contemplated the construction of a line from North Bay, Ontario, to the west coast, in competition with the CPR.

Negotiations with the dominion government, however, produced a transcontinental scheme which, in the view of Sir Wilfrid Laurier, would "penetrate the north land and give Canada breadth as well as length." At the time immigration to Canada was at its peak, agriculture expanded in the prairies and the existing single track of the CPR between Winnipeg, Manitoba, and Fort William, Ontario, was hard-pressed to handle the great wheat shipments efficiently and without delays.

The new line was to be built in two divisions in collaboration with the government, which agreed to construct the eastern portion from Moncton, New Brunswick, to Winnipeg, Manitoba, via northern Ontario and Quebec. This division, known as the National Transcontinental, was to be operated by the Grand Trunk upon completion. The Grand

Trunk Pacific, a subsidiary of the Grand Trunk, was to build the western division of the line from Winnipeg to Prince Rupert, British Columbia. The combined length of the two divisions was to be 4,708 miles.

The venture, enthusiastically begun and carried out, ended in catastrophe. World War I dried up the flow of immigrants, and boom days came to a sudden end. Economic depression gripped the world. Canadian railways in general, and the Grand Trunk in particular, were deeply in debt. The latter was unable to operate the National Transcontinental when it was completed in 1915, and the line became the responsibility of the Canadian government. The Grand Trunk Pacific went bankrupt and was taken over by the Minister of Railways as receiver in 1919. Four years later, on January 30, 1923, its once powerful but always financially troubled parent, the Grand Trunk Railway of Canada, ceased to exist. On that day the Grand Trunk's 4,776 miles of track became part of Canadian National Railways, today the largest railway system in the world.

Grain elevator of the Grand Trunk Railway, Toronto, Ontario

PAC

Railways of the Maritimes

First Intercolonial Railway train to Dalhousie, New Brunswick, 1884

Before the advent of steam, Nova Scotia and New Brunswick were more closely linked with far-off nations than with the interior of British North America. Those were the days of the great sailing vessels, plying the oceans and calling at the ports of the world. But not long after news reached the colonies of the first practical steam railway operating in Great Britain, forward-looking Maritimers began advocating railway connections with Lower Canada.

In 1836 a delegation was sent to England to seek financial assistance from the imperial government for such a project. A bill was passed in the legislature of New Brunswick, incorporating the St. Andrews and Quebec Railroad Company. Soon after, surveys were started on a route between St. Andrews, New Brunswick (then a prominent trading centre on the Bay of Fundy) and Quebec City in Lower Canada.

The proposed line might well have become one of the pioneer railways of British North America, had it not been for a boundary dispute between New Brunswick and the state of Maine. When it appeared that a railway would be built, the United States pressed for settlement of the boundary question. The outcome was that most of the territory through which the line had been projected was ceded to the United States by the terms of the Ashburton Treaty of 1842, which fixed the boundary. Immediate plans for the railway were shelved. Military considerations had altered the picture. The State of Maine was now wedged between New Brunswick and Lower Canada. St. Andrews was situated too close to the border to be chosen as the terminus of a railway which would be a vital link between the British colonies in case of war. Not until 1856 did St. Andrews become the terminus of a line known as the New Brunswick and Canada Railway, which was pushing its way northward but never developed into more than a local venture.

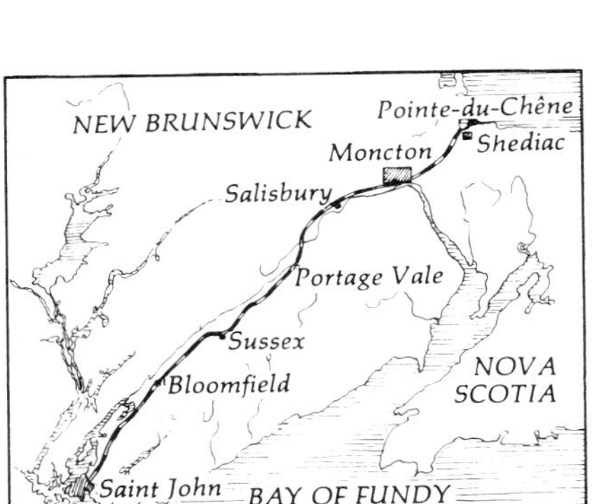

European and North American Railway

Meanwhile, the question of an intercolonial railway which would weld together British possessions in North America and promote settlement and trade remained uppermost in the minds of politicians on both sides of the Atlantic. More than thirty years were to pass, however, before this railway became a reality.

In 1846 Gladstone, then Secretary of State for the Colonies, commissioned Captain Pipon and Lieutenant Henderson, of the Royal Engineers, to survey three possible routes for a line from Halifax to Quebec City. Captain Pipon drowned in the Restigouche River while trying to rescue one of his men. He was succeeded by Major Robinson who handed in his report in August of 1848. This recommended a 635-mile route from Halifax to Truro over the Cobequid Mountains, thence along the Gulf to the Miramichi, across that river and via the Nepisiguit to the Bay of Chaleur and to the Matapedia, up the Matapedia to the St. Lawrence and up that river to Rivière du Loup and Point Levis. The cost was estimated at five million pounds.

The negotiations which followed between the British government and the provinces brought little progress. In 1850 the Honourable Joseph Howe of Nova Scotia went to England to solicit the support of the Colonial Secretary for the railway scheme. It took five months and a direct appeal to the British public before he succeeded in obtaining the promise of an imperial guarantee of the necessary loan. Upon his return to Halifax, Howe addressed a public meeting and prophesied: "Many in this room will live to hear the whistle of the steam engine in the passes of the Rocky Mountains, and to make the journey to the Pacific in five or six days."

Representatives from New Brunswick, Nova Scotia and the State of Maine in the meantime gathered for a railway convention in

Portland, Maine, to further a project known as the European and North American Railway. Designed as an overland link from Halifax, Nova Scotia, to Saint John, New Brunswick, where it would connect with a line from Portland, the European and North American Railway was to shorten the travelling distance between Europe and the principal cities along the eastern seaboard of the United States.

Discussions between railroad-minded parties failed to bring about immediate results regarding the European and North American. Realization of the intercolonial scheme seemed doomed, when the imperial guarantee was withheld on a plea of misunderstanding. Tired of waiting, New Brunswick and Nova Scotia turned their attention to local railways and proceeded to build separately, each province within its own boundaries.

EUROPEAN AND NORTH AMERICAN RAILWAY

New Brunswick led the way by beginning to construct the European and North American Railway. The ground-breaking for what was to become the first "Common Carrier" in the Atlantic Provinces was celebrated in a grand manner at Saint John on September 14, 1853. At dawn militia artillery fired a thunderous salute from Fort Howe. Later that day a two-mile long parade wound its way through the streets with a colourful assortment of floats, representing a variety of trades. Ships' carpenters showed off models of famous ships, with teams of horses pulling a large, fully rigged clipper ship and a steamer; tailors displayed "Adam and Eve in the Garden;" printers, who had a printing press mounted on their float, distributed copies of a farewell song to stage coach drivers about to be put out of business:

Soon will cease your occupation,
when the rail cars take the station.
When you see the steam horse start,
go and burn your lazy cart.

An estimated twenty thousand citizens and visitors assembled in the vicinity of what was later to be called "Celebration Street." There they watched the Lieutenant-Governor's wife, Lady Head, turn the first sod. The wooden wheelbarrow used to cart away the soil is still preserved. A masterpiece of craftsmanship, it is carved in the shape of a lion, his forepaws grasping the wheel, his tail forming the handle.

The first train on the European and North American Railway is believed to have been hauled by the Portland-built locomotive *Saint John*. It ran over three and a half miles of finished track on March 17, 1857, covering the distance in twelve minutes. The entire line from Saint John on the Bay of Fundy, via today's busy railroad centre of Moncton, to Shediac and Point du Chêne on the Northumberland Strait in the Gulf of St. Lawrence (a distance of approximately 108 miles) was opened for traffic in 1860.

American railways at the time extended as far as Bangor, Maine. Stage coaches provided transportation between Bangor and Saint John, while steamships plied between Saint John and Portland where travellers could board a train to Montreal.

The first two locomotives purchased by the European and North American Railway came from the Boston Locomotive Works. They were the *Hercules*, brought to Shediac in the summer of 1854, and the *Samson*, like the *Hercules* a 4-4-0 type engine.

CNR

CNR

European & North American Railway: spade and barrow used for sod-turning ceremonies, Saint John, New Brunswick, 1853

European & North American Railway: engraving on spade for sod-turning ceremonies, 1853

CNR

European & North American Railway: Point du Chêne wharf, Shediac Harbour, 1873

The *Hercules* faithfully served the railway for many years. Eventually she was sold and hauled from the railhead to Bateman's Brook. It took three weeks, sixteen teams of horses and a yoke of oxen for the veteran engine to complete her last journey. At Bateman's Brook she spent her declining years as a steam boiler in a sawmill.

Several of the locomotives running on the Saint John-Shediac line originated from the works of Fleming & Humbert at Saint John. This company, established in 1835 under the name of Phoenix Foundry, was one of the earliest locomotive builders in Canada.

THE NOVA SCOTIA RAILWAY

The Province of Nova Scotia, in the meantime, did not remain idle. Joseph Howe was appointed chairman of a Board of Railway Commissioners set up to supervise the building of the Nova Scotia Railway.

Requests for tenders on construction work for the first section from Halifax around the Bedford Basin to Nine Mile River appeared in the Halifax papers on May 3, 1854. Advertisements also asked for a thousand men and sixty teams of horses and carts. On June 8th, a general holiday in Halifax, an enthusiastic crowd gathered at the Governor's Farm at Richmond, north of the city, hoping to witness the ground-breaking ceremony for the railway. But the rumours proved false, the first sod was not turned until June 13th, in the presence of the Lieutenant-Governor and a number of local dignitaries. As soon as official ceremonies were over, work on the Nova Scotia Railway got under way. The contract had been awarded to Messrs. Cameron, Fraser and Turnbull of Pictou.

The first train ran from Halifax to Fairview on January 20, 1855. It was headed by the *Mayflower*, a 4-4-0 type 22-ton woodburner, built at Bridgewater, Massachusetts. Both sides of the track were lined with spectators, many of whom had never seen a locomotive before.

On June 8th of the same year the Nova Scotia Railway was opened to the public from Richmond to Sackville. Horse-drawn buses at that time provided regular service between Richmond Station and Province House at Halifax.

Soon after, work began on a 32-mile branch line from Windsor Junction to Windsor, and the *Nova Scotian* told its Halifax readers that they would soon be able to

> *daily sip the rich cream of the Windsor dykes in their coffee— and butter, yellow and fresh, the product of the Falmouth*

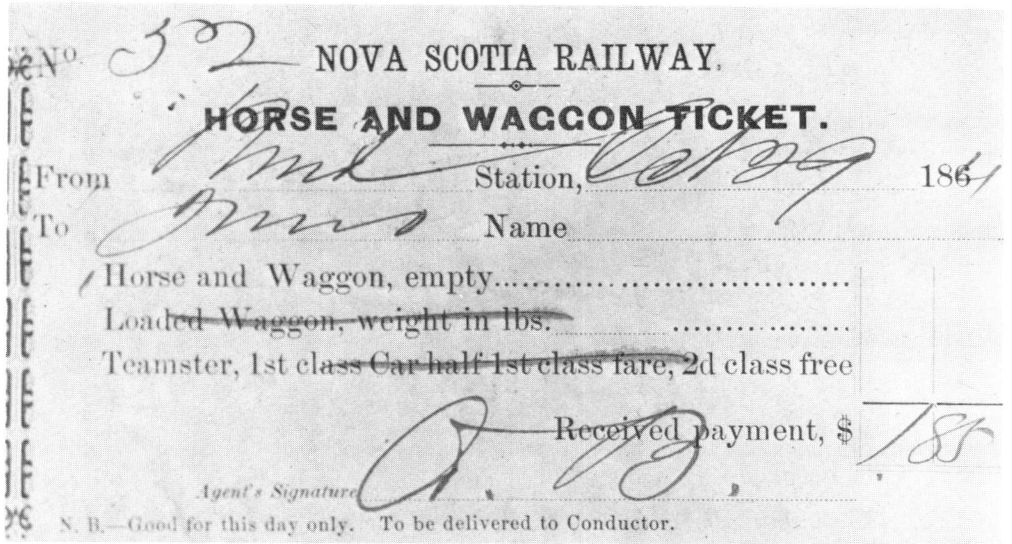

No.

NOVA SCOTIA RAILWAY.

HORSE AND WAGGON TICKET.

From ______ Station, 186_

To ______ Name

Horse and Waggon, empty ....

Loaded Waggon, weight in lbs. ....

Teamster, 1st class Car half 1st class fare, 2d class free

Received payment, $

Agent's Signature

N. B.—Good for this day only. To be delivered to Conductor.

CNR

The Nova Scotia Railway pioneered the "piggyback"

NOVA SCOTIA RAILWAY.

TARIFF.

PASSENGERS.

Morning and Evening Trains run daily between Halifax and Grand Lake, and a Mid-day Train to and from Bedford.

Miles.	Stations.	1st Train.	2d Train.	3d Train.	Fares. 1st Class. s.	d.	2d Class. s.	d.
	Up Trains.	A. M.	NOON.	P. M.				
	Halifax, depart	7.30	12. 0	3. 0				
3¼	Four Mile House	7.40	12 10	3.10	0	7½	0	5
8	Bedford	8. 0	12 30	3.30	1	3	0	10
10¼	Scott Road	8.10		3.40	1	10	1	3
13¼	Windsor Junction				2	3	1	6
20	Fletcher's	8 40		4.10	3	4	2	3
22½	Grand Lake, arrive	8.50		4.20	3	9	2	6
	Down Trains.	A. M.	P. M.	P. M.				
	Grand Lake, depart	9.25		5.10				
2½	Fletcher's	9.35		5.20	0	5	0	4
9¼	Windsor Junction				1	6	1	0
11¾	Scott Road	10.15		5.50	1	11	1	3
14½	Bedford	10.25	1.45	6.	2	6	1	9
19¼	Four Mile House	10.45	2. 5	6.20	3	2	2	8
22½	Halifax, arrive	10.55	2.15	6.30	3	9	2	6

Excursion Tickets—for use same day, up and down—a rate and a half. Tickets for Children under 12 years of age, half price. Passengers not providing themselves with tickets before entering the Cars, will be required to pay 7½d. extra. Special Trains provided on reasonable notice, and Passenger Cars hired to parties or families at diminished rates.

HORSES & CARRIAGES.

	Bedford. s.	d.	Grand Lake s.	d.
1 Horse and empty Carriage	1	10½	2	9
1 do. Carriage and load	2	6	3	9
Driver in Horse Car	0	7½	1	0
2 Horses and empty Carriage	3	1½	4	8
2 do. Carriage and load	3	9	5	6
Driver in Horse Car	0	7½	1	0
3 Horses and empty Carriage	4	4½	6	8
3 do. Carriage and load	5	[illegible]	[illegible]	6
Driver in Horse Car	0	7½	1	0
4 Horses and empty Carriage	5	0	7	6
4 do. Carriage and load	6	3	9	6
Driver in Horse Car	0	7½	1	0
Saddle or other Horse	1	6	2	3

MISCELLANEOUS.

Small Parcels and Packages according to size and value.	Bedford. s.	d.	Grand Lake s.	d.
Barrels, each	0	4	0	7½
Hhds. and Puns., 80 to 120 galls.	1	3	2	0
Bags of 2 bushels	0	3	0	5
Do. of 3 do.	0	4	0	7½
Bundles, equal in size to a barrel	0	4	0	7½
Heavy Articles, by weight, per ton, per mile	0	3		
Furniture, per ten cubic feet, at	0	5	0	7½
Dry Fish, in bundles of 1 cwt	0	2	0	4
Parcels under 50 lbs. or bulk of half-barrel size	0	3	0	4
Cordwood, per cord	2	6	3	9
Bark	2	6	3	9
Lumber and Scantling, per M.	2	0	3	0
Screwed Hay, per ton	2	6	3	9
Shingles, per 4 bundles	1	0	1	6
Timber, per ton, per mile	0	3		
Do. per M. soft	2	9	4	0
Do. per M. hard	3	3	5	0
Calves and Pigs, each	0	4	0	6
Sheep	0	3	0	4
Neat Cattle, single	0	10	1	3
Do. when more than one	0	8	1	0

The rates between Bedford and Grand Lake are the same as those between Halifax and Bedford. Freight taken in quantity by agreement. Freight to be labelled or marked legibly in all cases—unless so marked, transportation will be at freighter's risk. No responsibility assumed by carriers, unless contents of packages or parcels are distinctly and legibly marked upon them ☞ No Horses, Carriages, or other freight received within ten minutes before the starting of Trains In all cases the Cars to be loaded and discharged at expense of freighters—and not loaded above the stancheons.

N. S. Railway Office, Feb. 2, 1857.

JOSEPH HOWE, Chairman.

PAC

Nova Scotia Railway timetable, 1857

CNR

Early Nova Scotia Railway steam engine at Windsor Junction. Built at Sydney Mines in 1886

marshes, will be brought from beyond the Ardoise Hills each morning to grace our breakfast tables.

The same paper reported the first serious accident on the Nova Scotia Railway under the date of Sept. 19, 1855:

> *THE LOCOMOTIVE OFF THE TRACK!—The community was startled on Friday with the intelligence that the locomotive with the first returning morning train from Sackville had been thrown off the track near one of the crossings in front of the coloured village. We proceeded immediately to the scene of disaster and there lay the Iron Horse, but a few minutes before so fleet and powerful, now bruised and prostrate. The Locomotive, tender, and two of the baggage wagons had been precipitated over the embankment, a depth of nearly 20 feet, and were all more or less crushed and injured by the fall. There were about 30 passengers in the cars who had a most providential escape, one of the wagons having luckily been thrown into an oblique position which completely checked the progress of the passenger carriages and prevented their being hurled down the steep.*
>
> *The accident was caused by a horse. The animal had forced his way up a small ravine running at right angles with the railway, which was inadequately fenced, and stood upon the track. The train was travelling at the average speed, the usual lookout was kept, and the steam whistle, we understand, was vigorously plied to frighten the unwelcome intruder off the moment he was seen, but in vain. The brakes were promptly put in requisition, but there was too much way on the engine to stop and there was nothing for it but a collision which literally tore the horse to pieces and, as we have seen, threw the Locomotive, tender and wagons off the rails.*

Traffic had to be suspended temporarily as the *Mayflower* required extensive repairs, and two new engines on order from Glasgow, Scotland, were still in transit at sea. With the arrival of the two locomotives, named the *Joseph Howe* and the *Sir Gaspard Le Marchant*, the railway's rolling stock consisted of three engines, four passenger coaches, three sheep and cattle trucks, four platform trucks, seven ballast trucks and a snowplough.

Today, railway flat cars, transporting highway trailers, are taken for granted. Hardly anyone remembers that it was the old Nova Scotia Railway back in 1855 which first pioneered the "piggyback" practice in Canada by carrying stage coaches and farmers' wagons. The charge was twelve cents a mile for every hundred pounds for the vehicle, and two cents a mile for the horse. The driver went along free. Farmers were happy with the service since it allowed them to bring their produce to the city by rail, unload at the station and drive their wagons to market. For the railway, however, the service turned into a headache rather than a profitable business. Heavily loaded trains were hard on the roadbed and required large amounts of fuel and tallow. With an ever increasing number of people taking advantage of the piggyback service, delays up and down the line became the order of the day. To top it all, the railway was frequently sued for damages when hayloads caught fire from locomotive sparks, and eventually the company was forced to discontinue the popular service.

CNR

Sandford Fleming

Work on the Windsor branch as well as on the Truro extension proceeded slowly. There was a labour shortage because of the Crimean War, and the cost of materials had risen beyond estimates. Riots in 1856, instigated by Irish railway workers unsympathetic both to the war and the government's war policy, repeatedly brought construction to a halt. Not until June 8, 1858 was the Windsor branch line opened for traffic. On December 15th of that year regular service was also inaugurated between Halifax and Truro, a distance of sixty-two miles.

Construction on the extension from Truro to Pictou Landing started in 1864, but work bogged down the following year. Consequently, completion of the line was put into the hands of Sandford Fleming, the Scottish-born railway engineer, who, as we have seen, later became known to the world as the father of standard time. Fleming promised to have the Pictou branch operational by the spring of 1867. It was of great importance to the government of Nova Scotia that the deadline be met, as the prospect of Confederation was beginning to take shape. Given a free hand to proceed as he saw fit, Fleming introduced a number of techniques hitherto unknown in railway building. In order to continue working in the winter, he had temporary roofs placed over such construction projects as bridge and tunnel approaches. To fill in ravines which otherwise would have had to be bridged, he imported two American steam shovels.

The line was completed in 1867, one month before Confederation. Contemporary railway engineers described it as "the finest half hundred miles of railway in British North America." To celebrate completion of the job, Fleming staged a picnic for everyone concerned with the project. A heavy downpour did not dampen his spirits. He ordered the picnic tables moved into a large railway culvert between Lorne and Glengarry, and a good time was had by everyone.

INTERCOLONIAL RAILWAY

On July 1, 1867 the Provinces of Ontario, Quebec, New Brunswick and Nova Scotia joined in Confederation to become the Dominion of Canada. At the time, the railway map of the Maritime Provinces comprised the following:

New Brunswick: The European and North American Railway from Saint John on the Bay of Fundy to Shediac and Point du Chêne on the Strait of Northumberland, connecting the Bay of Fundy with the Gulf of St. Lawrence; the New Brunswick and Canada Railway running from St. Andrews to Woodstock with a branch line to St. Stephen.

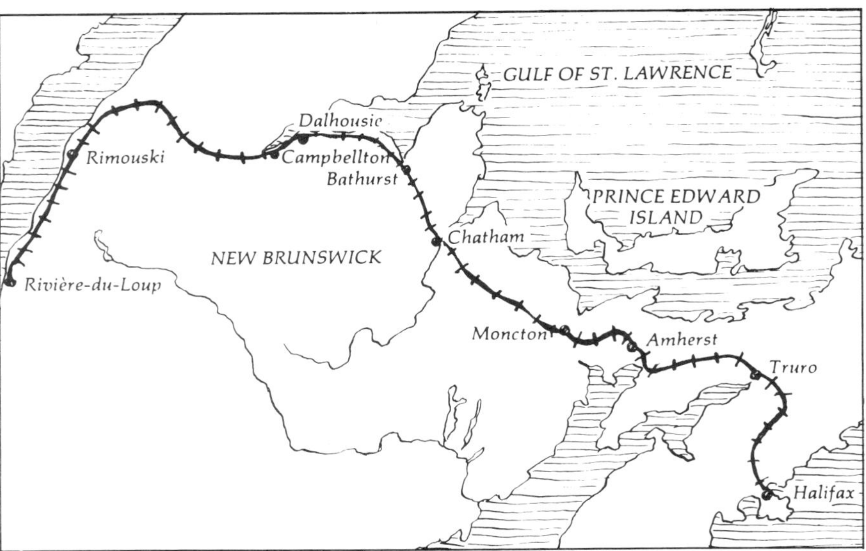

The Intercolonial Railway

Nova Scotia: The Nova Scotia Railway, a total of 145 miles, operated by the Nova Scotia government, and consisting of a line from Halifax to Truro, an extension to Pictou Landing and a branch line from Windsor Junction to Windsor.

While the principal waters surrounding both provinces were thus joined by rail, one of the main objectives of railway planners had not as yet been realized. The Maritime Provinces still had no rail connection with the remainder of Canada. But this was about to change.

Construction of a railway, linking the Atlantic port of Halifax with Quebec, had been one of the conditions under which Confederation was effected. Under the terms of the British North America Act, the dominion government was to assume ownership of existing railway lines which had been built by the two Maritime Provinces. Work on the Intercolonial Railway, the official name of the government project, was to start within six months after Confederation.

The European and North American and the Nova Scotia Railways formed the nucleus of the new Intercolonial. The first objective was the construction of a connection between Truro, Nova Scotia and Rivière du Loup, Quebec, which was then the eastern terminus of the Grand Trunk Railway of Canada. Sandford Fleming was appointed engineer-in-chief. Surveys for the line had been carried out under his direction since the winter of 1864 when he and his assistants had set out on snowshoes, carrying their supplies and instruments on dog sleds. Mapping out three possible routes, they had probed the wilderness by day and slept in lumbermen's cabins or out in the open at night. In 1865 Fleming presented his report on the Frontier, the Central and the Bay Chaleur routes. Although the northerly Bay Chaleur Route was the longest and most expensive, it was selected, mainly for military reasons, because it was the farthest removed from the international boundary. Much like the route surveyed by Major Robinson in the 1840s, it followed the Bay of Chaleur to Campbellton, then through the Matapedia Valley into the St. Lawrence Valley and along the river to Rivière du Loup, Quebec.

A railway builder far ahead of his time, Sandford Fleming had his own ideas, and many a heated controversy developed between the engineer-in-chief and the authorities during the course of construction. Determined as he was, Fleming usually succeeded in getting his way, but it meant fighting a running battle with officials who held the

Toronto Public Libraries

Accident on the Intercolonial Railway

purse strings. One such argument involved the material to be used in bridge construction along the Intercolonial line. Short-sighted railway commissioners determined that bridges should be constructed of wood. Fleming insisted that they be built of iron to diminish fire hazards. He went as far as the Privy Council and the drawn-out dispute ended on May 12, 1872, when an Order in Council was passed stipulating that all bridges on the Intercolonial, excepting three small ones, were to have their superstructures of iron. Fleming also recommended the use of steel rather than iron rails on the Intercolonial.

Initial orders were placed by the Railway Commissioners with locomotive builders in Glasgow, Scotland, as well as Kingston, Ontario and Halifax, for a total of forty new engines. Contracts also went out at the time for 250 box and 150 platform cars.

The first link in the Intercolonial system was opened for traffic on November 9, 1872, between Truro and Amherst in Nova Scotia. The section from Rivière du Loup eastward to Ste Flavie, now Mont Joli, Quebec, was completed in August 1874. By the end of the following year the line became operational between Moncton and Campbellton in New Brunswick and the remaining gap between Campbellton and Ste Flavie was closed by the summer of 1876. On July 1 of that year Sandford Fleming declared the Intercolonial Railway ready for traffic. He was not boasting when he said that this railroad was built "second to none, either on this continent or in Europe." The construction of the Intercolonial necessitated no spectacular engineering feats, such as the Canadian Pacific was later called upon to perform on its way west. Nevertheless, the route presented many difficult problems which had to be solved by skilful practical men.

In 1879 the Intercolonial acquired the Grand Trunk line from Rivière du Loup to Chaudière. Ten years later the Drummond County Railway, a private line from Chaudière to Ste Rosalie, Quebec, was integrated into the system. Running rights over the Grand Trunk into Montreal established a through route from Halifax and Saint John to Canada's largest city. From New Glasgow to Mulgrave, Nova Scotia, the Intercolonial operated the Eastern Extension, and in 1890 the line reached Sydney on Cape Breton Island, with ferry service being provided across the Strait of Canso between Mulgrave and Point Tupper. Today a causeway from Aulds Cove to Port Hastings links Cape Breton Island with the mainland.

Aug 1869

INTERCOLONIAL RAILWAY.

THE Commissioners appointed to construct the Intercolonial Railway, give

PUBLIC NOTICE

that they are now prepared to receive Tenders for five further Sections of the line.

CONTRACT No. 8

will be in the Province of Quebec, and extend from the Easterly end of Contract No. 5, at Rimouski, to a point near the Metis River, about 20½ miles in length.

CONTRACT No. 9

Will be in the Province of New Brunswick, and extend from the Easterly end of Contract No. 6, towards the Town of Bathurst, about 20½ miles in length.

CONTRACT No. 10

Will be in the Province of New Brunswick, and extend from the centre of the Chaplin Island road, near the Court House, at New Castle, towards Bathurst, about 20 miles in length.

CONTRACT No. 11

Will be in the Province of Nova Scotia, and will extend from the Easterly end of the Eastern Extension Railway to the Westerly end of Section No. 4, (including the bridge across the Missiquash River, except the western abutment) about 3½ miles in length.

CONTRACT No. 12

Will be in the Province of Nova Scotia and extend from the Easterly end of Contract No. 7, at Folly Lake, to a junction with the existing Railway at Truro, about 24½ miles in length.

Contracts Nos. 8, 9 and 10, to be completely finished before the first day of July, 1871.—Contract No. 11 to be completely finished by 1st July, 1870.

That portion of Contract No. 12, east of Folly River to Truro to be finished and ready for laying the track by the 1st day of October, 1870, from Folly River to a point opposite the Londonderry Iron Works by the 1st January, 1871, and the remaining portion of said contract by the 1st day of July, 1871.

Plans and profiles, with specifications and terms of contract, will be exhibited at the offices of the Commissioners in Ottawa, Rimouski, Dalhousie, St. John, Halifax, Toronto and Quebec, on and after the 13th September next, and sealed Tenders addressed to the Commissioners of the Intercolonial Railway, will be received at their office in Ottawa up to 7 o'clock p. m., on the 18th October, 1869.

Sureties for the completion of the contract will be required to sign the Tender.

Commissioners' Office,
Ottawa, 3rd August, 1869.

A. WALSH,
ED. B. CHANDLER,
C. J. BRYDGES,
A. W. McLELAN, } Commissioners.

86-8t

Toronto Public Libraries

New Intercolonial Railway Station, Halifax

Prince Edward Island had not joined Confederation in 1867. In contrast to the other Atlantic provinces, it was slow to catch the railroad fever that had long gripped the rest of the country. It was not until 1871 that Islanders commenced construction of a narrow-gauge railway, planned to run the entire length of the island, with branch lines to Tignish and Souris. A fixed price per mile was agreed upon with the contractor, but the actual number of miles was not specified. Today the numerous curves of the railway, meandering across Prince Edward Island, may add to its picturesque appeal for the traveller, but at the time of construction the winding road meant increased mileage and higher costs. Shoddy workmanship and local politics also added to the price, and before too long the government of Prince Edward Island found itself in financial difficulties over the railway, serious enough to cause it to take another look at the possibility of joining Confederation. When the dominion government promised to complete the troubled railroad, Prince Edward Island entered Confederation as Canada's smallest province in 1873.

The line between Charlottetown and Tignish was opened two years later. The first regular train service was inaugurated on December 31, 1875, only to be tied up immediately by a severe snow storm. Continuous service over the line did not resume until the following spring. During the summer steam vessels linked the island with the mainland. In the early days communications were kept open in the winter by boats, mounted on runners, which were pushed across the ice of the Northumberland Strait. Passengers paid reduced fares if they were willing to hang onto the ropes at the side of the boat and help shove and pull it across the treacherous surface. The icebreaker *Northern Light* ushered in a new era when it first went into service between Pictou, Nova Scotia, and Georgetown, Prince Edward Island, in the winter of 1876-77. A far cry from to-day's powerful icebreaking ferries, it was not unusual for these early boats to get caught in the ice. The *Stanley*, a successor of the *Northern Light*, was once held up in the ice for sixty days.

Built for political rather than commercial reasons, the Intercolonial Railway never became a financially successful venture; but it fulfilled its purpose as an instrument of Confederation by welding the provinces together.

In the Maritimes the Intercolonial drastically changed a way of life. It provided work for hundreds and helped to develop hamlets and towns along its route. Values of farm land and timber stands greatly increased in the areas it served. Coal mines were kept busy supplying fuel to run the trains. Shops and sawmills were established in communities along the line. Produce and goods were transported faster and more economically across country, and as money began to circulate more people prospered. As for the railway's deficits, the taxpayer footed the bill.

In 1918 the Intercolonial Railway came under control of the Canadian Government Railways, a corporate enterprise created to operate government-owned railway lines. Today the Intercolonial forms part of the Atlantic Region of the Canadian National Railways.

CNR

Intercolonial Railway: old windmill water tank pump, Oxford, Nova Scotia

THE DOMINION ATLANTIC RAILWAY

The Dominion Atlantic Railway displayed on its locomotive tenders a crest picturing Longfellow's famous "Evangeline" with the inscription: LAND OF EVANGELINE ROUTE. The railway, which was formed by the consolidation of two companies before the turn of the nineteenth century, traces its origin back to 1866. That year the Windsor and Annapolis Railway was incorporated to build a line from Windsor, Nova Scotia, through the fertile Annapolis Valley, to the town of Annapolis, site of Samuel de Champlain's Habitation, the first white settlement built on Canadian soil. For construction work the company purchased two pioneer engines of the Nova Scotia Railway, the *Joseph Howe* and the *Sir Gaspard Le Marchant*. Six new locomotives were ordered from Bristol, England, for the opening of the road. The first three which arrived were christened *Evangeline, Gabriel* and *Gaspereau.*

Part of the Windsor and Annapolis Railway was ready for traffic in June of 1869. The first regular train between Annapolis and Grand Pré was scheduled for August 19, 1869 and a grand celebration marked the opening of the line. Houses along the route were decorated with flags and bunting. Governor-General Lord Lisgar and a party of over one hundred distinguished guests arrived in a procession of stage coaches at Grand Pré where evergreen arches greeted them with the inscription "Welcome to the Land of Gabriel and Evangeline." At the station waited the inaugural train, headed by the gleaming *Evangeline,* to take the visitors to Kentville where from the other end of the line a similar train, carrying more dignitaries and officials, was due to arrive about the same time. A banquet laid out at the railway's machine shop turned into a long-drawn-out affair of speeches and toasts, giving the train crews ample time to celebrate happily on their own. As a result, the general manager of the railway spent more of his time trying to round up his men for the return trip than attending to his guests.

In the fall, soon after the opening of the line, a heavy gale caused dykes to break at Grand Pré. The roadbed of the railway was swept away and tracks were badly damaged over a wide area. Nevertheless, by the end of 1869 the Windsor and Annapolis Railway was completed.

Life in the pioneer days, of course, was more leisurely than now. One story tells of a passenger on the Windsor and Annapolis Railway leaving the train to milk a cow grazing along the tracks and bringing the milk to a crying baby in the coach while the train crew waited patiently. At Annapolis, in those days, the company employed a white horse for shunting the cars.

The first pay sheet of the Windsor-Annapolis Railway records the income of stationmasters at large stations as $400.00 annually, with only half that much being paid to the men in smaller places. Engine drivers earned about $32.00 in half a month, while firemen made $18.75 in the same period.

After the company had been granted the use of the Intercolonial's Windsor branch and had obtained running rights from Windsor Junction to Halifax, travellers could board a through train at Annapolis for Halifax. The first such train arrived at Halifax on New Year's Day 1872.

That winter went down in local history as one of the worst on record. A snow storm tied up roads throughout the province. No trains were moving for over two weeks. Farmers, who had brought their produce to Halifax as usual, were stranded with nothing to do but shovel snow for the railway.

Some of the old-time travellers remember the Windsor Junction stop. It seems that a number of goats in that area were not too happy with the pasture and made it a regular habit to board the train, wander through the cars and pester the passengers for food.

Dominion Atlantic Railway, Engine No. 1 — Queen Mab

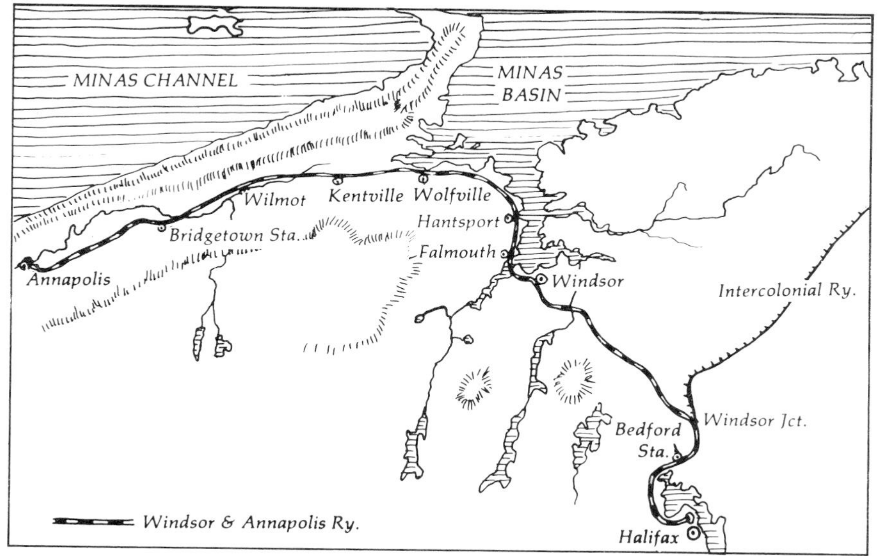

In the summer of 1875 the original 5′ 6″ gauge of the line was changed to what is now the standard gauge of 4′ 8$^1/_2$″, the work being performed with astonishing speed and efficiency in only ten hours.

In 1892 the Windsor and Annapolis purchased the Cornwallis Valley Railway, a line which had been incorporated to build from Kentville to Kingsport, Nova Scotia. In the meantime, the Western Counties Railway had been authorized to build from Annapolis to Yarmouth, Nova Scotia. The first sod for this enterprise was turned at Yarmouth on September 22, 1873, but the line was built only as far as Digby, where, among other problems, quicksand was encountered. In the years to follow, the stretch between Digby and Annapolis, known as the "Missing Link," had to be covered either by stage coach or by steamer, until the gap was closed in 1891, when the so-called "Missing Link Railway" was completed by the dominion government and handed over to the Western Counties Railway for operation. On July 27, 1891 the first through train ran from Yarmouth to Annapolis where it connected with the Windsor and Annapolis train for Halifax.

In 1893 the Western Counties Railway changed its name to the Yarmouth and Annapolis. A year later it was consolidated with the Windsor and Annapolis, and a new company emerged under the corporate name of Dominion Atlantic Railway. In 1905 this company purchased the Midland Railway, thus adding to its mileage a line from Windsor to Truro, Nova Scotia. An order of the Privy Council, dated January 3, 1912, approved the lease of the Dominion Atlantic to the Canadian Pacific Railway for a period of 999 years.

THE CHIGNECTO SHIP RAILWAY

This is the story of a railway which, if it had been completed, would probably have been the only one of its kind in the world. In any event it would have been the first railway anywhere built for the sole purpose of transporting vessels over land from one body of water to another, in this case across the Isthmus of Chignecto, a narrow strip of land joining Nova Scotia with the rest of Canada.

Started in 1888, the project was abandoned three years later for lack of money and government support, although three-quarters of the work had already been done. Machinery and locomotives ordered by the Marine Railway Company were ready for delivery and twelve miles of track, out of a total of seventeen miles required, were completed.

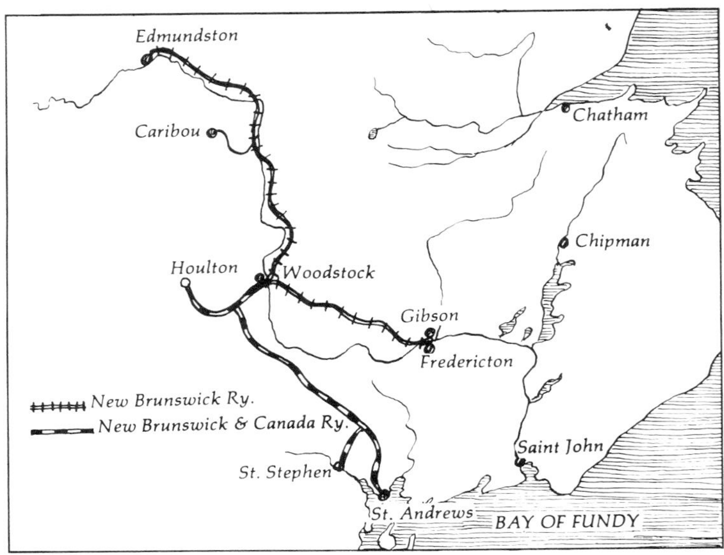

Had the railway become a reality, it would undoubtedly at the time have profoundly influenced the economy of the Maritimes by bringing about a considerable increase in trade between Canadian and New England ports. Vessels from the St. Lawrence bound for United States ports on the Atlantic coast via the Canso Strait could have saved at least 300 miles, and those bound for Saint John would have had their route shortened by something like five hundred miles.

The Ship Railway, projected to connect the Gulf of St. Lawrence with the Bay of Fundy, was one man's magnificent dream. He was a New Brunswicker by the name of Henry G. C. Ketchum. A graduate in civil engineering of King's College at his native Fredericton, he spent a number of years in Brazil working on the São Paulo Railway. After his return home, he was engaged in the construction of the railway from Moncton, New Brunswick, to Amherst, Nova Scotia, and later became chief engineer on the New Brunswick Railway between Fredericton and Edmundston.

Talk about building a canal across the Isthmus of Chignecto was very much the order of the day, and had been ever since the area was first settled. But nothing had been done until Henry Ketchum came up with his idea of constructing a marine railway. His figures proved that a railway for the conveyance of ships across the isthmus would be far less expensive than a canal, the construction of which might have cost anywhere from six to eight million dollars. This was a vast sum of money in the 1870s. Furthermore, Ketchum argued, a canal would present serious problems because of tidal differences between the Gulf and the Bay of Fundy, and could not accommodate the paddle-wheel steamers which were then plying Canadian waters. Unfortunately, the plans he had prepared of his railway were burned in the great fire which swept Saint John in 1877, and for a while the ship railway seemed forgotten; but in 1881, after having made a survey at his own expense, Ketchum approached Sir Charles Tupper, then Minister of Railways and Canals in the federal government. He succeeded in winning his support for the venture along with the promise of a government subsidy. A year later the Chignecto Marine Transport Railway Company was incorporated and authorized to construct a ship railway across the Chignecto Isthmus from the mouth of the La Planche River to a point on Baie Verte. The county of Cumberland in Nova Scotia provided the necessary land at no cost.

Work got under way in the fall of 1888 and proceeded pretty much on schedule for nearly three years, despite countless difficulties and obstacles which often seemed insurmountable. As it progressed, the railway was justly called one of the greatest engineering feats of its time. A ship to be transported was to be hoisted by hydraulic power from the water onto the tracks and drawn by two heavy locomotives across the isthmus at a speed of between five and ten miles an hour. The vessel was to be carried on a cradle or gridiron, capable of handling a weight of one thousand tons. The rails of the track were the heaviest ever rolled up to that time.

By 1890 five hundred men employed on the line were pushing for its completion. Others were working around the clock on the two terminal basins. Just as the end of a tremendous effort seemed in sight, however, work ground to a halt. Nearly five million dollars had been spent on the project, but a depression in the world market and a change in government policy forced the company to abandon the railway. Rails and fastenings were eventually purchased by the dominion government for a little over $105,000 for use on the Intercolonial Railway. Bricks, once part of the company's powerhouse, were utilized to build a museum.

Few reminders of a once fabulous enterprise remain today. One of its lasting monuments is a stone bridge near Tidnish. But the river that was to flow beneath this bridge was never diverted. And the man whose dreams were shattered lies buried at Tidnish, not far from the site of his ship railway that never was to be.

NEW BRUNSWICK LINES

Alexander Gibson, a man known to many New Brunswickers of the nineteenth century simply as "Boss," was the lumber baron of the Nashwaak country in central New Brunswick. He owned not only vast timber stands in the area but also a cotton mill in Marysville, which was then the second largest such factory in Canada.

Naturally, as an industrialist and the owner of millions of feet of lumber ready to be marketed, he was anxious to have his district served by a railroad. Consequently, he surveyed a route from the village of Gibson, at the mouth of the Nashwaak, to Edmundston and a branch line to Woodstock. The outcome of this survey led to the incorporation of the New Brunswick Railway Company in 1870. Since this enterprise fell into the category of a colonization road, it received a provincial government subsidy and construction began at once.

The section from Gibson to Newburg Junction opened in 1873, and 112 miles between Woodstock and Edmundston were operational by 1878. Four years later Gibson sold his holdings in the company when he failed to convince the directors of the necessity of standardizing the original narrow gauge railway.

In the course of time the New Brunswick Railway acquired various lines in the province, among them the network of the New Brunswick and Canada Railway in 1883. This latter company had come into existence nine years earlier and comprised four small lines, some of them built during the pioneer years of the 1850s. The last line to be taken over by the New Brunswick Railway was the Fredericton Railway which had been opened on November 18, 1869, when the first passenger train steamed into the provincial capital.

Incidentally, it was a Fredericton man by the name of John Taylor who, in 1870, devised a simple but important innovation to be used by railways. By placing a steam coil between two thicknesses of glass he created the first non-frosting cab window for locomotives.

The entire system of the New Brunswick Railway became part of the Canadian Pacific which by 1890 was expanding its empire into the eastern provinces of Canada.

In the meantime, Alexander Gibson had turned to other goals. During the 1880s, he and Senator J. B. Snowball were building the Canada Eastern Railway between Fredericton and Chatham, New Brunswick, Gibson pushing his way eastward, while Snowball was building westward. The two rights of way joined near Doaktown. For the purpose of bringing his railway into Fredericton, Gibson formed a separate company to construct a bridge, the cornerstone of which was laid on June 20, 1887 by Lady Macdonald, wife of Sir John A. Macdonald, Canada's first prime minister. Gibson later inaugurated the locally famous Marysville Suburban, which for half a century, day in and day out, shuttled back and forth between Fredericton and Marysville.

Unfortunately Gibson and Snowball, the two partners in the railroad business, did not always see eye to eye. When, at one of the annual company meetings, Senator Snowball was voted into the presi-

Intercolonial Railway yard, Moncton, New Brunswick, 1877

dency, replacing Alexander Gibson, the latter vowed no longer to patronize his own railway. For an entire year he shipped every last bail of cotton from his Marysville mill to Fredericton by horse and dray, rather than by rail. To end the dispute, Snowball sold his holdings to Alexander "Boss" Gibson, and the Canada Eastern, in the course of time, became part of the Intercolonial Railway.

Rails into Ottawa

In 1854 Bytown, a lumbering community of ten thousand in the Ottawa Valley, was incorporated as a city and changed its name to Ottawa. It was in the spring of the same year that the railway came to the future capital of Canada. The first train steamed into the newly erected station at the corner of Sussex and McTaggart Streets. Soon after a regular service was inaugurated between Ottawa and Prescott on the St. Lawrence River, with connections via Ogdensburg to Montreal.

Chartered in 1850 as the Bytown and Prescott Railway and later renamed the Ottawa and Prescott, the pioneer line had been surveyed by Canadian railway engineer Walter Shanley who walked over four possible routes, a combined distance of two hundred miles, before recommending the route via Kemptville. As the length of the railway was less than seventy-five miles, it did not qualify for government assistance. Lack of funds forced the company to buy steel rails from a Welsh foundry in exchange for mortgage bonds. With eight locomotives and one hundred and thirty-one cars on order from Boston, Bytown voters were asked to authorize a substantial municipal loan to the company.

The first passenger train over the partially completed line ran on June 21, 1854 between Prescott and Spencerville. In contrast to the usual celebrations accompanying an event of this nature, the occasion undoubtedly was a somewhat "sober" affair since the passengers consisted mainly of Prescott's "Sons of Temperance," who were travelling to Spencerville to attend a rally. The opening of the Kemptville-Prescott stretch of the line, however, turned out to be a lot livelier. As one contemporary observer tells it: "Passengers slaked their thirsty souls with the water of the St. Lawrence, improved mightily with Gillman's Brandy."

As the railway inched its way closer to Bytown, the company was so hard-pressed for money that for the last few miles to New Edinburgh (across the Rideau River from Bytown) the contractor had to make do with hardwood rails capped with strips of iron. On Christmas Day 1854, the first train, pulled by the locomotive *Oxford*, arrived at New Edinburgh. Passengers were ferried across the Rideau River and walked to Sussex Street Station, where a banquet was held to celebrate the inauguration of the line. By the following spring a bridge across the river was completed and Ottawa could be entered by train.

The Carillon and Grenville Railway carried passengers past the Long Sault Rapids on the Ottawa River

For a number of years the Ottawa and Prescott Railway (in connection with a steam ferry from Prescott to Ogdensburg) remained

PAC

The engine Ottawa *of the Ottawa and Prescott Railway. Picture taken in Ottawa in 1861*

CP

PAC

Ottawa Station at the corner of Sussex and McTaggart Streets, winter 1855

the only rail link between the flourishing lumbering centre and the hungry markets in the United States. But the railway company's profits were so low that the management was forced to issue a large number of promissory notes. These notes, from five dollars upward, circulated for a while in the area just like ordinary paper currency.

Ottawa's railway was the first in Upper Canada to adopt today's standard gauge of 4′ 8½″. Financial assistance was offered by the Grand Trunk, had the Ottawa line been willing to conform to the latter's gauge of 5′ 6″, but the owners of the Ottawa and Prescott preferred to remain independent.

Among the earliest locomotives of the Ottawa and Prescott Railway were the *St. Lawrence*, a 32-ton woodburner, and the *Ottawa*, described by railroaders of the day as a "splendid engine." An iron steam-ferry, built in the States, served between Prescott and Ogdensburg.

In 1867, the old Ottawa and Prescott (under different ownership by then) was renamed the St. Lawrence and Ottawa Railway. On this company's locomotive roster of 1873 appeared the *Lucy Dalton* which later, as CPR engine No. 9, became famous for being the first locomotive seen in the North Bay district. By the mid-seventies, some of the St. Lawrence and Ottawa engines switched from wood to coal.

Going back to the early days in the Ottawa district, another railway had been chartered in 1853. It was the Brockville and Ottawa, projected to link the town of Brockville with the vast timber resources of the Ottawa Valley. By the beginning of 1859 it reached Smiths Falls, a distance of twenty-eight miles from Brockville. The first train steamed into Smiths Falls on January 25th, after several hours' delay. Apparently a coupling had broken en route and the engine had run out of water. The crew searched the ditches along the right of way for water and found enough to get her going, but one of the coaches had to be pulled into the station by a rope. Before steel had reached the Ottawa River at Sandy Point, the company was bankrupt. The Canada Central, which had been chartered to build from Quebec to Lake Huron, eventually acquired control of the Brockville and Ottawa, and a line was constructed from Carleton Place to Ottawa.

On December 3, 1877 the Northern Colonization Railway or, as it was then called, the Quebec, Montreal, Ottawa and Occident,

Little Portland engine built about 1873 for the St. Lawrence and Ottawa Railway, now the Prescott Branch of the CPR. (Taken over in 1882)

PAC

reached Hull across the river from the capital city. This line, using Atlantic-type engines with eight-foot drivers, catered to members of parliament and senators of the day by providing luxurious "palace cars" on its less-than-two-hour run between Montreal and Ottawa.

Yet another railway came to Ottawa in September 1882. It was the Canada Atlantic, founded by John Rudolphus Booth, a wealthy Ottawa sawmill operator and owner of one of the finest pulp and paper plants on the continent. Born on a farm in Waterloo, Quebec, Booth had come to Ottawa in 1852 with only a few dollars in his pocket. In less than a quarter-century he had become the "Lumber King" of Ontario. He stimulated Ottawa's growth not only as a forceful businessman but also as a far-sighted railroad builder. His Canada Atlantic Railway was chartered to "construct a line from Ottawa to the St. Lawrence River at or near Coteau Landing, thence to the town of St. Johns or to some place on the international boundary, and to construct bridges over the St. Lawrence River and the Beauharnois Canal."

As the railroad progressed hundreds found employment, and factories along its route were busily producing supplies and materials. In 1890 the railway reached the international boundary via the Coteau Bridge, connecting there with the Central Vermont and linking Ottawa directly with Boston on the Atlantic coast. But John Booth also planned to connect the capital with the vast hinterland to the north. In 1891 the Ottawa and Parry Sound Railway, one of Booth's projects, was amalgamated with the Ottawa, Arnprior and Renfrew Railway under the name of Ottawa, Arnprior and Parry Sound Railway. In 1896 the road was completed from Ottawa to Depot Harbour on Georgian Bay, an island terminus across from Parry Sound. Three years later the company became part of the Canada Atlantic. Booth built a grain elevator at Depot Harbour and put steamships on the Great Lakes to carry his grain shipments. By connecting with the Central Vermont, he provided himself with an ocean outlet for his vast and flourishing lumber business.

In the meantime the Canadian Pacific Railway was spreading its band of steel across the country as Canada's first transcontinental line. The Canada Central Railway had come under the CPR's control; so had the Quebec, Montreal, Ottawa and Occident. In 1884 the successor of Ottawa's first railway line, the St. Lawrence and Ottawa, was also leased to the Canadian Pacific. The railway of John R. Booth, however, remained independent until February 1914, at which time it was amalgamated with the Grand Trunk Railway Company of Canada.

Advertisement of the St. Lawrence and Ottawa Railway in the Carleton Place Herald, *1871*

The St. Lawrence and Ottawa

RAILWAY CO.

The Shortest, Quickest, & Best Route from Belleville to Ottawa.

ASK FOR TICKETS BY PRESCOTT JUNCTION.

Winter Arrangement, 1871-2.

FOUR Passenger Trains will run daily on this Line making CERTAIN CONNECTIONS with those on the GRAND TRUNK, the VERMONT CENTRAL and the ROME and WATERTOWN RAILWAYS for all points East, West and South.

Comfortable Sofa Cars

On the Train connecting with the Grand Trunk Night Expresses, available only to passengers with first class through tickets. Charges for berths 50 cents each.

20 Minutes Allowed for Refreshments at Prescott Junction.

☞Through Tickets for sale at all the principal Stations on the Grand Trunk. Baggage checked through.

FREIGHT NOTICE.

A FLOATING ELEVATOR always in readiness during navigation at Prescott Wharf, where Storage for Grain, Flour, Pork, &c., can be had.

A Change Gauge Car Pit

Is provided in the Junction Freight Shed by means of which freight loaded on Change Gauge Cars COMES THROUGH TO OTTAWA WITHOUT TRANSHIPMENT.

THOS. REYNOLDS,
Managing Director.

R. LUTTRELL,
Superintendent, Prescott.

Ottawa, Nov. 17, 1871.

More about Ontario Lines

Toronto Public Libraries

LONDON AND PORT STANLEY RAILWAY

Opening of the Toronto and Nipissing Railway: Uxbridge Station c. 1871

In 1965 Canadian National Railways acquired the property of a company which, in one hundred and twelve years of existence, had never changed its name—a rare occurrence in the history of railways. Although leased and operated by various other companies, the line had neither been amalgamated with any of the large systems nor had its original charter ever been cancelled. The railway was no more than twenty-four miles in length, but it was important enough to compel both the Great Western and the Grand Trunk to cut freight rates in order to meet its competition.

Opened for traffic in 1856, the London and Port Stanley Railway, as the line was called, was owned jointly by the city of London, the town of St. Thomas and the counties of Elgin and Middlesex. To Londoners in particular, the railway was a source of pride and joy. For the return fare of a quarter, they were able to indulge frequently in the pleasure of taking a train to Port Stanley and spending a fun-filled day at the Lake Erie beach. When the picnic train (which consisted mostly of box or flat cars furnished with wooden seats) picked up additional passengers at St. Thomas, it was as a rule so crowded that one newspaper reporter of the 1870s declared: "No more excursions for me, by Jove!" The unfortunate chap had to stand all the way "without the opportunity even of resting his hand on the back of a seat, and then, in his hurry to get out, his hat fell off and he saw it trampled under a dozen feet."

From the start, the London and Port Stanley Railway was plagued by financial troubles. Construction costs were considerably higher than estimated. Hopes of turning the line into "an artery of trade between Canada and the United States" never materialized. Revenues from excursion traffic and transport of coal were not sufficient to improve the monetary situation, and the city of London was repeatedly called upon to come to the railway's rescue. But, costly as it was, no Londoner back in those days would have considered selling his railway. It was only after much heated debate that the city fathers in 1874 agreed to lease the line to the Great Western Railway for a period of twenty years.

Subsequently, the Grand Trunk acquired the Great Western and, with it, the London and Port Stanley lease. Anxious to add the leased line to the properties of their growing empire, Grand Trunk officials decided to buy it from the city. When they found that their request was most emphatically refused, they gave orders to remove immediately the Grand Trunk's entire rolling stock and equipment, including even the water barrels on the wooden bridges, and clean out all stations and shops along the line. The next morning nothing was left but the rails.

After this incident the Michigan Central stepped into the picture, operating the railway on a month-to-month basis. In 1893 the London and Port Stanley was leased for twenty years, this time to the Cleveland, Port Stanley and London Transportation and Railway Company. When the latter was purchased by the Père Marquette Railway in 1903, the old London and Port Stanley, being part of the deal, once more acquired a new master. But when the lease expired, London got its railway back. By 1914 the London Railway Commission had been formed to convert the London and Port Stanley into an electric road and to manage it for 99 years.

BUFFALO AND LAKE HURON RAILWAY

The first tracks to reach Lake Huron proper in 1858 were those of the Buffalo, Brantford and Goderich Railway. The line owed its existence to a group of Brantford merchants, who were determined to remedy a situation which had been brought about by the Great Western's decision to bypass Brantford by several miles to the north. They formed the Brantford and Buffalo Joint Stock Railway Company and engaged the well-known American engineer William Wallace to survey a route from Brantford to Fort Erie, opposite Buffalo. The city of Buffalo bought seventy thousand dollar's worth of stock in the venture.

By 1852 surveys were completed to the port of Goderich on Lake Huron, and in November of that year the name of the company was changed to the Buffalo, Brantford and Goderich Railway. The road was opened from Fort Erie to Brantford on January 6, 1854. Three trains, which steamed into Brantford that afternoon, were greeted with cheers and the firing of cannon. At an elaborate civic dinner, guests from Buffalo outnumbered the local residents. A display of fireworks, customary at such celebrations, was enjoyed in the evening, followed by a "Grand Railway Ball" held on the second floor of the depot machine shop.

In March of the same year the Buffalo and Goderich connected at Paris, Ontario, with the Great Western Railway. But by the time the road was nearing Stratford towards the end of 1854, the company was embarrassingly short of money. It owed a considerable sum to its contractors who in turn were unable to pay their workers. Trouble came in Janauary of the following year when about thirty of the workers lost their patience and tore up the tracks near Ridgeway, Ontario, west of Fort Erie. One man was killed during the ensuing fight. A subsequent attempt to raise money in England for the remainder of the road to Goderich failed. On May 16, 1856 the property of the Buffalo, Brantford and Goderich was taken over by the newly incorporated Buffalo and Lake Huron Railway Company which carried on the projected line.

More problems arose in the fall of 1856. Grand Trunk contractors ripped out the Buffalo and Lake Huron rails at Stratford where they found them blocking their path. Company officials ordered work crews to tear up the Grand Trunk track in reprisal and re-lay their own. But a Grand Trunk locomotive bringing two carloads of armed navvies appeared on the scene. The men were drunk and ready for a fight, which fortunately was averted. Eventually, the two railway companies worked out an amicable settlement of the matter.

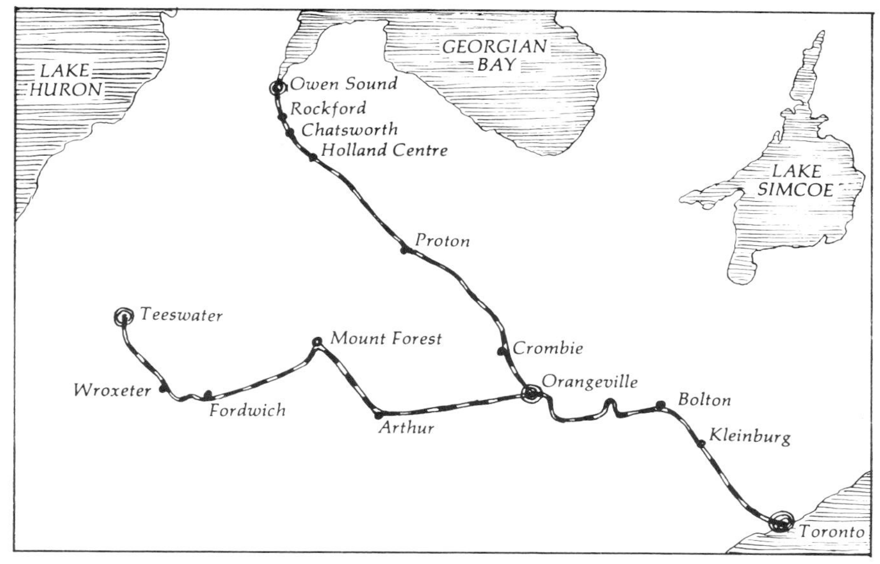

Toronto, Grey and Bruce Railway

The first through train from Fort Erie reached Goderich on June 28, 1858. With harbour facilities completed, Goderich soon became a busy shipping port in the summer months. The Grand Trunk, aware of the line's potential, began buying Buffalo and Lake Huron stock, and by the end of 1869 was ready to take over the road.

The London *Times* reported that "the formal embodiment of an agreement was now being prepared by the solicitors of the two companies and would be submitted to the proprietors at meetings specially called for that purpose." The writer commented:

> *The terms now proposed might be considered far short of what this company believed they were fairly entitled to, still, in view of existing circumstances, and to avoid, if possible, a harassing and costly struggle too likely to follow their rejection, the Board unanimously advised their acceptance. The injuries which the two companies had it in their power to inflict upon each other would be poor compensation for the indefinite postponement of any share in the prosperity which appeared again about to dawn upon Canadian enterprise.*

TORONTO, GREY AND BRUCE RAILWAY

A plaque erected on the station lawn at Orangeville, Ontario, reads as follows:

> *TORONTO, GREY AND BRUCE RAILWAY*
>
> *This pioneer railway received its charter in 1868 and the first sod was turned at Weston on October 5, 1869, by Prince Arthur, third son of Queen Victoria. The main line from Toronto to Owen Sound via Orangeville was completed in 1873 under the direction of chief engineer Edward Wragge, and a branch line to Teeswater was added the following year. The first train went into operation on the Southern section in April 1871. The original choice of narrow-gauge track proved ill-advised, and standard gauge was adopted 1881-83. The line was leased to the Ontario and Quebec Railway in 1883 and absorbed by the CPR in 1884.*

In a nutshell, this is the story of a railway which, like many of its contemporaries, opened up new districts, created a link and promoted trade between outlying settlements and the markets of larger cities. First advocated by Orangeville merchants as a tramway through the Credit River Valley to Brampton, the proposal aroused the interest of a group of prominent Toronto businessmen and, with their backing, the Toronto, Grey and Bruce Railway came into being.

Although the project was approved in the provincial legislature by only a small majority, the public was wholeheartedly in favour of the proposed line. The city of Toronto granted a quarter of a million dollars in support of the venture. In addition, Torontonians subscribed 320,000 dollars' worth of shares.

The charter authorized the company to build from Toronto through Orangeville to Mount Forest, thence to the border of Bruce County, and from there to Southampton on Lake Huron, with a branch line to Kincardine. It further provided for a branch to be constructed from Mount Forest, or Durham, northward to Owen Sound on Georgian Bay. Due to financial difficulties, the company had to abandon their plan of building to Lake Huron, and instead an extension was constructed via Wroxeter to Teeswater. The route to Owen Sound, known as the Grey Extension, was also altered when the County of Grey refused a grant to aid in the construction of the branch from Mount Forest to Owen Sound. Communities north of Orangeville had offered bonuses and it was therefore decided to build from Orangeville direct to Owen Sound.

Men wanted on the Cobourg and Peterboro Railroad. Note the wages offered in 1854

Travelling Register.

COBOURG AND PETERBORO Railroad.

WANTED on the C & P. R. R, Five Hundred Labourers and Two Hundred good hands for track laying, to whom the following wages will be paid—Labourers per day, 5s cy, Track Layers from 5s to 10s per day.

Any men coming by any of the Lake Ontario Steamers will have their passage money returned, providing they remain on the works one month and produce a certificate from the Captain.

JOHN FOWLER,
Contractor.

Cobourg, May, 1854. 1273-12t

For the sod-turning ceremony in Weston on October 5, 1869, a large crowd assembled, the overwhelming majority being ladies eager to get a glimpse of the royal visitor, Prince Arthur (later, as the Duke of Connaught, to be Governor-General of Canada). According to the Toronto *Globe* reporter, special constables "armed with batons and a bunch of ribbons in their button holes," had quite a job pleading with the "fair creatures" to stand back and behave. For a while the surging crowd threatened to disrupt procedures, and not until the constables got angry and "flourished their batons" was the young prince able to descend from the platform. With a hand-crafted silver spade he lifted a small pre-cut square of turf, marked with a miniature Union Jack, and placed it on a maple wheelbarrow. Not quite satisfied with his job, however, he went back to dig a second, larger spadeful of dirt on his own, but the sod was tough and, using only his arms, he had quite a struggle. "Take your foot to it, Your Highness, and you will make more of it," someone said in a whisper. "His Highness took the hint," so the paper says, "and soon cut it in right good style, pitched it into the barrow and wheeled it to the end of the tramway amid the most enthusiastic cheers from the crowd, and thus quite royally the first sod of the Toronto, Grey and Bruce Railway was turned."

Construction began without delay. By agreement with the Grand Trunk, a third rail was laid along the latter company's right of way from Weston to Queen Street in Toronto to accommodate the Toronto, Grey and Bruce, which had chosen a 3′ 6″ gauge despite the fact that the so-called "narrow gauge" was seldom used in Canada and had yet to prove its merit. With the exception of three, all other lines at that time were running on the broad gauge of 5′ 6″. There were plenty of arguments for and against the proposed narrow gauge, but the ultimate decision seemed more than justified by the initial savings on labour and material during construction. Narrower rights of way, lighter rails, less ballast and smaller equipment, all contributed to lower costs. But what the promoters of the narrow gauge did not foresee was the ever increasing demand on their railway following the opening of new areas. Soon their low-capacity cars could no longer handle the resulting increase in freight, and produce began piling up in the communities they served. Heavy snow in the winter added to the hazards of narrow cuttings. Delays meant losses in revenue. Eventually, between 1881 and 1883, the gauge was changed to conform with other lines which had since converted their broad gauge tracks to the now prevailing standard gauge of 4′ $8^{1}/_{2}$″.

The first train on the Toronto, Grey and Bruce left from Toronto for Alton Station on April 10, 1871. One week later, the rails reached Orangeville, where citizens welcomed the gaily decorated locomotives, *Kincardine* and *A. R. McMaster*. In December of the same year the railway was opened to Mount Forest.

The *Globe* of January 10, 1870 relates the financial burden the construction of the railway threatened to impose on farmers in Grey County:

> *The increased demand of the Toronto, Grey and Bruce Railroad of $400,000 from the county of Grey, will not meet with general acceptance. In the present temper and circumstances of the people, if the Directors insist on this sum, it is tantamount to shelving indefinitely, the railroad questions as far as the North Riding of Grey is concerned. The people are, really, not able to give $400,000 of a bonus. For two years the crops have been short, the midge has fairly made a lodgement in the district, and the heavy crops that a new county always gives for the first ten years, while the new land is being cleared, are now ceasing. The farmers, (who are an intelligent, sober, honest class of men, on the whole,) see all this clearly, and hesitate to mortgage their lands to that figure on the promises. One township, for instance, in the neighbourhood, that was assessed by the Provisional Directors for $38,000 has just 500 ratepayers, one-third of whom have not as yet got their deeds. The railroad would cost each ratepayer on an average $76, or one year's average rental. This might be borne; but one glance can satisfy any man that nearly double this sum, or $140, which would be the sum, were $400,000 the bonus, would be hard to grant, and should not in fact be asked. This is not a favourable season to submit to the votes of the people here a question involving an increase of taxation, inasmuch as there is considerable depression all over; but so great is the desire to have railway communication with Toronto that a reasonable demand will now be granted in all probability, for if the bonus must be $400,000 the vote should be postponed till a new harvest is seen.*

The Grey Extension to Owen Sound was not completed until the summer of 1873, when a special train, headed by the engine *Owen Sound*, carried company directors over the entire distance from Toronto to the terminus on Georgian Bay.

Passenger trains on the narrow gauge, as a rule, travelled at speeds not exceeding sixteen miles per hour. A trip from Toronto to Owen Sound therefore would take nine hours on a "slow" train. The "fast" train, leaving at 8:00 in the morning, made it in seven hours and thirty-five minutes. But in 1876, the engine *Mono* carrying freight papers to be placed aboard a departing steamer, set a record for the Toronto, Grey and Bruce, when she succeeded in making the run in 168 minutes.

The most unusual engine of the line, undoubtedly, was the *Caledon*, one of the few Fairlie-type locomotives to operate in Canada. She was a double-header, with two boilers joined back to back by one fire-box dividing the centrally located cab. Although in frequent need of repair, the *Caledon* worked as a freight engine for nearly a decade. She was scrapped in 1881.

COBOURG AND PETERBOROUGH RAILWAY

A tiny settlement founded by United Empire Loyalists on the shore of Lake Ontario in the late 1790s, only thirty years later Cobourg was a thriving village on the verge of becoming a town. Among its leading citizens at that time was William Weller, owner of the famous Royal Mail Coach Lines, who along with other progressively minded men advocated a railway for Cobourg.

Under one of the two earliest railway charters granted in Canada, the Cobourg Railroad Company was incorporated in 1834 with power to construct a "double or single iron or wooden rail road" from Cobourg northward to Rice Lake. Although the provincial government generously pledged a subscription of £10,000 in the venture, promoters were unable to raise sufficient capital. Consequently the project was shelved until 1846, at which time it was resurrected as the Cobourg and Rice Lake Plank Road and Ferry Company. Built by Samuel Gore, this eleven-mile stretch of plank road barely survived the frosts of the first two winters before it had to be abandoned.

Then came the Cobourg and Peterborough Railway, incorporated in 1852. It turned out to be haunted by a series of misfortunes right from the start. Following a portion of the old plank road, it ran to Harwood, and headed via Tick Island across the widest part of Rice Lake to Hiawatha and then on to Peterborough.

The first sod for the Cobourg and Peterborough was turned at Cobourg on February 9, 1853. Before the rails reached the south shore of Rice Lake, a cholera outbreak took a heavy toll among construction workers, mostly German immigrants, who had been hired to work on the line for a dollar a day.

The Rice Lake Bridge, although largely completed by the end of 1853, was severely damaged by shoving ice that winter, and the opening of the bridge had to be postponed until the following November. A pile structure from Harwood to Tick Island, the bridge changed to thirty-three 80-foot truss spans, with a 120-foot-long swing section in the navigation channel, and continued to the north shore again as a pile trestle. Just under three miles in length, it was likely the longest bridge on the North American continent at the time but undoubtedly also one of the most ill-fated. No sooner had trains begun to operate over the completed line to Peterborough, when on January 1, 1855 ice jams in the lake pushed the north pile bridge towards the Peterborough shore, the truss span towards Tick Island, and the southern trestle towards the Cobourg shore. Near the island there was a gap of seven feet. Repairing and maintaining the Rice Lake Bridge soon turned out to be considerably more of a problem than its original construction.

To add to the railway's troubles, contractor Samuel Zimmerman demanded a price far in excess of his original estimate. At first he refused to turn the road over to the company, but when the directors eventually ran out of ready cash he let them have the line in a partially completed condition. The sum of £10,000, borrowed by the railway from the Marriage Licence Fund of Canada West, was merely a drop in the bucket in trying to bring the road up to standard. Nearly every winter the bridge was damaged and required extensive repairs; the line's operating costs were staggering.

Although the Cobourg-Peterborough Railway contributed a great deal to Peterborough's flourishing export trade, the town of Peterborough had never invested a single cent in the enterprise nor did it intend to come to the railway's rescue at any time. The fact was, Peterborough already possessed a perfectly reliable connection to Lake Ontario via the Port Hope, Lindsay and Beaverton Railway, which was not encumbered with such hazards as a flimsy railroad bridge most of the time out of commission. Consequently Cobourgers alone ended up spending nearly one million dollars on their troubled railroad which was no more than thirty miles long. Their pride received another blow in the fall of 1860, when the Prince of Wales, who visited the area, was not permitted to cross Rice Lake on the infamous railway bridge.

During the following winter fate struck its mortal blow. The bridge disintegrated and floated down the lake. There were those who claimed that some extra help must have been supplied by agents of the rival Port Hope Railway. In any event, the bridge was never again rebuilt, although plans for its reopening did not die for many years to come.

In 1866 the company merged with the Marmora Iron Works under the new corporate name of Cobourg, Peterborough and Marmora Railway and Mining Company. The following year it opened a spur line from the Trent River to the Marmora Iron Mines at Blairton. Ore shipments from the mines were transferred onto barges at the Trent River and transported to Harwood, whence they continued by rail over the old tracks to Cobourg. Traffic on the Blairton branch, however, came to a halt in the late seventies when no more new ore deposits were found.

The long-abandoned Rice Lake-Peterborough section of the road was leased to the Grand Junction Railway which required entry into Peterborough, and in 1886 the railway was sold to a Mr. Pearse. Reorganized the following year as the Cobourg, Blairton and Marmora Railway and Mining Company, the road eventually came under the control of the Grand Trunk Railway of Canada.

Toronto and Nipissing Railway: Locomotive Shedden, *a double-header, c. 1870*

Toronto Public Libraries

MIDLAND RAILWAY OF CANADA

The Midland Railway of Canada evolved from the 1846 charter of the Peterborough and Port Hope Railway Company which changed its name in 1854 to the Port Hope, Lindsay and Beaverton. From 1869 on, it was known as the Midland Railway of Canada. The Port Hope-Lindsay section opened for traffic in December 1857, and a branch from Millbrook to Peterborough was built the following year. After the reorganization in 1869, construction continued and a branch was also built from Peterborough to Lakefield. By 1879 the line was completed from Lindsay via Beaverton and Orillia to Midland.

Under an agreement with the respective companies, signed in 1882, the Midland Railway of Canada and five lines adjacent to its property were consolidated into one system to be known as the Midland Railway of Canada. Involved in this merger with the Midland were the following lines: the Toronto and Nipissing Railway; the Toronto and Ottawa Railway; the Whitby, Port Perry and Lindsay Railway; the Victoria Railway and the Grand Junction Railway. The entire Midland system was leased to the Grand Trunk in 1884.

THE TORONTO AND NIPISSING RAILWAY

The Toronto and Nipissing Railway was incorporated in 1868 by William Gooderham, a prominent Toronto flour miller and distiller, who firmly believed that pioneer railways should be built in the most economical manner. He therefore chose the narrow gauge of 3′ 6″ for his line, which meant a large comparative saving in construction and maintenance. Whenever he travelled over his railway, he instructed the engineer "to slow down the train" in order to minimize wear and tear on the light forty-pound rails. His "colonization railway" into the Ontario Northland was owned in the beginning solely by its shareholders and borrowed no more money than was absolutely necessary. Gooderham wanted the railway to serve not money-lenders but the small communities along its way. Of course, he also knew that traffic with northern Ontario would contribute greatly to the growth and prosperity of Toronto, and just what this might amount to becomes apparent from an 1867 article in the Toronto *Globe*:

> *The projectors of this line contemplate connecting Toronto with the waters of Gull River, thereby giving access to a region now completely shut out from a profitable market — a region in which vast tracts of splendid pine and hardwood abound, and which to the matter of lumber and cordwood alone would, it is believed, yield enough to freight the road for 50 years. If to this a passenger and general traffic be added, an idea can be formed of the immense business which could be done by such a road, and the incalculable good which it would effect in opening up a very valuable tract of back country, and giving the farmers in Victoria, North Ontario and York easy access to the markets in this city, where they can always command better rates for their products than at any other point in the Province of Ontario. Of the district thus sought to be opened up it may be mentioned that there are twelve townships in Victoria through*

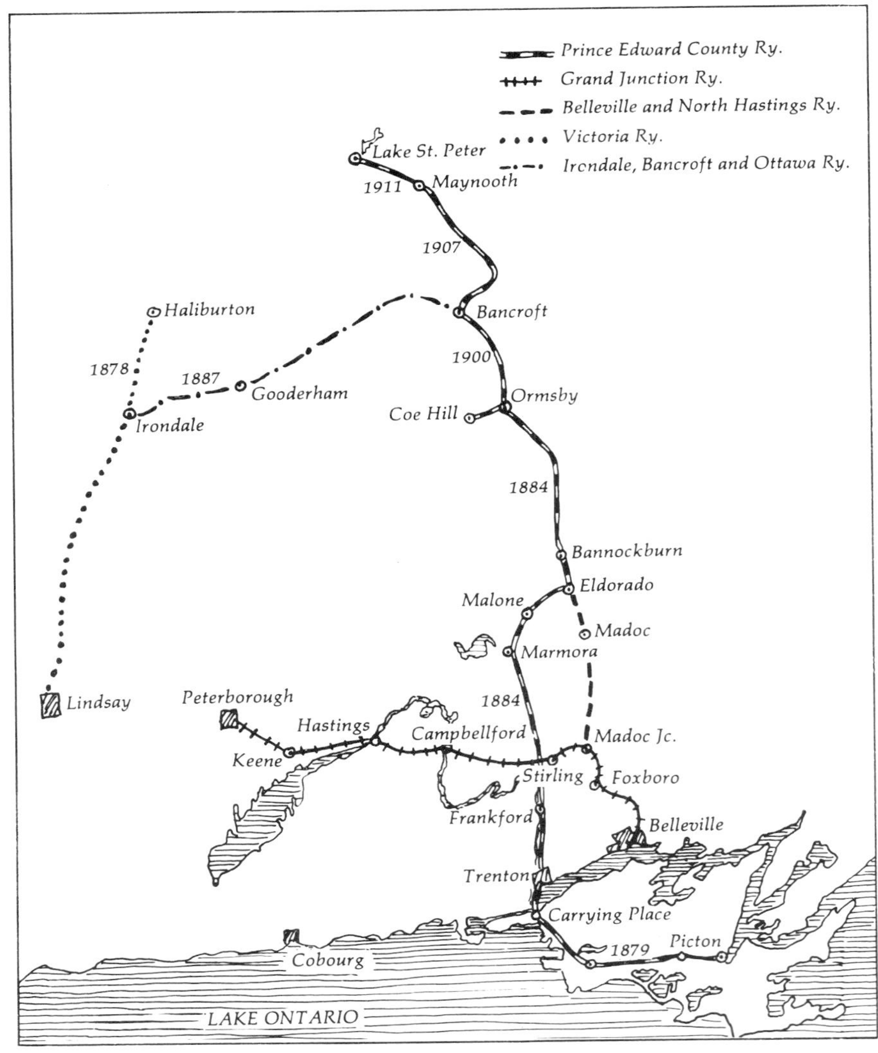

which no road passes, which hardly contains a settler, although much of the land is excellent for farming purposes and heavily timbered.

The line was opened to Uxbridge in 1871 and in November of the following year it reached Coboconk. An extension, built in 1877 by the Lake Simcoe Junction Railway from Stouffville to Jackson's Point on Lake Simcoe, was leased by the Toronto and Nipissing.

In its early years of operation, the railway transported a hundred thousand tons of freight annually. Narrow gauge freight cars had to be carefully balanced when being loaded, to avoid derailment of the train on curves. Trips into town were a popular feature for farmers along the route in the summertime. The train crew called them "Watermelon Excursions" because passengers always brought along a huge supply of home-grown watermelons for their refreshment.

One of the first engines owned by the Toronto and Nipissing was the *Gooderham and Worts,* named after the firm of distillers which is now the oldest of its kind still existing in Canada. The locomotive, although weighing only twenty-five tons, was capable of speeds of up to forty miles an hour, probably much to Mr. Gooderham's dismay.

The line also acquired a Fairlie engine, one of the few double-headers to run on Canadian roads. Named the *Shedden,* after the president of the company, she was a sister engine of the *Caledon* owned by the Toronto, Grey and Bruce. A contemporary described her as a "two-headed fiery dragon, belching sparks and billows of black smoke from both her huge smoke stacks, as she raced or crawled through forests and farms of the backwoods country." One day, in the early 1880s, near Coboconk, the *Shedden* blew up in a boiler explosion, killing a railway crew of seven men.

THE TORONTO AND OTTAWA RAILWAY

The Toronto and Ottawa, chartered in 1877, with powers to build from Toronto to Ottawa via Peterborough, was never constructed. Its charter was used by the Midland group merely to build three missing links in their system, totalling approximately thirty miles: one from Bridgewater Junction to Bridgewater, another from Blackwater Junction to Manilla Junction and, finally, one line from Peterborough to Omemee Junction.

WHITBY, PORT PERRY AND LINDSAY RAILWAY

A line from Whitby to Port Perry was opened in 1871 under the charter of the Port Whitby and Port Perry Railway. The little locally promoted line of less than twenty miles never seemed overloaded with traffic, but its supporters, hoping for better days, changed the name to Whitby and Port Perry Extension Railway and dreamed for a while of building a transcontinental line to the Pacific. They eventually made it to Lindsay, a distance of just over twenty-six miles from Port Perry, under the new corporate name of Whitby, Port Perry and Lindsay Railway Company.

It seems that early in its career the company ran into trouble with Canadian customs officials. In a letter to the editor of the Belleville *Intelligencer* of May 13, 1872 an irate citizen described the incident as follows:

> *The Whitby and Port Perry Railway Company purchased a Locomotive and Tender for their line, the price to be paid was $11,500 on delivery. This was to include duty. That duty is 15 per cent. It would amount of course on $10,000 to $1,500 leaving $10,000 as the figure at which the Locomotive should have been entered. It was entered at the frontier at $4,590!!!*
>
> *As the Collector and his second officer had received a second hand affair from some Portland Company twelve months ago*

> *which had been entered at $5,000, their doubts were raised as to the smaller entry for articles quite new and very superior indeed, to the old rattle-trap of last year. They therefore seized the Locomotive.*

VICTORIA RAILWAY

Formerly known as the Lindsay, Fenelon Falls and Ottawa River Valley Railway, the Victoria Railway was built from Lindsay to Kinmount and thence to Haliburton, reaching the latter in 1878. Reporting to his readers on the last weeks of feverish construction activities before completion, the Haliburton correspondent of the Bobcaygeon newspaper wrote: "Haliburton is very lively. Lots of fun. We have as many as twelve navvies in the lockup in one week."

GRAND JUNCTION RAILWAY

Leading citizens of Peterborough, Belleville and Cobourg were the promoters of the Grand Junction Railroad back in 1852. They planned a loop line from Belleville, at the Bay of Quinte, to Peterborough, and from there back to Lake Ontario at Toronto. When they found that the Grand Trunk Railway also had this route in mind, they amalgamated with the latter in 1854, and then waited sixteen years for the line to be built.

With the consent of the Grand Trunk, the original charter was revived under the corporate name of Grand Junction Railway. Under the date of September 1, 1870 the Belleville *Intelligencer* published a notice that

> *Application will be made to the Legislature of the Province of Ontario, at its next sittings, for an Act to legalize and confirm any and all By-Laws passed by any of the Municipalities through which the line of the Grand Junction Railway passes, granting bonuses to the said Company to assist in the construction of their Railway.*

At the meetings held by the promoters of the road in the County of Peterborough, it became apparent that some people there felt that the enterprise would greatly benefit Belleville, being the eastern terminus, but that the road would not be of equal value to the municipalities of the western section. Nevertheless, according to the *Intelligencer* reporter, they "concluded that it is of sufficient importance to themselves and all sections of the country through which it would necessarily run, that it is their duty as well as their interest to lend the enterprise their cordial support and give it substantial municipal assistance."

As for Belleville, the overwhelming majority of property holders gave their sanction to the scheme, and the town's fathers voted the sum of $100,000 to aid in the construction of the Grand Junction Railway. Abandoning the loop line idea, the company built the road from Belleville to Peterborough via Stirling, Campbellford, Hastings and Keene, and the first GJR train reached the outskirts of Peterborough on January 1, 1880.

During the same year, the Grand Junction Railway Company acquired a twenty-two-mile narrow gauge road built by the Belleville and North Hastings Railway from Madoc Junction to Eldorado, site of the first gold mine in Ontario. Although a relatively short line, the Belleville and North Hastings, conceived to develop the mineral resources of the north country, had long been a dream in the area. When the first sod was turned on September 9, 1875, near White Lake in Huntingdon Township, some eight miles from the point of the planned intersection with the Grand Junction, hundreds turned out to celebrate the important event.

PRINCE EDWARD COUNTY RAILWAY

Prince Edward County is a picturesque peninsula, surrounded by the navigable waters of Lake Ontario. There seemed little need for building a railway since the county's major export, grain, was going south across the lake and most communities were within a short distance from docking facilities for lake steamers. For many years people in the county were contemplating the merits of a railway, but nothing came of it until 1873, when the Prince Edward County Railway was incorporated to construct a line from any point on the Grand Trunk, between Trenton and Brighton, to the Prince Edward County town of Picton. There were those who called its supporters "railway maniacs" and predicted financial disaster for the county.

The first sod was turned that same year, but there the matter rested. Rails which had been purchased started to rust while the company's stockholders at county council were arguing over grants and the date of the line's completion. Finally a Toronto financier stepped into the picture, and the railway was opened in the fall of 1879 with the first train arriving in Picton or, to be exact, reaching the town's limits. Pictonians had been so eager to get their railway under way that they had neglected to specify the exact location of their terminal to the builder. Consequently, the contractor located the station at Sandy Hook, just on the town's boundary line, thus fulfilling his part of the bargain, which was to bring the railway to Picton.

In order to get the station moved into town, council had to hand £1,000 to the McMullen Brothers, who had since purchased the railway. Moving day was June 7, 1881. A large crowd waited at the future station site on Lake Street, where a ceremony was to be held. The station house (loaded in two parts on flat cars) rolled along the short stretch of the extended track until it came to a sudden stop at Mr. Crandall's barn. Either the barn stood too close to the track or the station house stuck out too far! In any event, there seemed little else to do, but to saw off the corner of the barn. It was several hours before the train was able to proceed.

The first two locomotives of the line were the *Trenton* and the *Picton*. The original rolling stock also included two coaches, two flat and two baggage cars.

With the railway new industries came to town. Picton saw the erection of its first canning factory in 1882, and in October of that year the first car load of canned goods, which was consigned to Winnipeg, left Picton via railway on the initial lap of its journey. Near the station, a large apple evaporator went into operation. To protect the apples from frost during transit, they were packed in barrels and shipped in box cars heated by small wood stoves.

The McMullen brothers, who renamed the line the Central Ontario, built an extension from Trenton to the iron mines at Coe Hill in 1884.

For a while the railway's business flourished. Docks at Weller's Bay bustled with activity as iron ore arrived by train from the north and was loaded onto large vessels, which carried it across the lake en route to the smelters in the United States. But all too soon the mining boom came to an end. Never really prosperous at any time, the Central Ontario was sold in the early 1890s to an American businessman. Eventually the line, which by 1907 had been extended to the lumbering community of Bancroft, was integrated into the Canadian Northern system, which in turn became part of Canadian National Railways. During the 1950s the CNR extended the tracks of the old Central Ontario to Picton Bay, and ore trains once more began rolling over its tracks.

Crossing the Continent

On a cold November morning in 1885 a special train arrived in the lonely Eagle Pass, some twenty miles west of Revelstoke, British Columbia. It carried a group of distinguished gentlemen who had come to witness the driving of the last spike, which would join the Dominion of Canada by rail from coast to coast.

The spot where the tracks of the first transcontinental railway from east and west were about to meet was christened Craigellachie, originally two Gaelic words meaning "stand fast." A fitting name indeed, for it was the steadfastness of determined men in the face of the seemingly impossible which had built the railway over treacherous terrain and forbidding mountain ranges, accomplishing one of the greatest engineering feats in history. When financial crises had threatened to halt the gigantic project, these men had mortgaged their personal possessions in order to carry on. It was their stubborn perseverance which saw the Canadian Pacific Railway to its completion. Not only did they conquer a thousand obstacles in their path, but they succeeded in building the line in less than five years, or half the time they had been allotted.

Among those present at Craigellachie for the ceremony were William Van Horne, whose genius had masterminded the enormous task of constructing the CPR; Sandford Fleming, once the engineer-in-chief in charge of surveying; Major Rogers, discoverer of the Rogers Pass through which the railway crosses the Selkirk Range; and Donald A. Smith, senior director of the Canadian Pacific, later to become Lord Strathcona in recognition of his services to Canada.

There were no flags waving and no bands playing that morning; only the sound of a sledge hammer driving the last spike into the rail. The spike, unlike most spikes used for ceremonial purposes, was made from neither gold nor silver. Van Horne, who in his day had seen many a railway company go bankrupt, considered a fancy spike bad luck, and he had ordered a plain iron spike to be used. Being a man of action, he did not believe in long speeches either. "All I can say is that the work has been well done in every way," was the famous fifteen-word address he delivered on the morning of November 7th at Craigellachie.

The contemporary correspondent of the Toronto *Globe* gave the following account of the historic event:

> *Tracklaying was commenced at six this morning on the last half mile of the Canadian Pacific Railway. At nine o'clock the last rails had been brought forward and measured for cutting to*

CP

make connections. One rail was cut and placed and the other left until Mr. Van Horne and party arrived. Major Rogers made several blows with a heavy sledge on the last rail, helping to cut it. One hundred and twenty feet of rails were then taken up and left on lorries to be placed when Mr. Van Horne arrived. Immediately after the train made its appearance with the railway magnates aboard. The train drew close to the ends of the track and the party came forward. At twenty-two minutes past nine everything was in readiness to complete the connection. Hon. Donald A. Smith took the maul in hand to drive the last spike, and after missing it a few times he drove it home amid cheers from all present.

After congratulations had been exchanged on the completion of the great work, the Van Horne special train passed over the connecting link, and departed for Port Moody, the present terminus of the road, where they will take a special steamer for Victoria. Chips from the last tie were carried away as mementoes by those present.

(In the summer of 1970 the authors visited the Eagle Pass and found that someone had taken, probably as a memento, the bronze plaque from the cairn erected near the track to commemorate the driving of the last spike.)

The following day, on November 8th, the "special" consisting of the engine, a baggage car and the official cars *Metapedia* and *Saskatchewan*, arrived at Port Moody and thus became the first railway train ever to travel across Canada from the Atlantic to the Pacific.

Canadian Illustrated News, *October 1878*

Canadian Pacific Railway.

To CAPITALISTS and CONTRACTORS.

THE GOVERNMENT OF CANADA will receive proposals for constructing and working a line of Railway extending from the Province of Ontario to the waters of the Pacific Ocean, the distance being about 2,000 miles.

Memorandum of information for parties proposing to tender will be forwarded on application as underneath. Engineers' reports, maps of the country to be traversed, profiles of the surveyed line, specifications of preliminary works, copies of the Act of Parliament of Canada under which it is proposed the Railway is to be constructed, descriptions of the natural features of the country and its agricultural and mineral resources, and other information, may be seen on application to this Department, or to the Engineer-in-Chief at the Canadian Government Offices, 31 Queen Victoria Street, E.C., London.

Sealed Tenders, marked "Tenders for Pacific Railway," will be received, addressed to the undersigned, until the

First Day of January next.

F. BRAUN,

Secretary.

Public Works Department,
Ottawa, 24th October, 1878.

The Canadian Pacific Railway has been called the "Wedding Band of Confederation." It was the promise of a transcontinental railway link with eastern Canada that persuaded British Columbia to join Confederation in 1871. Sir John A. Macdonald, first prime minister of the dominion, at the time pointed out:

It is quite evident to me that the United States Government is resolved to do all it can, short of war, to get possession of the western territory, and we must take immediate and vigorous steps to counteract it. One of the first things to be done is to show unmistakably our resolve to build the Pacific Railway.

The government of Canada, as a Confederation promise, agreed to "the commencement simultaneously, within two years from the date of the union, of the construction of a railway from the Pacific towards the Rocky Mountains and from such point as may be selected, east of the Rocky Mountains towards the Pacific, to connect the seaboard of British Columbia with the railway system of Canada; and further to secure the completion of such railway within ten years from the date of union."

Prior to British Columbia entering Confederation, Prince Rupert's Land and the Northwest Territories had been purchased by the government from the Hudson's Bay Company, and in 1870 Manitoba had become a province of Canada. But without a railway linking the east and the west, the "Dominion from Sea to Sea" existed only in name. Isolated from one another by thousands of miles, its far-flung people did not feel they belonged to one single nation. At the time of the Red River Rebellion, which preceded Manitoba's entry into Confederation, troops from the east took nearly three months to reach Fort Garry where the trouble had started.

John A. Macdonald succeeded in interesting some of the country's leading businessmen in the construction of a transcontinental railway. Parliament resolved that "public aid to be given to secure that undertaking should consist of such subsidy in money, or other aid, not unduly pressing on the industry and resources of the Dominion." As a result, two companies were incorporated in 1872. One, known as the Canada Pacific Company, headed by Sir Hugh Allan of the Allan Steamship Lines, represented a group of Montreal financiers. The other, named Inter-Oceanic Railway Company, was made up of influential Toronto businessmen. Repeated government efforts to amalgamate the two companies proved fruitless. The Toronto group alleged that the Canada Pacific was influenced by American interests and would not do justice to the needs of Canada.

Eighteen hundred and seventy-two was an election year. After the Macdonald government had been returned to power, the matter of the railway appeared to be solved by the formation of a new company which was to receive fifty million acres of public land and thirty million dollars in government assistance. Sir Hugh Allan, president of the company, promised to exclude American interests and repay investors. And then broke the so-called "Pacific Scandal" that toppled the government. Somehow the opposition got wind of the fact that the Tories had received the amount of $160,000 for election expenses from Allan, and the Liberal leader moved a vote of non-confidence. Consequently Sir John A. Macdonald resigned on November 5, 1873, and the railway scheme floundered.

Alexander Mackenzie, who formed the next government, publicly stated that "all the power of man and all the money of Europe" would not be able to construct the transcontinental railway within the span of ten years as had been promised by his predecessor. Furthermore, politicians argued, a railway traversing the sparsely populated western provinces "would not pay for its axle grease."

Little or no progress was made during Mackenzie's regime, and aside from two small sections of railway between the Red River and the Great Lakes, no significant construction took place. The first contract was let for a line from Selkirk to Emerson, Manitoba, where it would connect with an American railroad; the second for a 45-mile stretch along Dawson Road from Fort William, Ontario, to Lake Shebandowan. The first sod on what was eventually to become the Canadian Pacific Railway was turned on June 1, 1875, on the bank of the Kaministikwia River at Fort William.

In the fall of 1877 the first steam locomotive came to the prairies. The now famous little engine, known as the *Countess of Dufferin,* is still preserved and on public display in central Winnipeg. When she arrived on October 9, 1877 at St. Boniface, Manitoba, on a Red River barge towed by the steamer *Selkirk,* she was greeted by church bells and a gun salute. The 4-4-0 type engine had been built in 1872 by the Baldwin Engine Works of Philadelphia, and had served on an American line until she was purchased in 1877 by a contractor named Joseph Whitehead, who intended to put her on construction work in the Red River district.

Unfortunately, the locomotive did not arrive in time to take part in the ceremony of driving the first track spike in the Pembina Branch of the future CPR at St. Boniface. The Governor-General and Lady Dufferin had been invited to perform the historic task, and it had been hoped that the Countess would start the first locomotive which was named in her honour. The ceremony took place on September 29, 1877, and the *Manitoba Free Press* reported:

> *At twenty minutes past eleven the vice-regal party arrived at the station grounds, where they were met by Mr. Jas. H. Rowan, Chief Engineer of the Manitoba District, C.P.R., who conducted their Excellencies to the spot for the auspicious event.*

The famous Canadian Pacific Railway Locomotive No. 1, Countess of Dufferin, *Winnipeg, Manitoba*

Rails had been laid on the track for some distance, ready for spiking, and on the particular tie selected — one of huge proportions, and very cleanly planed were inscribed the following words:

CANADIAN PACIFIC RAILWAY
THE FIRST TWO SPIKES DRIVEN BY
THEIR EXCELLENCIES THE GOVERNOR-GENERAL AND THE COUNTESS OF
DUFFERIN 29TH SEPTEMBER, 1877.

Of course, the Canadian Pacific Railway as such was not incorporated until several years later, but the name was already widely used of the future transcontinental line, for the present being constructed piecemeal fashion by the Canadian government. Although contractor Whitehead painted "C.P.R. NO. 1" on his locomotive, the *Countess of Dufferin* did not actually come into the possession of the Canadian Pacific Railway until 1882, when she was purchased for the sum of $5,800 and was given road number 151. But to the countless railway fans who visit her year by year she will always remain the "first lady" of the Canadian Pacific.

Under Alexander Mackenzie, the government adopted a policy of linking Canada's splendid waterways and extending the rails gradually, as funds became available and local traffic began bringing revenue. Private financiers, willing to take on the task of building the transcontinental railway and risk their capital on a rather doubtful venture of such magnitude, were hard to find. Many believed it to be sheer madness to attempt construction of a line through the rugged wilderness north of Lake Superior and to try crossing such formidable barriers as the Rockies, the Selkirks and the Coastal Range on the way to the sea. Even if it could be completed, a railway through the endless, lonely prairies could never be made to pay for the great expenditure required to build it. The entire population of the northwest at the time was less than 170,000. There were some financiers who might have invested in the line, had the government consented to abandon an all-Canadian route and build part of the railway through the United States.

Sir Richard Cartwright, a prominent figure in Canadian public life, declared the undertaking was beyond the country's means and the building of the Pacific Railway would place upon every man's farm a mortgage "so heavy that it would require two or three generations to clear it off."

In England an article appeared in *Truth* stating that the "Canadian Pacific, if it is ever finished, will run through a country about as forbidding as any on earth." British Columbia was described as "a barren, cold mountain country that never should have been inhabited." Of Manitoba, the writer said, "Men and cattle are frozen to death in numbers that would rather startle the intending settler if he knew. . ."

But there were others undaunted by the gloomy pictures painted. John A. Macdonald was returned to power in 1878 on the strength of his so-called "National Policy" advocating railways, new settlement and protective tariffs for the development of the country and its resources. British Columbia, which threatened to withdraw from Confederation if railway construction in the province did not begin by the spring of 1879, was assured that the government intended to live up to its obligation. Immediate completion of the line from Port Arthur to Winnipeg was promised and, further west, contracts were let to Andrew Onderdonk, a young American railway builder, to construct the CPR through the Thompson and Fraser canyons from Savona's Ferry to Yale, and thence to Port Moody, British Columbia.

George Stephen, first president of the Canadian Pacific Railway

Work on the Yale-Savona section began in the fall of 1880, when Onderdonk's engine No. 1 arrived by barge near Yale at the head of navigation on the Fraser River. Known as the *Yale*, this locomotive was the first to operate on the CPR main line, west of the Rockies. Employed immediately in construction work, she was joined the following spring by the *Emory*, nicknamed "Curly." A little later the *New Westminster* arrived on the scene.

Labourers being scarce in the west, Onderdonk brought several thousand Chinese coolies to Canada to work on the line. Many of them later settled in British Columbia and today account in part for the large Chinese population of that province.

By the end of the ten-year deadline originally agreed upon for completion of the rail link with eastern Canada, only 264 miles of main line had actually been constructed. The government, faced with financial disaster if called upon to continue sinking millions of dollars in the railway, decided to turn the project over to private enterprise. Men willing to take the risk had at last been found. An agreement was signed on October 21, 1880 by the government of Canada and the members of a syndicate consisting of Messrs. George Stephen and Duncan McIntyre (Montreal), James J. Hill (St. Paul, Minnesota), John S. Kennedy (New York), Kohn, Reinach & Co. (Paris, France) and Morton, Rose & Co. (London, England). Sir Charles Tupper, Minister of Railways, signed on behalf of the government.

On February 15, 1881 the House of Commons passed an Act which incorporated the Canadian Pacific Railway Company. The letters patent of the CPR bear the date of February 16th, the day the Act was given royal assent. The government, under the terms of the agreement, pledged a subsidy of twenty-five million dollars to the railway in addition to a grant of twenty-five million acres of land. The lines, contracted for by the government from Port Arthur to Emerson and from Savona to Port Moody, were to be turned over to the CPR upon completion. The company was to be exempt from taxes on the land for a period of twenty years. No charters were to be granted for twenty years to any competitor seeking to build within fifteen miles of the international boundary. The company in return promised to build the line within ten years. On February 17th the CPR's first directors' meeting was held. George Stephen became the company's first president, Duncan McIntyre its first vice-president.

George Stephen, a former president of the Bank of Montreal, invested in the CPR venture despite the warnings of his associates that it would be "the ruin of us all." Some seriously believed he had lost his senses. But "railways" were his specialty. Together with his cousin, Donald Alexander Smith, who was the head of the Hudson's Bay Company, and a group of equally daring financiers which included James Hill, a Canadian residing in St. Paul, Minnesota, he had earlier purchased the bankrupt St. Paul, Minneapolis and Manitoba Railway. Recognizing the line's potential, the new owners had reorganized it and within a short time they all had become millionaires.

For political reasons dating back to the "Pacific Scandal" in the House of Commons, Donald A. Smith did not become a member of the Canadian Pacific Syndicate. But he was one of the driving forces behind the railway, and when a financial crisis loomed, it was he and George Stephen who risked their personal fortunes by endorsing promissory notes and putting up their properties and furniture as collateral for loans to help keep the railway alive.

The first shareholders' meeting of the Canadian Pacific Railway Company was held at No. 18 Parliament Street in London, England, on March 29, 1881. During a second meeting, two days later, it was decided to acquire the Canada Central Railway, a line which included the Brockville and Ottawa Railway and would bring the eastern terminus of the CPR to Ottawa and Brockville.

In April of that year the company took over the government-built lines from Selkirk to Pembina and from Selkirk to Cross Lake. A month later construction was started by the CPR and the first train steamed into Winnipeg over the Red River Bridge on August 26, 1881.

On New Year's Eve that year William Cornelius Van Horne arrived in Winnipeg to set up his headquarters as the general manager of the Canadian Pacific. He had been hired on the recommendation of James Hill and had left his post as superintendent of the Chicago, Milwaukee and St. Paul Railroad to take charge of the Canadian road. A big burly man and a huge eater who claimed to have a "Dinner Horn, Pendant, upon a Kitchen Door" as his coat of arms, Van Horne was hard as a hammer, full of boundless energy and drive. When he gave orders, he saw to it that they were carried out. The Indians soon called him "Great Chief of the Railway," and history remembers him as a human dynamo.

As his right hand in charge of purchasing the supplies for the entire road, Van Horne picked the best man he could find, Thomas G. Shaughnessy, who had been on his Milwaukee staff. Thanks to precise planning, originating in Shaughnessy's Montreal office, supply trains soon began arriving at regular intervals at the railhead, each carrying sufficient rails, ties and hardware to build one mile of track.

With Van Horne showing up unexpectedly at construction camps along the line on a flat car, in a wagon or in a caboose at any time of day or night, urging the men to work "faster," an average of 3½ miles of track were laid per day.

Tons of food were required to feed the army of five thousand railway workers and seventeen hundred teams of horses, which Van Horne employed on the prairies during the summer of 1882. Shaughnessy saw to it that supplies never ran out, and only the best cooks were hired for the construction camps.

Van Horne had promised to build five hundred miles of track before freeze-up. By October trains were running from Winnipeg to Regina, a settlement known in those days as "Pile o' Bones," for buffalo bones were the only freight it had to offer and immense piles were stacked along the railway track. Before the end of the year the railhead was located near Medicine Hat, Alberta.

CP

CP

The lifetime pass given to Chief Crowfoot by Van Horne

Natives of the Blackfeet tribe in the area put on their war paint and prepared to massacre construction gangs who dared to drive their steel into the Indian reservation. A quick-witted missionary, known as Father Lacombe, came to the rescue. He took it upon himself to promise the Indians government compensation for their land and succeeded in winning Chief Crowfoot's confidence. Bloodshed was averted, and track laying continued toward Calgary, no more than a little cattle town at the time. The first train arrived in Calgary on August 10, 1883, cheered by settlers and Mounted Police. Father Lacombe and Chief Crowfoot were invited to a dinner served in Van Horne's private car *Saskatchewan*. Company directors attending included President Stephen, who relinquished his post for one hour to Father Lacombe in recognition of the missionary's service to the Canadian Pacific. Both Father Lacombe and Chief Crowfoot were given life passes on the CPR. The Indian Chief proudly wore his pass in a leather folder around his neck until he died.

Meanwhile the Canadian Pacific acquired the Western Division of the Quebec, Montreal, Ottawa and Occidental Railway, thus gaining entry into Montreal. In the spring of 1883 the company leased the Ontario and Quebec Railway, the Credit Valley Railway and the Toronto, Grey and Bruce. With its steamer connection to Owen Sound the latter provided the link with the east over which men and material could be moved to construction sites north of Lake Superior. The northern route along the lake shore had been adopted by the CPR, in spite of vigorous objections from James Hill, who advocated the route via Sault Ste. Marie, Ontario, and St. Paul, Minnesota, as a link to the prairie section of the Canadian Pacific. As a result, he resigned from the board of directors and vowed to build his own railway empire south of the border in competition with the CPR, which is exactly what he did.

Van Horne himself referred to part of the Lake Superior section as "200 miles of engineering impossibilities." But nothing was impossible for a man like him. He hired twelve thousand labourers, offering them "two dollars and upwards a day," and put five thousand horses to work. From Chicago he imported a track-laying machine to speed up work in the mosquito-infested swamp areas. To blast his way through the solid rock of northwestern Ontario, he built three dynamite factories on the spot which produced over a million dollars' worth of

Joseph Whitehead, *4-4-0 eight-wheeler type, Engine No. 2 belonging to contractor Joseph Whitehead; built in 1872, sold to the CPR as No. 144 in 1883, scrapped in 1902*

explosives for the job. Three miles of track in the Jack Fish Bay district cost the staggering sum of $1,200,000. On their way west towards Rat Portage (now Kenora), construction workers watched seven successive sets of tracks being swallowed by the muskeg.

By that time the company's coffers were becoming empty. A twenty-million-dollar government loan was merely a drop in the bucket but for a while it kept the railway going. "It may be that we must succumb," Donald Smith said when finances were at their lowest ebb, "but not while we individually have a dollar." Van Horne continued his feverish pace and offered to complete the line five years ahead of schedule.

As the operational mileage of the CPR increased, the general manager devoted some of his energy to creating profitable traffic. In Winnipeg and at the Lakehead he erected grain elevators. West of Winnipeg, which was already beginning to flourish, he laid the groundwork for ten model farms, and at Lake-of-the-Woods he constructed a flour mill that eventually grew into one of the largest in the world. Foreseeing the days of busy tourist traffic on the Canadian Pacific, he made plans for a string of railway hotels. The famous Banff Springs Hotel had its origin in a mountain chalet erected for the CPR's passengers in 1887. Château Frontenac, long since a landmark of Quebec City and known to travellers from all over the world, was opened in 1893.

In the Rockies, survey crews and location parties had been at work since 1883, and the railhead reached the summit near Lake Louise by the end of that year. From there the steel moved westward through Kicking Horse Pass and down "Big Hill" where tracks, descending from the top of the pass to Field, British Columbia, dropped 237½ feet per mile. The Big Hill was once the terror of engine drivers, particularly in the winter when temperatures sometimes hit 40 degrees below zero and slippery tracks added to the hazards of runaway engines. Rockslides and avalanches were another danger the rugged railroader of the Mountain Division feared. To haul a fifteen-car train up the hill required four heavy locomotives, two to push and two to pull. The first work train going down Big Hill after the track had been completed "ran away" and plunged into the Kicking Horse River below.

While construction workers pushed ahead, Major Rogers searched for a pass through the Selkirks. When news of his discovery reached the camp at the mouth of the Beaver River, engineers found that the route he had chosen would cut seventy-seven miles from the original survey.

The railway's conquest of nature's giant barriers called not only for engineering skill and equipment far beyond the imagination of the nineteenth century, but for expenditure which reached staggering proportions. Many a mile of road in the mountains consumed more than half a million dollars. By the spring of 1885 money had run out again. On the verge of bankruptcy, the builders of the Canadian Pacific Railway faced their darkest hour. Had it not been for the second Northwest Rebellion of Louis Riel, the government might not have come to the company's rescue. When the uprising in the west threatened to get out of control, Van Horne offered to move militia and supplies from the east to the trouble spot. He brought a small army over his railway from Ottawa to the prairies in less than five days, using marches and water transportation where the track had not yet been completed.

Thus convinced of the railways' great value to the nation, parliament on July 20, 1885, passed the Canadian Pacific Bill granting a

A publicity photograph depicting the NWMP guarding the railhead of the new CPR line into the prairies

five-million-dollar loan and sanctioning a thirty-five-million-dollar bond issue. George Stephen went to London and found to his amazement that the head of the financial house of Baring Brothers, Lord Revelstoke, who once had been vigorously opposed to the Canadian Pacific venture, offered him 91 cents on the dollar for the entire issue. A grateful CPR later named its division point at the Columbia River crossing "Revelstoke."

On September 26, 1885 Andrew Onderdonk discharged his men after having laid the last rail in the Eagle Pass, completing his contract for the western end of the line. Eleven days later the tracks from the west and the east were joined at Craigellachie, and all that remained to be done was to put the Rocky Mountain section into shape for the inauguration of regular traffic. The CPR now operated over 4,300 miles of road, nearly 2,900 miles of which comprised the main line.

At 8 o'clock on the evening of June 28, 1886 the first regular through train, the *Pacific Express*, left Montreal's Dalhousie Square Station for Port Moody, British Columbia, which was the western terminus of the line. Included in the train was the dining car *Holyrood* and two sleeping cars, *Yokohama* and *Honolulu*. Exactly as scheduled, the train arrived in Port Moody at 12 o'clock noon of July 4th after a transcontinental journey of five and a half days. The same journey by other means of transportation ten or fifteen years earlier would have taken half a year.

On August 6, 1886 a ten-car train from Port Moody steamed into Montreal after having carried the first shipment of tea from Yokohama, Japan, over the CPR to eastern markets. The brig *W. B. Flint* which had brought the cargo across the Pacific was the forerunner of a large fleet of steamships operating under the Canadian Pacific. Regular service was inaugurated by the company with Yokohama and Hong Kong in 1887, using three chartered steamships. Four years later the first three ships of the CPR's famous "Empress" fleet were built and put on the Pacific route. By the early 1900s the company's ships also plied the Atlantic and, as its builders once had promised, the CPR was "spanning the world."

Newspapers of the 1880s carried glowing descriptions of the breathtaking wonders seen for the first time by passengers from eastern Canada, travelling along the scenic mountain route of the Canadian Pacific. Undoubtedly, one of the most vivid accounts of the journey was that of Lady Macdonald, wife of Canada's prime minister, in an article published in England in *Murray's Magazine*, under the title "By Car and by Cowcatcher." The passengers on the luxurious scenic-domed transcontinental train, *The Canadian*, inaugurated on April 24, 1955, could not have enjoyed their trip through the Rockies more thoroughly than did the nation's First Lady back in 1886.

That year Lady Macdonald accompanied her husband on his first transcontinental trip to the Pacific coast. In the Rockies she insisted on exchanging the comfortable luxury and relative safety of their private car for a seat in front of the engine. A candle-box was securely bolted to the dusty platform of the cowcatcher, and there she sat, with a robe over her knees, riding through Kicking Horse Pass, down hazardous Big Hill, through Rogers Pass and on to Port Moody, six hundred miles in all, getting a "front seat" view of the scenery, happy and oblivious to the danger of her precarious position. In her article in *Murray's Magazine* Lady Macdonald described the experience:

> *With a mighty snort, a terribly big throb, and a shrieking whistle No. 374 moves slowly forward. The very small*

Manitoba Archives

Arrival of the first transcontinental train at the CPR Station, Winnipeg, Manitoba, 1886

population of Laggan have all come out to see. They stand in the sunshine and shade their eyes as the stately engine moves on. "It is an awful thing to do!" I hear a voice say, as the little group lean forward; and for a moment I feel a thrill that is very like fear; but it is gone at once, and I can think of nothing but the novelty, the excitement, and the fun of this mad ride in glorious sunshine and intoxicating air, with magnificent mountains before and around me, their lofty peaks smiling down on us, and never a frown on their grand faces!

The pace quickens gradually, surely, swiftly, and then we are rushing up to the summit. We soon stand on the "Grand Divide"—5,300 feet above sea level—between the two great oceans. I look and lo! the water flowing eastward towards the Atlantic side, turns in a moment as the Divide is passed, and pours westward down the Pacific slope!

Another moment and a strange silence has fallen around us. With steam shut off and brakes down, the 60-ton engine by its own weight and impetus alone, glides into the pass of the Kicking Horse River, and begins a descent of 2,800 feet in twelve miles. We rush onward through the vast valley stretching before us, bristling with lofty forests, dark and deep, that, clinging to the mountain side, are reared up to the sky. The river, widening, grows white with dashing foam, and rushes downwards with tremendous force.

No doubt, the engine driver was less enthusiastic over the trip, being responsible for his passenger's safety. Just what the prime minister's feelings were about his wife's adventurous escapade, is not altogether known. Certainly Lady Macdonald was enthralled, particularly by the ride through Rogers Pass to Revelstoke.

Every possible sense of fear is lost in wonder and delight. Perhaps no part of the line is more extraordinary an evincing of daring engineering skill, than the pass, where the road bed curves in loops, over trestle bridges of immense height, at the same time rapidly descending. In six miles of actual travelling the train only advances two and a half miles, so numerous are the windings necessary to get through the canyon.

The wild and magnificent canyon of Illecillewaet is now leading us to Revelstoke, at the second crossing of the Columbia. The

The Canadian Pacific Railway Co'y

EMIGRATION TO MANITOBA
AND THE
CANADIAN NORTHWEST.

Sale of Lands.

To encourage the rapid settlement of the Country, the Canadian Pacific Railway Company will be prepared, until further notice, to sell lands required for Agricultural purposes at the low price of $2.50 an acre, payable by instalments, and will further make an allowance by way of rebate from this price, of $1.25 for every acre of such lands brought under cultivation within three to five years following the date of purchase, according to the nature and extent of the other improvements made thereon.

The lands thus offered for sale, will not comprise Mineral, Coal or Wood lands, or tracts for town sites and Railway purposes.

Contracts at special rates will be made for lands required for cattle raising and other purposes not involving immediate cultivation.

Intending settlers and their effects, on reaching the Company's Railway, will be forwarded thereon to their place of destination on very liberal terms.

Further particulars will be furnished on application at the offices of **The Canadian Pacific Railway Company**, at Montreal and Winnipeg.

By order of the Board,
CHS. DRINKWATER,
Secretary.
Montreal, April 30th, 1881. 104

CP

Red Sucker Trestle, north shore of Lake Superior, Ontario

river rising in Rogers Pass, pours through a ground defile in wild rapids, enclosed by majestic mountains, surging past rocky and gravelly shores . . .

Told by the train crew that one of the many tunnels in the Gold Range was "wet" because of spring water pouring out of the arching rocks onto the train, Lady Macdonald, by then the sole occupant of the cowcatcher, "with praiseworthy economy," took off her hat, tucked it safely under her wrap and put up her umbrella. Emerging from the tunnel, she encountered a group of young English sportsmen standing near the roadside:

They have evidently just climbed the bank guns in hand, leaving a large canoe with two Indian paddlers on the lake below . . . Just imagine the feeling with which these well-regulated young men beheld a lady, bareheaded and with an umbrella, seated in front of an engine at the mouth of a tunnel in the Gold Range of British Columbia! I am sorely afraid I laughed outright at the blank amazement of their rosy faces, and longed to tell them what fun it was; but not being introduced, you know, I contented myself with acknowledging their presence by a solemn little bow which was quite irresistible under the circumstances!

Neither the discomfort of wind and dust nor the danger of rock slides and wild animals could induce Lady Macdonald to leave her seat and seek the shelter of her car. Even at night she steadfastly remained on her "observation deck":

We soon find darkness closing around us in that tremendous canyon which for twenty-seven miles holds the Fraser in its depths, and along the side of which we now travel, mere specks in the vast solitude of mountain precipice, black with wild rugged rocks, and awful with immense shadows. The train proceeds slowly. A lookout man sits with me on the buffer beam and the Comptroller, unmindful of his interesting young family at home in Ottawa, with an admirable sense of duty, shares with us the risks of that night ride along the Fraser . . .

Arriving at Port Moody, the CPR's most unusual passenger had but one regret: "I must bid good-by to candle-box and cowcatcher, and content myself with an easy chair on the deck of a steamer bound for Victoria."

CP

Canadian Pacific Railway Station in Regina, built in 1882

Arrival of the first transcontinental passenger train in Vancouver, May 23, 1887

CP

Engine 285. The first engine built by CPR (opposite)

During the spring of 1887 the main line of the CPR was extended a little over twelve miles from Port Moody, along Burrard Inlet to the new town of Vancouver, which had been incorporated a year earlier and shortly after that had burned to the ground. It was the construction of the CPR which led to the birth of Vancouver, today Canada's largest seaport and her third largest city. On an inspection tour in 1884 William Van Horne had chosen the site as the future terminus. He named it "Vancouver," and predicted that there would be "a great city right there with steel tracks carrying endless trains of freight and passengers, an all-year port with fleets of vessels trading with the world." When the first transcontinental train reached Vancouver on May 23, 1887, the town had a population of about two thousand. CPR engine No. 374 which pulled the train into the station is still preserved and on display at Kitsilano Beach in Vancouver.

With its transcontinental line completed, the Canadian Pacific turned its attention to colonization, and the story of the railway soon became synonymous with the development of the west. In the prairies,

Canadian Pacific train on high level bridge, Lethbridge, Alberta

CP

land from the company's government grant was offered to settlers for $2.50 an acre. Half of this was to be refunded on every acre that was ploughed. At one time Van Horne set up a "real estate office" in the hollow of a big tree. An even more intensive campaign was launched in Europe, and a tide of immigrants arriving in Canadian ports followed the steel westward and settled the land. Over the years vast sums of money have been spent by the Canadian Pacific on land settlement, colonization and irrigation projects to create prosperous agricultural areas. When the bumper wheat crop of 1891 ripened in the prairies and hands were urgently needed to bring in the harvest, the CPR advertised for labourers. To enable them "to reach the bountiful harvest of Manitoba and the North-West," it offered low one-way rates of fifteen dollars from any station in Ontario. Thousands answered the call. The first of the Harvest Specials from eastern Canada left for the west on July 28, 1891, and the practice continued as a yearly feature until the 1920s. A trip in the CPR's early colonist cars with their wooden seats may not have been the most luxurious one but, if it were a harvest excursion, it was bound to be a gay and boisterous stag party on wheels.

Two years after its incorporation the Canadian Pacific embarked on a vast programme of building locomotives, many of which were noted for their outstanding quality and long-time service. The first locomotive built by the CPR left the company's Delorimier Avenue shops at Montreal in 1883 and began a period of service that was to last for thirty-seven years. A standard 4-4-0 type, bearing the number 285, she became the forerunner of many highly reliable engines produced by the company.

The oldest of the CPR's eight-wheelers, until recently at the Canadian Rail Transportation Museum in Delson near Montreal, was locomotive No. 144. Built in 1886, she worked first out of Winnipeg and in her later years was transferred to the Chipman-Norton branch in New Brunswick. By the time she retired she had chalked up a record of nearly one and a quarter million miles. And speaking of long-time service, No. 29, another of the CP locomotives on display at Delson, worked from 1887 to 1960. On November 6th of that year she operated a special from Montreal to St. Lin, Quebec, in connection with the seventy-fifth anniversary celebration of the Canadian Pacific. It was the last train on the CPR to be pulled by a steam locomotive.

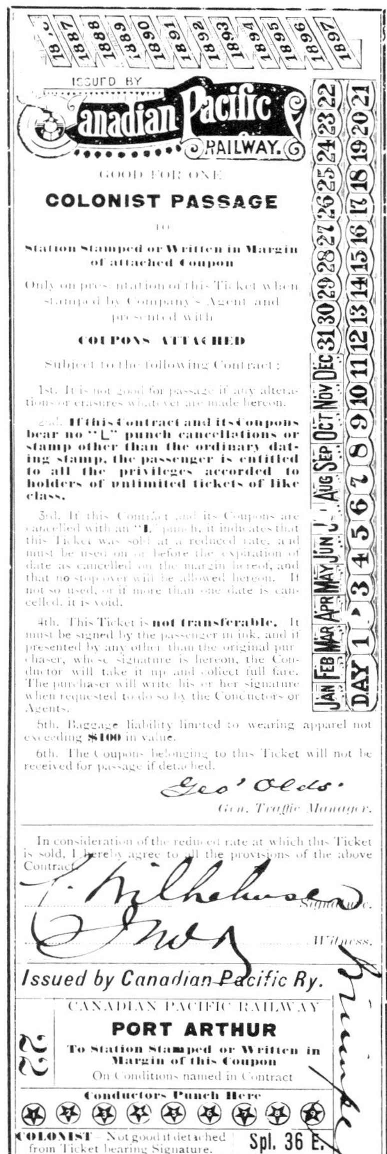

1886 1887 1888 1889 1890 1891 1892 1893 1894 1895 1896 1897

ISSUED BY
Canadian Pacific Railway.

GOOD FOR ONE

COLONIST PASSAGE

TO

Station Stamped or Written in Margin of attached Coupon

Only on presentation of this Ticket when stamped by Company's Agent and presented with

COUPONS ATTACHED

Subject to the following Contract:

1st. It is not good for passage if any alterations or erasures whatever are made hereon.

2nd. **If this Contract and its Coupons bear no "L" punch cancellations or stamp other than the ordinary dating stamp, the passenger is entitled to all the privileges accorded to holders of unlimited tickets of like class.**

3rd. If this Contract and its Coupons are cancelled with an "**L**" punch, it indicates that this Ticket was sold at a reduced rate, and must be used on or before the expiration of date as cancelled on the margin hereof, and that no stop-over will be allowed hereon. If not so used, or if more than one date is cancelled, it is void.

4th. This Ticket is **not transferable.** It must be signed by the passenger in ink, and if presented by any other than the original purchaser, whose signature is hereon, the Conductor will take it up and collect full fare. The purchaser will write his or her signature when requested to do so by the Conductors or Agents.

5th. Baggage liability limited to wearing apparel not exceeding **$100** in value.

6th. The Coupons belonging to this Ticket will not be received for passage if detached.

Geo' Olds.
Gen. Traffic Manager.

JAN FEB MAR APR MAY JUN JUL AUG SEP OCT NOV DEC
DAY 1 2 3 4 5 6 7 8 9 10 11 12 13 14 15 16 17 18 19 20 21 22 23 24 25 26 27 28 29 30 31

In consideration of the reduced rate at which this Ticket is sold, I hereby agree to all the provisions of the above Contract.

.... *Signature.*

.... *Witness.*

Issued by Canadian Pacific Ry.

CANADIAN PACIFIC RAILWAY
22
PORT ARTHUR
To Station Stamped or Written in Margin of this Coupon
On Conditions named in Contract
Conductors Punch Here
COLONIST Not good if detached from Ticket bearing Signature. Spl. 36 E.
Via C.P. Main Line.

Engine No. 30 was another eight-wheeler built in 1887. After sixty years of continuous service, hauling freight and passengers in the Province of Quebec, she was still in beautiful shape and climaxed her career by becoming a "film star" in the movie *Canadian Pacific* made in the Banff area.

During the first four decades of railroading 4-4-0 locomotives were the most widely used type of engine on Canadian roads. By the end of 1887 the CPR owned close to four hundred of these eight-wheelers. In the late 1880s the company introduced the "Mogul" type, a locomotive with a 2-6-0 wheel arrangement. Found too light for transcontinental service, this locomotive proved not very successful and as a result only about fifty of this type were operating on CPR lines.

Considerably more popular were the so-called ten-wheelers, the first of which was built by the Canadian Pacific in 1889. These engines performed extremely well, both in passenger and freight service, and the company owned nearly one thousand of the 4-6-0 type locomotives over the years.

CP

CPR Engine 434, Mogul type 2-6-0

CPR Engine 29, a 4-4-0 type built in 1882

CP

CPR Engine 894, ten-wheeler type 4-6-0

Towards the latter part of the century competition with other lines focused attention on speed, particularly in the Montreal-Ottawa district. As a result, three Atlantic engines were built with a 4-4-2 wheel arrangement. They are CPR Nos. 209, 210 and 211, well-known among steam buffs for their extra high driving wheels which measured 84″ in diameter.

To cope with heavy passenger service and provide the motive power required to move trains weighing five hundred tons or more, the 4-6-2 Pacific-type engines were developed. The Pacifics with their large cylinders, greater boiler capacity and tremendous tractive power, are considered the first successful modern steam locomotives. They not only were sleeker in appearance than some of the earlier types, but one of the most important factors was a considerable weight reduction which allowed their use on restricted trackage.

Among the CPR's less popular locomotives were six Mallets, each actually "two engines in one," with a pair of cylinders on each side and six driving wheels on either end. They were eventually modified to the 2-10-0 type and put into service as transfer engines in Montreal. Then there were the Decapod and the Santa Fe locomotives, built for freight service but discontinued because of their complicated machinery and high cost of maintenance.

A more outstanding freight engine was the "Mikado," developed in 1912 mainly for the purpose of hauling heavy trains on the steep gradients of the Rockies. It performed so successfully that the company continued to order them for more than thirty years. In all, over three hundred of the Mikados served on CPR lines.

The "Hudson" has the distinction of having been one of the most handsomely designed locomotives of the Canadian Pacific. As a high-speed passenger engine she easily surpassed all her predecessors. Her 4-6-4 wheel arrangement permitted a larger boiler with an increased heating surface and consequently increased steam pressure. Heavier than the Pacific, she also possessed a greater tractive effort.

The first of the Hudsons was built for the CPR by the Montreal Locomotive Works in 1929. Having acquired ten of these engines, the company immediately ordered another ten, identical in every respect except for No. 2811 and No. 2813; these were equipped with boosters increasing their weight, making it possible to exert an additional twelve thousand pounds of tractive effort. Later Hudsons, purchased by the CPR, became more streamlined. They were outfitted

CP

CPR Engine 2623, Pacific type 4-6-2

with glossy black jackets to enclose some of their external fittings and featured cowelled stacks with smoke deflectors, recessed headlights and solid pilots. The sides of the boiler and the tender were painted in a deep red with lettering done in gold.

Hudson locomotive No. 2850 became famous for operating the royal train over Canadian Pacific rails, from Quebec City to Vancouver, during the visit of King George VI and Queen Elizabeth in 1939. The pilot train over the 3,224-mile-long line was also headed by a Hudson locomotive and both engines performed flawlessly throughout the entire trip, which required no less than twenty-five changes of engine crews. In commemoration of the historic journey, Hudson locomotives Nos. 2820 to 2864 have since been known as "Royal Hudsons," and each of these engines displays a crown on the front end of the running board. Number 2850 was exhibited later at the World's Fair in New York. Today one of the Royal Hudsons, No. 2860, is on display in Vancouver.

One of the fastest steam engines on the Canadian Pacific road was the "Jubilee" with a 4-4-4 wheel arrangement. She was designed for the purpose of operating relatively short, but extremely fast trains on CPR lines traversing regions with a widely scattered population. Five of the Jubilees were built in 1936 by the Montreal Locomotive Works, and twenty similar engines were ordered from the Canadian Locomotive Company in Kingston, Ontario, the following year.

Then there were two "Northern" 4-8-4 type engines developed by the CPR's Angus Shops in Montreal during the late 1920s. They were the first locomotives in Canada to have one-piece cast-steel frames. Cylinder and valve chest casings, as well as the tender frames, were also single castings resulting in additional strength. Designers had attempted to produce a low-weight engine with high tractive effort. Although some of their new design features influenced the development of future steam locomotives, the two Northern engines, No. 3100 and No. 3101 remained the only ones of this type on the Canadian Pacific roster. Both were employed between Montreal and Toronto until diesels took over the reign on this stretch in 1954. Converted to oil-burners, the two Northern engines finished out their service years in the West.

The first of the CPR's well-known "Selkirks" was No. 5900, built by the Montreal Locomotive Works in 1929 for service on heavy gradients of troublesome mountain trackage. The Selkirks soon set new standards of performance in the Rockies. Consequently, an experi-

CP

CPR Engine 2850, Royal Hudson type 4-6-4

mental engine, No. 8000 on the roster, similar in most aspects to the successful 5900s, but operating with multiple boiler pressure, was tested in Revelstoke, British Columbia. Although No. 8000 proved to be more economical than conventional locomotives, she was eventually scrapped, as the complexity of her boiler system made repairs difficult and costly.

The last new steam locomotive acquired by the Canadian Pacific was a Selkirk. Number 5935, built by the Montreal Locomotive Works and delivered to the CPR on March 12, 1949, is now among the exhibits in the museum at Delson, Quebec. Accompanied by Mikado engine No. 5468, she travelled from Calgary, Alberta, to Delson not under her own steam but as part of a freight train hauled by the diesels.

Back in 1882 locomotives and rolling stock of the Canadian Pacific totalled seven hundred and fifty one. Half a century later the number had risen to nearly one hundred and four thousand.

If the four thousand bridges on the CPR line could have been placed end to end, they would have measured close to seventy miles in length. Among the most famous bridges in the world is the CPR's viaduct on the Crow's Nest Branch at Lethbridge, Alberta, completed in 1909. It is 5,327 feet long, and with a maximum height of 314 feet above the water of the Belly River, it is the highest railway bridge in Canada.

The two spiral tunnels between the Great Divide and Field, British Columbia, opened for traffic in August of 1909, were engineering marvels of the day. Eliminating the Big Hill with its emergency switches, spur lines and frequent "run-away" engines, the Upper Tunnel cuts into Cathedral Mountain, turns 234 degrees and, passing underneath itself, re-emerges again 48 feet lower down. The track then turns east and, after crossing the Kicking Horse River, enters the Lower Tunnel which spirals into Mount Ogden, turns 232 degrees and exits 45 feet below its entrance. Passengers in the last car are able to see the engine of their train emerge before they enter the tunnel.

On December 9, 1916 the CPR inaugurated the longest railroad tunnel in Canada. Replacing the old route over Rogers Pass in the Selkirks, the double-tracked Connaught Tunnel, slightly over 5 miles in length, shortened the rail distance by 4½ miles, lowered the summit by 540 feet, and eliminated curvature, equalling seven full circles, along with four miles of snowsheds.

CP

CPR Engine 3004, Jubilee type 4-4-4

In 1888 William Cornelius Van Horne, the giant who had pushed the steel of the Canadian Pacific westward with dynamic force and an iron will, succeeded George Stephen (later Lord Mount Stephen) as the second president of the CPR. He continued to create local traffic, planned new lines and acquired branch and feeder lines to add to the CPR's rapidly growing system. Knighted in 1894, he held the presidency for eleven years before becoming chairman of the board of directors, a post from which he resigned in 1910. He died at Montreal on September 11, 1915.

Thomas George Shaughnessy, purchasing agent during the construction years, and a man whose extraordinary business ability was internationally acknowledged, earned himself the title of "King of Railway Presidents" as the third to rule the empire of the Canadian Pacific. He was created a Knight Bachelor in 1901, a K.C.V.O. in 1907, and raised to the peerage as Baron Shaughnessy of Montreal in 1916.

His successor, Edward Wentworth Beatty, a brilliant lawyer, was the first Canadian-born president of the CPR. Educated at Toronto University and Osgoode Hall, he was called to the Ontario Bar in 1901 and appointed Assistant in the Law Department of the Canadian Pacific the same year. In 1918 at the age of forty-one, he was elected president of the CPR. In 1924 he succeeded Lord Shaughnessy as chairman of the board, retaining both his new office and the presidency until 1942. The men who have since served as presidents and chairmen of the Canadian Pacific include D. C. Coleman, W. M. Neal, W. A. Mather and Norris Roy Crump.

"Buck" Crump, as he was known to his boyhood friends, grew up beside the CPR tracks with the railroad fever in his blood. He quit school at fifteen to work for the CPR on the Revelstoke repair track for forty cents an hour, became a machinist's apprentice at Field, British Columbia, and later transferred to the CPR's shops at Winnipeg, where he worked to pay for a university course to become a mechanical engineer. In the years to follow as he climbed the ladder, learning every aspect of how a railway works, he travelled many times back and forth over the CPR's 21,000 miles, and personally met a large number of the company's 87,000 employees whom he was to head at the age of fifty. In 1961 he became chairman of the board, a post he still retains, after having resigned as president in 1964.

R. A. Emerson became the next president of the Canadian Pacific. He in turn was succeeded in 1966 by Ian David Sinclair, a

CP CP

distinguished lawyer from Winnipeg, who now guides the destiny of the world's largest privately owned railway company, which long since has grown into much more than a transportation network of vast proportions.

Today CP is one of Canada's biggest single industries. Its business interests and investments include steamships on the oceans of the world, the largest trucking company in the country, an international airline, palatial hotels in large centres from coast to coast, real estate holdings and mining enterprises. Some of its non-railway ventures are now grouped under the ownership of a subsidiary known as Canadian Pacific Investments Limited.

The railway, which early critics had feared "would never pay for its axle-grease," paid back government loans, received during construction, in full with interest less than a year after the line was completed. Throughout the years of its eventful history the company rarely failed to make a profit. With assets worth billions of dollars, Canada's first transcontinental line today spans the world by air, sea and rail.

CP

CPR 5921, Selkirk type 2-10-4, freight and passenger engine (top left)

Mountainous terrain caused difficulties in construction and operation. The locomotive of a modern CP freight train (below) emerges from one of the spiral tunnels before the back of the train has entered at the upper level

The TransCanada Limited leaving Windsor Station, May 1919

CP

Canadian Northern Railway

The Canadian Northern Railway, which by the end of 1915 operated Canada's second transcontinental line and possessed a system consisting of nearly ten thousand miles of track, started out as a small pioneer railway on the prairies.

The story of its astonishing growth is the story of two enterprising Canadians, William Mackenzie and Donald Mann, who back in 1896 purchased the charter of the Lake Manitoba Railway and Canal Company. With it they acquired a federal land subsidy of 6,000 acres per mile for a total of 125 miles. Having persuaded the Manitoba government to grant exemption from taxes and guarantee the company's bonds to the tune of $8,000 per mile, they set out at once to build a railway from Gladstone, northwest of Portage la Prairie, via Dauphin to Lake Winnipegosis, a distance of approximately 125 miles.

In December of 1896 trains began running between Gladstone and Dauphin, and soon after the first timetable of the Lake Manitoba Railway and Canal Co., effective January 3, 1897, announced service between Gladstone and Sifton, sixteen miles north of Dauphin. Running rights had been secured over the Manitoba and North-Western Railway between Gladstone and Portage la Prairie on the main line of the Canadian Pacific, and mixed trains were operating twice a week. The line was run with a total of fourteen employees and a minimum of equipment, most of it purchased secondhand.

Born in the little Ontario town of Kirkfield near Balsam Lake, William Mackenzie had been a schoolteacher and occasional storekeeper before getting involved in building railways, first for the Grand Trunk and later for the Canadian Pacific. Donald Mann was born on a farm at Acton near Toronto and worked in the lumber camps of northern Ontario and Michigan before becoming a railroad contractor in Winnipeg, where he constructed sections of the Canadian Pacific, which at the time was pushing its steel towards Calgary. The two men joined forces in the 1880s forming a partnership in the railway contracting business. Encouraged by the expanding grain traffic on the prairies and the financial success of the Canadian Pacific, they decided to build a railway of their own.

Brimming with energy and ambition, Mackenzie and Mann had no sooner finished their railway to Winnipegosis than they embarked on their next project, a line from Winnipeg eastward to Port Arthur at the Lakehead, in competition with the Canadian Pacific. With immigrants from the British Isles and other European countries flocking to the Prairie Provinces, the two promoters found it not hard to convince provincial and federal governments of the urgent need for new railways to serve the colonists. Farmers, dissatisfied with existing high freight rates, clamoured for additional outlets to break the monopoly still enjoyed by the Canadian Pacific.

Construction of the Winnipeg-Port Arthur line was to be carried out under the charters of the Manitoba and South-Eastern and the Ontario and Rainy River railways. The first stretch between St. Boniface and Marchand, a little over forty-five miles, opened for traffic in November 1898. Relying mainly on poplar, jack pine or tamarack from the swamps for its freight, the regular train on this line was dubbed the "Muskeg Special" by the residents of Winnipeg who used much of the cordwood for their winter fuel. The initial rolling stock of the railway included two locomotives, two used passenger coaches, a number of secondhand flat cars and fifty new freight cars.

On the prairies, Mackenzie and Mann's interests eventually included the Winnipeg Great Northern to Lake Manitoba. Originally incorporated as the Winnipeg and Hudson's Bay Railway and Steamship Co., the charter carried a land grant of four million acres and a provincial guarantee for the company's bonds. In 1899 the Lake Manitoba Railway and Canal Co. and the Winnipeg Great Northern were amalgamated and became known as the Canadian Northern. Extending westward, the railway reached Saskatchewan in 1900.

A wave of expansion by then was sweeping the country, based on new immigration policies, followed in turn by a spectacular boom in the west to which the Canadian Northern Railway contributed a considerable share. It encouraged settlers to go farther north than ever before by opening up large areas of virgin land on the Great Plains, and gathering with its feeder lines the grain from the great farming districts which developed in the wake of railway construction. In the ensuing years it gave life and impetus to hundreds of small communities and shipping points in the territories it traversed.

In 1901 Mackenzie and Mann acquired approximately three hundred and fifty miles of track in the Brandon-Winnipeg-Emerson district, originally subsidiary lines of the Northern Pacific Railway, an American company which had been allowed by the Province of Manitoba to build in the area after the federal government had cancelled the monopoly rights of the Canadian Pacific. When the Northern

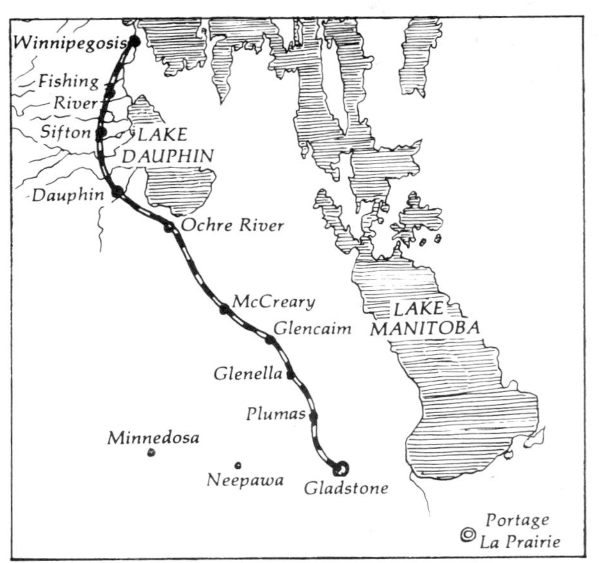

Lake Manitoba Railway and Canal Co.

Pacific went into bankruptcy, its Canadian lines were leased for 999 years by the Manitoba government, which in turn offered them to private companies for operation. Both the Canadian Pacific and the Canadian Northern were keenly interested in securing the attractive lease. Mackenzie and Mann won by agreeing to certain freight reductions on the line which forced the CPR eventually to follow suit—much to the prairie farmers' satisfaction. With the new addition, the Canadian Northern procured terminal facilities in Winnipeg and a connection with St. Boniface was ready in 1901.

A year later, on January 1st, the two railway builders jointly drove the last spike into the rails at Hanna's Point near Fort Frances on their Winnipeg-Lake Superior line. Several feet of rails were still missing when the official party arrived in a special three-car train from Port Arthur. After the ceremony the train proceeded to Winnipeg inaugurating traffic between the Lakehead and the fast-growing prairie centre. The cars, incidentally, had been supplied for the occasion by the Canadian Pacific to its younger rival. An important milestone had been reached in the development of the Canadian Northern. At Port Arthur an immense new grain elevator was waiting to receive the wheat from western Canada for transfer to Great Lakes vessels.

The Canadian Northern continued to grow both in mileage and scope, leasing and absorbing railways or constructing new links in its spreading system at an ever increasing pace. In the west it pushed into areas of central Saskatchewan where the climate once had been considered too severe for raising wheat by the scattered settlers who had tried their luck. But fast-ripening varieties, resistant to frost, were being introduced by agricultural experts and helped turn the province into one of the richest granaries in the world. Trains of the Canadian Northern brought in new colonists and supplies and carried away the bountiful harvests.

Edmonton, the capital of Alberta, on the main line of the Canadian Northern, was reached in 1905. Soon Mackenzie and Mann's roads of steel, consisting of feeder lines and connecting links, served such centres as Prince Albert, Moose Jaw, Regina, North Battleford and Calgary. Original construction was usually carried out in the cheapest possible way, to get traffic going and bring in revenues.

Meanwhile the two partners had begun to dream of even bigger things. Their holdings had long since grown into the third largest railway system in the country. But with their mileage increased their

appetite for more profitable ventures. As a preliminary step towards their ultimate goal, they acquired a short line in Ontario out of Parry Sound to a junction with the Canada Atlantic Railway. In Quebec they absorbed the Great Northern, and the Chateauguay and Northern railways which would give them the beginning of an Ottawa-Montreal-Quebec connection.

In Nova Scotia Mackenzie and Mann controlled among other interests the Halifax and Southwestern, and a railway was subsequently constructed between Halifax and Yarmouth via Lunenburg.

What the owners of the Canadian Northern had in mind, however, was not merely a patchwork of lines in various provinces, but a ribbon of steel tying together the major centres from coast to coast, with their prairie system supplying enough freight going east and west to make the new transcontinental road a paying proposition.

The fact that the eastern-based Grand Trunk also was anxious to share in the boom of the western provinces and contemplated the building of a transcontinental railway of its own, by no means frightened determined men like Mackenzie and Mann. Confident of success, they pursued their plans. Government attempts to reach some kind of compromise on the subject of a transcontinental railway between the Grand Trunk and the Canadian Northern failed. Suggestions to build a joint line eastward from the Lakehead may have vaguely appealed to the Canadian Northern, but it found no favour with Charles Melville Hays, the vigorous general manager of the Grand Trunk.

With Canada's population continuing to grow and her economy expanding in a spectacular way, the future looked bright. There were some who predicted that the country would have a hundred million people by 1920. In this general atmosphere of buoyant optimism, Ottawa was easily persuaded to sanction both the Grand Trunk and the Canadian Northern transcontinental schemes.

While Donald Mann was the highly competent construction expert of the team, William Mackenzie was the financial wizard, who had never yet failed to raise the necessary capital for their new expansion projects.

A master of political diplomacy, he succeeded in getting municipal, provincial and federal loans, subsidies and guarantees for large portions of the company's bonds. The extent of public assistance to the Canadian Northern eventually amounted to a total of a quarter-billion dollars. While government land grants had since been discontinued as a form of railway aid, many of the existing charters taken over in the course of time by the Canadian Northern carried valuable subsidies of this nature. Mackenzie and Mann's chief source of financing, however, was government-backed bond issues readily floated on the money markets of the world. As a result they were able to build their railway empire without investing anything but their organizational talents, their wide experience, skills, courage and resourcefulness. In return for their services as managers of the railway they retained the shares issued by the company, and thus they alone controlled their gigantic enterprise. For their contributions to Canada's development, both William Mackenzie and Donald Mann were subsequently knighted.

On its way westward from Edmonton the Canadian Northern pushed through the Yellowhead Pass, following Sir Sandford Fleming's route surveyed years earlier for the Canadian Pacific. Here the altitude was lower and construction easier than on Kicking Horse Pass, which had been chosen by the CPR to get through the Rocky Mountains. Heavy freight trains would require no extra engines in the mountains. Nevertheless, there were plenty of engineering difficulties to overcome, rock to be moved, streams and gorges to be bridged. Costs soared far beyond original estimates, particularly in the Fraser Canyon where cables had to be slung across the river to transport men and material in slings and buckets to the work site. Before reaching Vancouver, thirty-nine tunnels had to be constructed, most of them pierced through solid rock. Rain and melting snow in the spring swelled the rivers and often wiped out unfinished sections of the road.

In Ontario the Canadian Northern was building from Toronto via Belleville to Ottawa, and from Toronto via Parry Sound on Georgian Bay to the Northern Ontario mining centre of Sudbury. Tracks were being laid through the rocky wilderness of the Lake Superior section, with the line from Port Arthur eastward running in a northerly curve to Capreol and thence straight to Ottawa, avoiding the roundabout way via Toronto.

To complete the missing links in the transcontinental system, more money was urgently needed on top of lavish financial aid which had been received by the Canadian Northern throughout its years of feverish expansion activities. In 1913 Ottawa was asked once more for assistance, with the assurance that it would be the company's last appeal. But another more serious financial crisis arose a year later. Funds from the world's money markets had suddenly dried up with the threat of World War I. The flow of immigrants from Europe ceased temporarily, and so did the boom in western Canada.

If only to protect its prior investments in the Canadian Northern, the Canadian government once again had to come to the railway's rescue. After much debate a "positively last" guarantee of bonds was granted to the extent of forty-five million dollars. In return, the government received forty million dollars of the company's common stock and the right to appoint one director. But the following year the Canadian Northern returned, "begging" again.

Meanwhile, on January 23, 1915, at Basque, British Columbia, the last rail was laid on the Canadian Northern Railway, thus completing Canada's second transcontinental line from Atlantic tidewater at Quebec to Vancouver on the Pacific coast. In the fall of that year a fifteen-car train travelled over the entire route carrying seventy-eight senators and members of parliament and some thirty newspapermen from major centres in eastern Canada to the west coast, inaugurating regular transcontinental traffic on the line. On its way back east, the train stopped at Gladstone, Manitoba, where the Canadian Northern had had its humble beginning in 1896. An address was delivered to Sir William Mackenzie, president of the company, complimenting the builders on their achievement:

> *We had not conceived it possible [he said, according to the press reports] that a railway, possessing the standard of alignment, and gradient of your road could have been constructed across Canada within so short a period. The evenness of the roadbed and the facility with which one locomotive has hauled across the continent a train nearly one quarter of a mile in length, fully demonstrates the high standard of construction obtained throughout the line of travel.*

A final link in the Canadain Northern Railway system was completed in 1918, when the railway entered the heart of Canada's largest metropolis via a three-million-dollar double-tracked tunnel to the new Dorchester Steet Station of Montreal. Traffic was officially declared open as the first passenger train, carrying engineers, government and company officials, passed through the tunnel into the terminal on September 21st. But by now the Canadian Northern Railway was bankrupt. The hoped-for volume of traffic had never materialized. Heavy interest payments on past loans could no longer be met. New loans were not available. In the end, the government of Canada had no alternative but to take over the road. The *Mail and Empire* of September 12, 1918 reported:

> *Payment has been made by the Government for the 510,000 shares of common stock of the Canadian Northern Railway Company, of which Mackenzie and Mann and the Canadian Bank of Commerce were respectively owners and pledgees. Those shares were surrendered to the Goverment following the passage of the legislation of 1917, providing for the acquisition of 600,000 shares not then in the possession of the Government. A cheque for $8,500,000 payable to the order of Mackenzie and Mann and the Bank of Commerce jointly, has been issued accordingly by the Government.*

Thus the two men relinquished the empire that they had built. Aside from the railway, it consisted by then of hotels, telegraph and express companies, mining ventures and a fleet of ships on the Atlantic. Officially, the Canadian Northern ceased to exist when it became part of the government-owned Canadian National Railways system in January 1923. Sir William Mackenzie and Sir Donald Mann took their place in history as the last of the daring railway pioneers in Canada.

The Third Transcontinental

Expansion of railways was uppermost in the minds of Canadian politicians at the turn of the last century. And why not? It was a time of unprecedented growth, of rapid industrial and agricultural development combined with a buoyant prosperity never before experienced in the country. As the prime minister of the day, Sir Wilfrid Laurier, put it: the twentieth century belonged to Canada! It was his government which embarked on a grand-scale railway policy that eventually was to become the colossal "railway problem" of later years.

At the time, the Canadian Pacific was still the only transcontinental railway in Canada. Between Winnipeg and Fort William its single-track line, carrying the prairies' wheat crops to the Lakehead, was overloaded during the peak shipping season and frequent traffic delays were inevitable.

Aside from the infant Canadian Northern and the Northern Pacific (an American line which had ventured into Manitoba), there were no rival railways existing as yet in the west. Consequently, the Canadian Pacific, with its extensive prairie mileage, cashed in on the wheat boom, and freight rates were high. Western farmers as well as politicians began demanding construction of new railways in the hope that competition would lower rates.

The Grand Trunk Railway, which had passed up the opportunity of expanding westward in the 1880s, now enviously eyed the Canadian Pacific's success. The old pioneer line was undergoing a series of drastic changes in an all-out effort to spruce up its vast eastern-based network of lines and pull itself out of the financial doldrums that had been baffling it for years. Charles Melville Hays, formerly general manager of the Wabash Railroad in the United States and known as a highly capable administrator, had been appointed to do the job. Under his energetic management, large sections of the Grand Trunk were scheduled to be rebuilt. Sharp curves and heavy grades were being eliminated and bridges modernized. Double tracking of the main line was speeded up; outdated rolling stock was overhauled; orders for new equipment were placed; inefficient employees were dismissed.

In 1902 Mr. Hays announced the Grand Trunk's intention to extend its steel to the Pacific coast. The railway proposed to build westward from its northern Ontario terminal of North Bay. In the east, existing Grand Trunk lines to Montreal and Portland would provide the Atlantic outlet for the new transcontinental scheme.

Meanwhile, the western-based railway of Mackenzie and Mann, known as the Canadian Northern, had constructed a line from Winnipeg to Lake Superior and had obtained powers to extend eastward to Ottawa and Montreal, as well as westward via the Yellowhead Pass to Vancouver.

There were those who pointed out the folly of building simultaneously two new transcontinental lines, but suggestions to have the Grand Trunk and the Canadian Northern merge their schemes fell on deaf ears. Neither the Grand Trunk nor the Canadian Northern, although both eagerly seeking government assistance, was willing to curtail its ambitions. A compromise proposal, which would have had the Grand Trunk expand in the east and the Canadian Northern in the west, with a connecting link to be built and used jointly by both, was flatly rejected. In the end Ottawa decided to give its blessing to both railways, ignoring the fact that their tracks would run virtually side by side for hundreds of miles through the prairies.

Concern arose, however, that transcontinental traffic might be diverted to United States ports if the Grand Trunk were permitted to follow its original plans. To assure an all-Canadian route (with Halifax, Nova Scotia, or Saint John, New Brunswick, as the Atlantic outlets), Sir Wilfrid Laurier thought it advisable for the government to build and own the eastern portion of the line from Winnipeg to the Maritimes and lease it to the Grand Trunk upon completion. As for the western half of the transcontinental, it was to be solely a Grand Trunk enterprise with the dominion government guaranteeing most of the company's construction bond issue.

Accordingly, in 1903 the Grand Trunk Pacific Railway was chartered for the purpose of building the "western division" from Winnipeg to the Pacific coast within a period of five years. The Grand Trunk, as the parent company, was to hold not less than $24,900,000 of the common stock in its subsidiary in order to retain complete control. Both the western and the eastern portions of the transcontinental were to be built to the highest standards, with the Grand Trunk to approve specifications of the government line. The latter was to be operated by the Grand Trunk Pacific under a fifty-year lease, rental-free for the first seven years, and thereafter at three per cent of construction cost per annum.

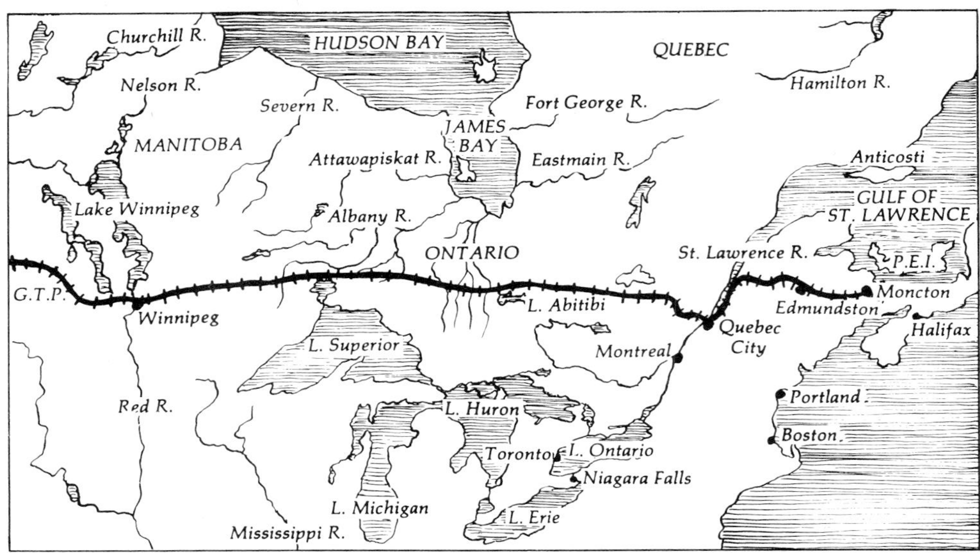

The National Transcontinental

A storm of controversy arose in the House of Commons when the terms of the agreement with the Grand Trunk Railway became known. The opposition offered alternative plans, the Minister of Railways, Andrew Blair, resigned in protest, and the railway policy turned into a fiery election issue. Nevertheless, the new transcontinental got under way, and all across the country construction was soon being pushed ahead from various points.

Known as the National Transcontinental, the government-owned part of the project commenced at Winnipeg, running eastward through Northern Ontario and Quebec to Quebec City, and thence via Edmundston to Moncton, New Brunswick, where it met the Intercolonial over which traffic continued to the Canadian ports of Saint John and Halifax. The reason for locating the line far to the north of existing settlement through a wilderness of forests, lakes, rivers and swamps, was to open up the hinterland to agriculture. At the time little thought seems to have been given to the rich timber and mineral resources of the northern territories, which ultimately proved of even greater importance.

The first survey crews, charged with finding the most favourable route for the railway, went out in the fall of 1904. The stretch of approximately 1,800 miles between Moncton and Winnipeg was divided into sections, and at the height of activities between forty-five and fifty survey parties were scattered in the field. All told, they explored some ten thousand miles, mostly in extremely difficult terrain, before the actual course of the National Transcontinental was plotted. Specifications called for the curvature of the line to be limited to a maximum of six degrees. Grades were not to exceed 0.6 per cent against westbound traffic and 0.4 per cent against the anticipated heavier eastbound movement. Consequently, surveyors had to spread out over a wide area in search of suitable terrain. As a rule, parties worked in pairs, advancing towards each other from two given points. In the largely trackless stretch of more than eight hundred miles between the upper St. Maurice in Quebec and Lake Nipigon in Ontario, the joining up of two survey crews was an accomplishment in itself. Due to the encounter of unexpected obstructions, such as large bodies of water or dense bush, frequent delays in the work schedule were unavoidable. Approaching parties occasionally missed each other entirely, overlapping by many miles. Communication in that event was reestablished by the discharge of ship's rockets after dark, a rescue device used also to guide lost crew members back into camp.

Although survey work was not completed in the desolate northern regions until 1908, construction of the National Transcontinental commenced in 1906 in Quebec as well as on the prairies. In New Brunswick work got under way the following year, and by the end of 1908 the entire line was under contract. By then an army of twenty-one thousand worked along the route. Supplies for individual sections were distributed from points on the Canadian Pacific, the Canadian Northern, the Intercolonial and other railways. Barges and ships were hastily being built, to be put into service on transportation routes over navigable rivers and lakes before freeze-up, to reach the remote districts which the railway was to traverse. Temporary tramways to circumvent rapids were constructed, and hundreds of miles of tote roads established, to move supplies during the winter. A narrow gauge railway, eighteen miles long, was built to bypass the chutes on the Nipigon River, while an eight-mile monorail, operated by horses, ran between the Abitibi and the Frederick House Rivers.

If surveyors had been faced with a host of obstacles, the construction gangs also encountered their share of problems, particularly in the unpredictable muskegs that were full of sinkholes and plenty of unforeseen hazards. In some places the seemingly bottomless swamps swallowed trainload after trainload of ballast before a foothold could be gained. At the Ontario-Quebec border, a thousand-foot steel bridge over the Okidodasik suddenly turned over on its side due to a clay slide. The treacherous soil throughout the clay belt accounted for numerous other mishaps. A succession of slides caused a concrete culvert near Lake Abitibi to break up and disappear in a sea of mud. Bridge builders resorted to structures with long spans, rather than attempt to put foundations in mid-stream where clay was found.

In the summer of 1909 the first section of the line was opened from Hervey Junction to Quebec City, a stretch of eighty miles. By 1912 much of the track had been laid eastward from Winnipeg and westward from Moncton, but large gaps in northern Quebec and Ontario still remained to be tackled. It was not until November 17, 1913 that the last spike on the National Transcontinental was driven in Ontario, completing the final section between Grant, west of Cochrane, and Nakina. The total construction cost had climbed to one hundred and sixty million dollars, a hundred million more than originally estimated.

In order to cash in on western traffic via the Great Lakes to eastern Canada at the earliest possible date, the Grand Trunk Pacific meanwhile had constructed a branch from Sioux Lookout, on the main line of the National Transcontinental, to Fort William, Ontario, thus providing a through-route for wheat shipments from Winnipeg to the Lakehead. Connection with North Bay and the Grand Trunk system was secured by the National Transcontinental with the acquisition of running rights from Cochrane, Ontario, over the Temiskaming and Northern Ontario Railway.

Although the railway was ready and sections of it were in operation, regular service throughout the line was not inaugurated until 1915. In accordance with the terms of the original contract, the Grand Trunk Pacific Railway was called upon to take over the operation of the National Transcontinental upon its completion. Charging that the line did not come up to the required construction standards, the Grand Trunk Pacific, however, refused to accept its obligation. Consequently, in 1915 the dominion government, using the newly created name of

"Canadian Government Railways," undertook to operate the line from Moncton, New Brunswick, to Winnipeg, Manitoba, and lease in perpetuity the Superior Junction (Sioux Lookout) branch to Fort William, which had been built by the Grand Trunk Pacific.

A car ferry at Quebec provided service across the St. Lawrence River during the early years of operation, while the famous Quebec Bridge, six miles above the city, was still under construction. Designed as the world's largest cantilever bridge, it had been started by the Quebec Bridge Company in conjunction with the Phoenix Bridge Company back in 1900. In the summer of 1907 a few defects in the structure were discovered and subsequently repaired. On August 29th of that year, as the bridge was nearing completion, the southern cantilever span suddenly collapsed under the weight of a load of steel, a locomotive and a crane. Seventy-five workers were killed.

An inquiry which followed blamed the disaster on an "error in judgment" by the engineers, and the dominion government took over the project. A new structure was designed and work resumed. By the summer of 1916 only the 640-foot centre span remained to be hoisted into place. On September 11th a large crowd of people watched it being towed on pontoons underneath the gap and slowly lifted up by powerful jacks. It had been raised about thirty feet above the water, when one of the hoist's castings split and the span plunged in the river. Ten men were killed in the second mishap. A new centre span was ordered immediately and safely put into place in 1917. On October 17th of that year the first train successfully crossed the structure. Quebec Bridge at last provided the final link on the Moncton-Winnipeg route.

Charles Melville Hays, the man in charge of the Grand Trunk Pacific, believed that only a first-class railway could successfully compete with its rivals. He therefore set out to build his railway to a standard far superior to any of the existing long-distance railways on the North American continent. Until then colonization lines traversing sparsely populated wilderness areas always had been constructed cheaply and fast, often with sharp curves and steep grades to avoid costly bridges and tunnels. For the Grand Trunk Pacific Hays specified low grades and minimum curvature, calculating that his steam engines would thus be capable of hauling heavier-than-usual pay loads. Lower operating and maintenance expenses of a perfect road, he argued, would more than compensate for the initial high cost of construction (which eventually amounted to over $140,000,000).

The route selected by the Grand Trunk Pacific ran from Winnipeg to Edmonton, thence via the Yellowhead Pass through the Rockies, and in a northwesterly direction towards the Pacific. Between Winnipeg and Brandon, Manitoba, the tracks parallelled those of the Canadian Pacific. For some two hundred and fifty miles between Edmonton, Alberta, and the Yellowhead Pass they invaded the territory of the Canadian Northern, and through the pass the two rival railways constructed their lines virtually side by side.

The first sod on the Grand Trunk Pacific's main line was turned without much fanfare at sunrise on August 29, 1905 by one of the foremen of McDonald, McMillan & Co., the contracting firm which had been awarded construction of the initial prairie section. The site was located in the vicinity of the little town of Carberry, some thirty miles east of Brandon, Manitoba. Work began almost at once and moved ahead with reasonable swiftness, despite an acute shortage of labourers which plagued contractors, particularly in the spring and at harvest time when farmers lured workers away from the railroad by offering higher wages. By the end of 1908 the roadbed was ready from Winnipeg, Manitoba, to Wolf Creek, Alberta, a distance of over nine hundred miles. The railhead however did not reach this point until two years later, because Canadian rail manufacturers had been unable to keep up with the demand for steel.

The first regular service on the main line of the Grand Trunk Pacific was established on the section between Winnipeg and Portage la Prairie on July 30, 1908. On August 13th of the following year the first Grand Trunk Pacific train arrived at Edmonton, Alberta.

Meanwhile, the company's Lake Superior branch still remained unfinished, although originally it had been intended to push it to completion well ahead of the main line. The government had promised to give priority to the Sioux Lookout-Winnipeg section of its National Transcontinental, in order to allow construction material for the Grand Trunk Pacific proper to be moved, economically and independently from competitors, over a through-route from the Lakehead to Winnipeg. Early revenues from this part of the line were expected to help pay for the costly mountain section of the Grand Trunk Pacific. As it turned out, the government as well as the Grand Trunk Pacific fell behind schedule and consequently the hoped-for benefits never materialized.

GRAND TRUNK

RAILWAY.

Special Settlers' Trains

—FOR—

MANITOBA

—AND THE—

Great North-West.

ONE of these Special Trains will leave Belleville each WEDNESDAY during the months of MARCH and APRIL.

An experienced Official will accompany each train.

REMEMBER—this is the SHORTEST route. Only First Class cars used.

LOWEST RATES.

For Tickets and all information apply to

U. E. THOMPSON,

Agt. Grand Trunk Ry.
Bridge St., Belleville.

3/21/82 d3m

Unlike most earlier railways, the Grand Trunk Pacific had received no land grants from the government. Nevertheless, the company ventured into the real estate business, purchasing land for its own use as well as for the purpose of reselling it to prospective settlers. In 1906 a wholly-owned subsidiary was incorporated under the name of Grand Trunk Pacific Town and Development Company. Following this, dozens of townsites were laid out on the prairies, spaced in an orderly fashion every ten or fifteen miles along the projected route of the railway. Surveyors marked out the streets and located the station house and all the major public buildings on their maps before moving on to the next site. Each town was planned to a prescribed pattern, which prompted a contemporary reporter to describe the Grand Trunk Pacific settlements as resembling "babies raised according to formula."

Among the first structures to be erected in the newly born towns were grain elevators, now familiar landmarks on the prairies. They began springing up like mushrooms overnight, as agents of elevator companies followed on the heels of surveyors and picked choice locations in anticipation of the need for storage facilities. By the time the first regular train puffed into the station of a Grand Trunk Pacific community, merchants, hotel keepers, bankers, doctors and lawyers had set up shop and, along with the other settlers in the area, were ready to celebrate the event.

For its western terminus the railway chose a lonely fiord at the mouth of the Skeena River, five hundred miles up the coast from Vancouver. A nation-wide contest was held in 1905 to find a name for the town which was to be built there by the Grand Trunk Pacific. The prize of $250 for the winning entry went to a Winnipeg school teacher who suggested "Prince Rupert," to commemorate the first governor of the Hudson's Bay Company. Actually, she had ignored one of the contest rules: that the name should not have more than ten letters.

The town was to be located on Ka-ien Island. Soon after the surveyors had left, several hundred prospective residents moved in, pitching tents and setting up shop. A newspaper, called the *Empire* was being printed in a large tent in the middle of what was marked as Center Street on the map. Early visitors described the place as resembling a "lumber camp" full of muskeg sinkholes and tree stumps. Prince Rupert town lots did not go up for sale until May of 1909. Fifteen hundred people showed up at the land auction held in Vancouver. The first two lots at the corner of Second Avenue and Second Street went for over ten thousand dollars.

The town of Prince Rupert was incorporated in 1910, and Grand Trunk Pacific engineers began building the new ocean port which possessed an excellent harbour, situated five hundred miles closer to the Orient and its riches than Vancouver. Charles Hays envisioned a great future for his railway's western terminus, with a steady flow of export and import cargoes passing over its docks, helping to make the Grand Trunk Pacific one of Canada's busiest traffic arteries.

After traversing the Rockies via the Yellowhead, which was considerably less rugged than Kicking Horse Pass which the Canadian Pacific had chosen earlier, the Grand Trunk Pacific followed the upper Fraser, the Nechako and Endako, and finally the Skeena River valley making its way through the Coastal Range to the sea. Thousands of miles of possible routes were explored before selecting the Skeena's course, which leads through the montains via a narrow canyon eighty miles in length. Fed by the spring thaw, the river races seaward at speeds of fifteen miles an hour, leaving the wild picturesque canyon through what is known as the "Hole in the Wall."

Later in the spring of 1908 work got under way on this western leg of the Grand Trunk Pacific. Construction material was shipped by rail via the Canadian Pacific to Vancouver, and thence by water to the mouth of the Skeena. From there to Hazelton, some 180 miles inland, the river served as the only supply route for the builders. To reach Hazelton, heavily laden stern-wheelers, winched through parts of the canyon by cables which were fastened to trees or rocks along the banks, often took as much as a week. Downstream they made the journey in less than twenty-four hours. Construction crews in the area virtually lived on salmon which filled the waters of the Skeena.

More than four million cubic yards of rock had to be excavated for the first one hundred miles of roadbed of the Skeena section. One rock-cut, some sixty-six hundred feet in length, required all of two years to complete. Drillers, placing explosive charges, often worked suspended on ropes, which were yanked up the side of the mountain wall out of danger once the fuse was lit. Units of so-called "station men" attacked a hundred feet of solid rock at a time. Being paid by the yardage they cut, they worked like slaves from daybreak till nightfall and sometimes after dark by the glimmer of their lamps. A few of the toughest men were able to make small fortunes in a short span of time.

In the eastern mountain section, meanwhile, contractors had to build a costly wagon road, 184 miles in length, to carry building and food supplies to their construction camps. Tunnelling through solid rock, clearing dense forests, bridging ravines and roaring rivers, the railway builders inched their way westward through some of the most difficult and adverse terrain. Among the enemies they faced were blizzards, rock slides, avalanches and floods. In the Jasper area they encountered a problem of a different nature. A beaver dam, causing flooding of the roadbed, brought things to a temporary standstill. No sooner had the work crew breached the dam than the beavers repaired it during the night. It became a contest as to who would hold out the longest. In the end the beavers won. The roadbed was completed, and the dam as well as the beaver lodges were left intact.

In the Fraser canyon, scows, skippered by daring rivermen, carried supplies for construction camps. Fifty men drowned in that area within a single year. By the end of 1912 the railhead was nearing the Yellowhead Pass.

On April 7, 1914 railway crews from the east and west met at Finmore, 374 miles inland from the Pacific. Here they drove the last spike which completed the Grand Trunk Pacific Railway. No special ceremony was held to mark the occasion. The following day the first through train from the east arrived at Prince Rupert, which had since been connected with the mainland by the Zenardi Bridge.

Charles Melville Hays, whose fondest dream had been to travel from Halifax to Prince Rupert over the rails of Canada's new transcontinental road, never saw its completion. In 1912 the man who had masterminded the Grand Trunk's western venture had boarded the ill-fated White Star liner *Titanic* on her maiden voyage from Europe. His wife and daughter were among the survivors in the last lifeboat to be lowered from the sinking ship. They never saw him again.

Had he lived, his hopes of seeing the Grand Trunk Pacific Railway prosper would have been shattered. Only three years after the railway was opened, ninety miles of steel rails between Wolf Creek and Jasper were ripped up and shipped to Europe for military use. The gap was closed by utilizing Canadian Northern Railway's trackage.

The Grand Trunk Pacific as well as its parent, the Grand Trunk, were in serious financial trouble. The latter had incurred liabilities to the tune of over one hundred and twenty-three million dollars in connection with its Pacific venture. The Grand Trunk Pacific was the first to totter. In 1919, barely five years after its completion, the line passed into the hands of the Minister of Railways as receiver. Eventually, integrated in the vast Canadian National Railways system, the scenic wilderness stretch between Red Pass Junction, in the shadow of mighty Mount Robson, to rail's end at Prince Rupert, became part of what is known as the Smithers Division of the Canadian National Railways. At Red Pass Junction the Canadian National Railways track forks and the main line heads in a southwesterly direction towards Vancouver.

The Grand Trunk Pacific town of Prince Rupert never gained significance as a shipping port until World War II, when it temporarily served as a huge military supply base. To lessen potential danger to densely populated areas and larger seaports to the south, thousands of tons of ammunition were routed via the Smithers Division to the lonely Pacific terminus. Prince Rupert turned into a beehive of activity as seventy-five thousand soldiers, bound for duty in Alaska, passed through its railway station, and trainloads of trucks, cranes, bulldozers and other supplies needed to build the Alaska Highway rolled off the ramps.

Near the town rises a mountain known as Mount Hays, commemorating the man who envisaged that Prince Rupert would some day take its place among the great cities of the continent.

Steel to the North

On September 10, 1929, the Minister of Railways and Canals made the following announcement in Ottawa: "The historic Hudson's Bay Company is to have the honour of taking out through Hudson Bay the first shipment of wheat to be exported via Churchill. . ."

It was in 1670 that Charles II of England had granted a charter to "The Governor and Company of Adventurers of England Trading into the Hudson's Bay," and for two and a half centuries the company's ships had plied the waters of the bay. Churchill, Manitoba, situated on the west coast of Hudson Bay, had its beginning when the Hudson's Bay Company built the first Fort Churchill near the mouth of the Churchill River in 1685.

The shipment referred to in the minister's statement consisted of a ton of prairie wheat, done up in two-pound sacks. It arrived by rail at the port of Churchill, where it was transferred to the company's steamer *Ungava* ready to sail for England. The wheat sacks were to be distributed overseas as souvenirs to commemorate the opening of the Hudson Bay Railway.

Although steel had reached the shore of Hudson Bay in the spring of 1929, the railway was still not fully completed. Nevertheless, newspaper editors and reporters picked up the story and spread it across the country. After all, the public had waited nearly half a century for the realization of a cherished dream — the "Bay Route." There was some official concern that the general enthusiasm was becoming "embarrassing to those responsible for the orderly progress of things" and, as it turned out, the Hudson Bay Railway was not opened for commercial use until 1931.

Conceived to provide the shortest possible shipping route from the wheat fields of Canada's prairies to the markets of Europe, by connecting Winnipeg with tidewater, the Bay Route was first proposed in the 1870s. Back then, its supporters were ridiculed for suggesting the construction of a railway line over the frozen muskeg and barren lands of Canada's north.

The railway's corporate history begins with the charter of the Winnipeg and Hudson's Bay Railway and Steamship Company, granted by act of parliament in 1880 to a group of prominent citizens of Manitoba. One of them was Hugh Sutherland, a fervent promoter of the project for many years to come. About the same time, another company, empowered to build a line to Hudson Bay, was formed by eastern interests under the name of Nelson Valley Railway and Transportation Company.

Churchill railway station

Manitoba Archives

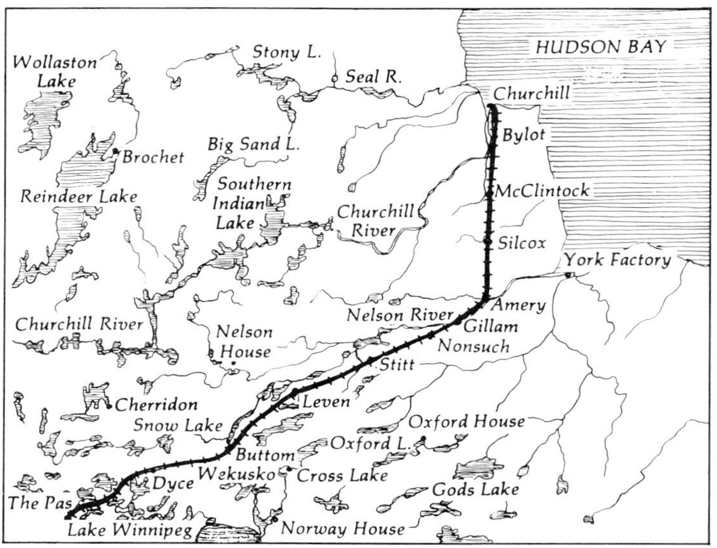

When it became apparent that neither of the two rivals were in a financial position to realize their ambitions, they decided to amalgamate. As a result, the Winnipeg and Hudson's Bay Railway and Steamship Company took over the properties of the eastern-based company. Despite a substantial land grant which the new charter carried, attempts to raise sufficient construction capital continued to fail. A steady decline in the price of wheat, a general slowdown of the economy, and the outbreak of the second Northwest Rebellion in 1885, were some of the reasons. Construction work that had been carried out terminated at Shoal Lake, and the stretch became known as "Hugh Sutherland's Forty Miles."

A man who never said die, Mr. Sutherland remained undaunted. For more than a decade he continued to explore every possible source of financial assistance for his Hudson Bay project, only to have success elude him every time it seemed within his reach. Eventually the company was reorganized and its name was changed to Winnipeg Great Northern Railway. The company's problems, however, remained the same.

Meanwhile, the two railway contractors William Mackenzie and Donald Mann had acquired the dormant Lake Manitoba Railway and Canal Company charter and started to build a line from Gladstone to Dauphin, Manitoba. With their eyes on expansion, they soon took over Hugh Sutherland's Winnipeg Great Northern, and the merger had produced the Canadian Northern Railway.

Construction on the Bay Route finally commenced under Mackenzie and Mann in 1906, with the line running northward from Hudson Bay Junction (now known as Hudson Bay) on the Canadian Northern. The latter by then had been extended westward to Prince Albert. By 1910 eighty-eight miles of the Hudson Bay line were operational. The northern terminus was The Pas, then little more than a trading post.

At The Pas the efforts of private enterprise ceased. Northward, the Bay project was to be continued by the government of Canada. Sir Wilfrid Laurier had made it an election promise to complete the railway. Survey parties had been sent out in 1909 to plot the route. Construction of a bridge across the Saskatchewan River at The Pas began in 1910 and a contract for 185 miles was let the following year to J. D. McArthur, a Winnipeg railway builder.

Preparations were barely under way when, following the defeat of the Liberals in the 1911 election, a new Minister of Railways ordered construction work on the line halted. The future of the Hudson Bay Railway seemed once again in doubt. Dedicated supporters of the project, however, outnumbered the sceptics in the House of Commons, and work was soon resumed.

Ever since plans for a railway to Hudson Bay had been first under discussion, the choice of its northern terminus had been a matter of considerable controversy. Some promoters favoured Churchill as the future seaport, others Port Nelson. The distance to Nelson was approximately eighty-six miles shorter, and consequently the latter was selected in 1912. Construction of harbour facilities began. By 1918 grading and bridging of the line to Port Nelson was nearing completion. The railhead had reached the Nelson River crossing at Kettle Rapids north of Gillam, when work once again was suspended. One of the prairie newspapers sarcastically remarked that "Gabriel the Trumpeter should be invited to drive the last spike on Judgment Day."

It seemed as if the Hudson Bay Railway was doomed after all. Considering the severe climatic conditions and the short shipping season of a Bay port, its successful operation was questioned by many. Why throw any more money away on "icebergs and polar bears," as one of the opponents of the project put it? Approximately one hundred miles of track between Mile 214 and the railhead at Gillam were subsequently abandoned. Operations and maintenance ended just south of Pikwitonei.

In 1924 impatient prairie representatives and other interested supporters formed the "On-to-the-Bay Association" at a mass meeting held in Winnipeg. As pressure mounted in the west, a new Liberal government in Ottawa two years later committed itself to the completion of the line. The government-operated Canadian National Railways which had recently been created to amalgamate tottering private enterprises, was asked to rehabilitate the trackage of the Hudson Bay Railway.

Once again doubts arose about the merits of Port Nelson becoming the terminus. New extensive surveys were carried out, with the result that Nelson and its already constructed facilities were abandoned in favour of Churchill which, in the opinion of experts, possessed a superior harbour. A route, turning north at Amery on the Nelson River and traversing extensive muskeg areas with underlying permafrost on its way to Churchill, was located.

At the close of 1928 the railway had progressed to within forty-eight miles of the new tidewater terminus where temporary docks had been built and permanent structures were well under way. During the winter, tracks were laid on the frozen muskeg. Men fought blizzards and subarctic temperatures, sometimes advancing only a matter of feet per day. Finally, on March 29, 1929, the chief engineer of the Canadian National Railways' Western Region received a telegram containing five simple words: "Steel laid to Churchill today."

A dream that had persisted for half a century at last had become reality. Covering the relatively short distance of 510 miles, from The Pas, Manitoba, to the deep-sea harbour in the heart of the continent, the Hudson Bay Railway linked the granaries of the western plains with the ocean and the world.

WHITE PASS AND YUKON RAILWAY

George Carmack and two Indian companions, Skookum Jim and Dawson Charlie, made history on August 17, 1896, when they gleaned a few grains of gold from the bottom of Bonanza Creek, a tributary of the Klondike River in the Yukon Territory. News of their discovery started the gold rush fever which spread like wildfire across the continent. Men left their homes and families by the thousands. They sold their shops and belongings to buy passages at Vancouver, Victoria or Seattle on one of the coastal ships going north to Skagway. From there they carried their supplies for forty miles on their backs, climbing the rugged White and Chilkoot Passes to the head of Lake Bennett, where they constructed makeshift boats and rafts for a five-hundred-mile trip to Dawson in the heart of the fabulous gold fields.

Dawson, in the short span of a couple of years, mushroomed from nothing into a city of twenty-five thousand with stores, banks, saloons, dance halls and the reputation of being the "Paris of the North." More than a hundred thousand fortune seekers swarmed through Dawson at the peak of the Klondike Gold Rush.

To solve the transportation problem from tidewater at Skagway over the difficult White Pass to Whitehorse on the Yukon River, two men contemplated the building of a railway. They were Sir Thomas Tancrede, representing English interests, and Michael J. Heney, a Canadian railway contractor known as "Big Mike." They met at Skagway and decided to tackle the tough job together.

Construction of the White Pass and Yukon Railway started in May of 1898, one day after men, horses and supplies had been landed at Skagway. Two months later the first passenger train in the far north began operating over a distance of four miles. The summit of White Pass was reached in February of 1899, with tracks ascending from Skagway 2,885 feet in twenty-one miles.

Meanwhile, construction gangs worked their way from Whitehorse south to Carcross, where the last spike was to complete the line on July 29, 1900. A group of dignitaries was on hand to help drive the golden spike home. One after another tried his hand at striking a blow. They finally succeeded in bending the spike so badly that they decided to give up and celebrate instead, leaving the job to the track foreman who finished it by using a regular spike.

Before the railway came, Carcross was known as Caribou Crossing. The story goes that the man who painted the station sign had cut the board too short, and rather than waste the lumber, he abbreviated the name of the place.

The route of the 111-mile narrow gauge White Pass and Yukon Railway from Skagway to Whitehorse today is one of the most beautiful in the world, but its construction was one of the most difficult ever attempted. Supply bases were a thousand miles from the site. Steamer service to Skagway was irregular and telegraph connections with the outside world were non-existent. Men, blasting and hacking their way through solid rock, hung suspended on ropes from the sheer cliffs of mountains. One of the rocks they had to dynamite was 120 feet high, 70 feet wide and 20 feet thick. Their main equipment, aside from powder, consisted of picks and shovels.

Over Dead Horse Gulch they built a 215-foot-high steel cantilever bridge, the most northerly of its kind in the world. To construct a 250-foot tunnel, sixteen miles from Skagway, they had to hoist their machinery up a steep trail over crags that barely offered a foothold for one man. Doors were placed at the portals of the tunnel, to be closed during the winter to prevent snow drifts from blocking the passage.

Regular service over the White Pass and Yukon started in August 1900. Before that time horse-drawn wagons conveyed passengers over incompleted sections of the line. From Whitehorse a fleet of sternwheelers carried the gold seekers over the Yukon River to Dawson City and the land of their dreams.

MANITOBA NORTHERN RAILWAY

On Saturday, September 22, 1928 a simple ceremony was held in the north of Manitoba near the Saskatchewan border. The Honourable John Bracken, premier of the Province of Manitoba, was present to drive a golden spike into the rail marking the completion of the Manitoba Northern Railway. The line, in the words of the *Mail and Empire* correspondent of the day, was "reaching into formerly considered inaccessible regions, tapping a great northland, stored with the riches of countless thousands of years."

Gold, silver, copper and zinc had been discovered in the area in 1914, but at the time many considered it too far away from civilization to be worthwhile. The place was named Flin Flon by the prospectors, after Professor Josiah Flintabbatey Flonatin, hero of a novel set near the Churchill River. With the Hudson Bay Mining and Smelting Company ready to operate the Flin Flon mine in 1927, a railway became an urgent necessity.

A townsite was established and a railway construction contract was let to the Dominion Construction Company and W. S. Tomlinson in December of that year. A bonus of a quarter of a million dollars was to be paid, if the line were completed within one year of the contract— a monumental task, considering the difficult terrain the railway was to traverse.

Known as the Flin Flon Subdivision of the Canadian National Railways, the 87-mile line was built through an area of solid rock, treacherous clay and vast stretches of muskeg in the record time of nine months. To accomplish the job, contractors resorted to methods never used before. For the first fifty or so miles from the junction of the road with the Hudson Bay Railway (four miles north of The Pas, Manitoba) to Cranberry Portage, steel was laid in the winter on frozen muskeg. Sink holes were blocked with ties and timber. The line was unballasted and its level varied by several feet, but thousands of tons of construction material could be hauled over its rails. Stations were built and shipped on flat cars to their appointed locations along the line.

Even before steel reached Cranberry Portage, a community of tents and log structures had sprung up overnight. Thanks to the railway, new enterprises and towns were being born where no one had ever dreamed they would arise.

NORTHERN ALBERTA RAILWAYS

The Northern Alberta Railways came into being in 1929 when three existing companies were merged and purchased jointly by the CNR and CPR.

The main line of the Northern Alberta had its beginning in the Edmonton, Dunvegan and British Columbia Railway, which was opened in 1916 between Edmonton and Spirit River, Alberta. Today the terminus of the line is Dawson Creek, a bustling shipping centre in the Peace River district of British Columbia at the southern end of the Alaska Highway, which extends in a northwesterly direction to Whitehorse in the Yukon and thence to Fairbanks in Alaska.

The railway's route runs from Edmonton north to the Athabaska River crossing, thence northwest along the south shore of Lesser Slave Lake via Grande Prairie to the city of Dawson Creek. It serves the Peace River Country, one of the finest agricultural areas of Canada.

A branch line from the Alberta town of McLennan at Mile 264 goes in a northwesterly direction, crossing the Peace River and terminating at Hines Creek. Before the merger, it was known as the

Central Canada Railway, a line which opened in 1916 between McLennan and Peace River.

Another branch of the Northern Alberta Railways extends from Carbondale, a short distance north of Edmonton, to Waterways, Alberta, where shipments to and from the North are transferred from rail to river transportation and vice versa. The line dates back to 1917, when it was built between Carbondale and Lac la Biche under the name of the Alberta Great Waterways Railway.

GREAT SLAVE RAILWAY

Construction of the 442-mile Great Slave Railway got under way in 1962. A branch line of the Canadian National system, it was to carry traffic into the desolate wilderness of the far northern regions.

The *Pioneer*, a giant automated tracklayer capable of operating in temperatures of fifty degrees below zero and serving as a power unit for a seven-car work train, was used to lay the track at a rate of up to twelve feet per minute. Thirty-four bridges, both steel and timber structures, were required along the route, the largest one being a two-thousand-foot steel bridge across the Meikle Valley.

The railway, which was completed in 1965, runs from Roma, Alberta, on the Northern Alberta Railways, northward to Hay River, a community on the south shore of Great Slave Lake in the Northwest Territories. Seven miles south of the terminus, a branch extends eastward to the zinc- and lead-rich mining region of Pine Point. Sawmills, grain elevators, and the occasional homestead of a stout-hearted pioneer, dot the railway's route through forests and farmlands and the bleak landscape of Canada's North.

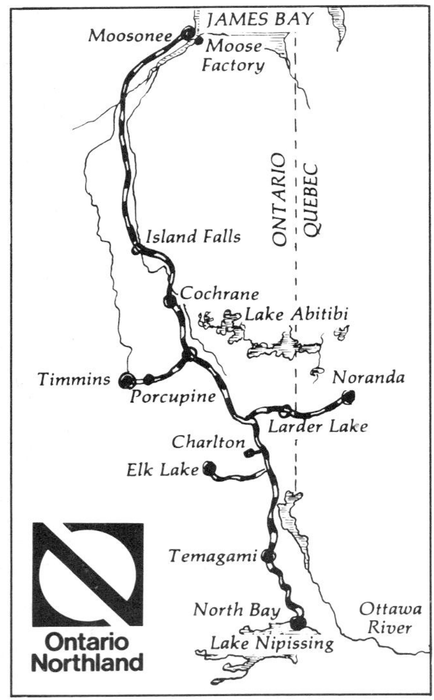

ONTARIO NORTHLAND RAILWAY

North of Cochrane, Ontario, roads are non-existent. The only means of transportation on land to Ontario's arctic tidewater is the railway to Moosonee on James Bay. It is the route of the famous *Polar Bear Express* which provides the life line for Ontario's northland. In winter a snowplough has to precede the train. In summer Indian children often jump aboard at stops to buy an ice cream cone at the dining car, because the *Polar Bear* is their rolling candy store and the only one they know.

The single-track line is part of the province-owned Ontario Northland Railway constructed through muskeg and the world's oldest rocks of the Canadian Shield. The first sod was turned in 1902 at North Bay after a contract had been let for construction to New Liskeard. Two years later, at Mile 103, workmen blasting through the rocks discovered fabulous silver deposits. Originally planned to serve a small group of settlers in the farming country of the clay belt at the head of Lake Temiskaming, the railway soon carried the wealth of the Cobalt silver mine camp.

As the railway builders pushed northward to Cochrane, more discoveries were made at Elk Lake, Timmins, Kirkland Lake and Noranda. Branch lines were built and towns and cities rapidly grew in the rich mining areas. Since the railway opened up the rugged wilderness of Ontario's north, hundreds of millions of dollars in gold, silver, copper and nickel have been mined along its route. With the railway gaining access to the great forest resources of the northland, giant pulp and paper mills supplied a steady flow of freight and in time the "paper trains" came to be called "Hundred Million Dollar Specials" because of their valuable cargo. In 1922 construction resumed in stages from Cochrane northward. Moosonee, on the shore of James Bay, was reached in 1932, when the last spike was driven in the road by Mr. Justice F. R.

An early photograph of The Northland *operating beween North Bay and Timmins*

Latchford who, thirty years earlier as Ontario's Commissioner of Public Works, had turned the first sod of the line.

The railway was once known as the Temiskaming and Northern Ontario. Some people said that its initials T & NO stood for "Time No Object." In 1946 the name was changed and the line became the Ontario Northland Railway.

ALGOMA CENTRAL AND HUDSON BAY RAILWAY

This railway was chartered in 1899 as the Algoma Central, to build from Sault Ste. Marie, Ontario, to a point on the CPR and to Michipicoten Harbour in the Algoma District. It was to link the Algoma Steel plant at Sault Ste. Marie with the iron mines at Michipicoten, site of one of the earliest mining operations in Ontario. Construction started in 1901, the same year that the company's name was changed to Algoma Central and Hudson Bay Railway. Today the company operates some 320 miles of main line as an integral part of the Algoma Steel Corporation.

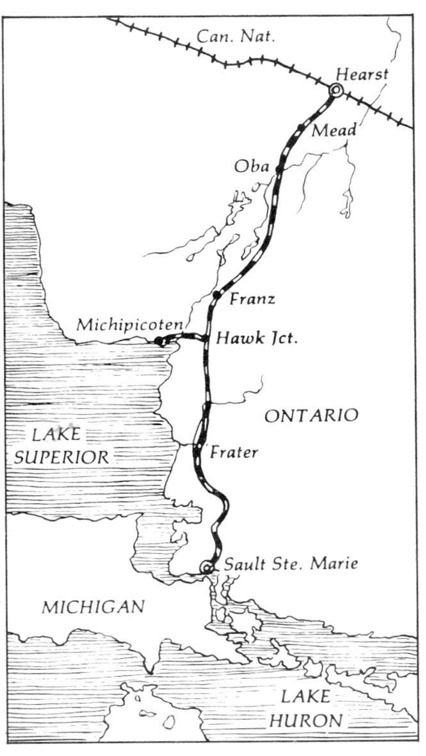

The Algoma Central and Hudson Bay Railway

THE QUEBEC, NORTH SHORE AND LABRADOR RAILWAY

Geological surveys back in the 1890s first reported iron-bearing deposits on the Labrador Peninsula. Later, attention focused on the Knob Lake region near the Quebec-Labrador border, where two hundred million tons of iron ore were estimated to lie buried within a two-mile radius. Subsequent activities of the Iron Ore Company of Canada in the area prompted the construction of a railway for the purpose of carrying the ore from the mines to tidewater, where it could be transferred to lake and ocean vessels.

The 358-mile line was started in 1950 as a subsidiary of the mining company. Aircraft, used to transport heavy equipment and supplies to initial working points, helped greatly to speed up construction. As the railway was expected to carry in excess of ten million tons of iron ore per year when completed, it was built to specifications of the highest standard. Rails weighing 132 pounds a yard were used in order to withstand the load of trains consisting of up to 125 ore cars hauled by four diesel-electric locomotives.

The single track railway, known as the Quebec, North Shore and Labrador, runs from Knob Lake, now the town of Schefferville, to Sept Iles, Quebec, on the north shore of the St. Lawrençe River. It was completed in 1954 at a cost of 120 million dollars.

Railways of British Columbia

Provincial Archives, Victoria, B.C.

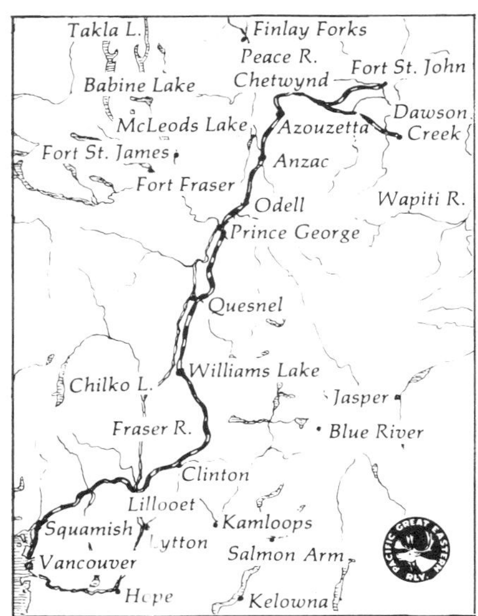

For over thirty years British Columbia's Pacific Great Eastern Railway began "nowhere" and went "nowhere." It consisted of 340 miles of rail, linking the hamlet of Squamish at the head of Howe Sound with the little community of Quesnel in the hinterland of the province. Projected to connect Vancouver with the main line of the Grand Trunk Pacific at Prince George, the PGE was dubbed "Prince George Eventually," "Please Go Easy" and "Past God's Endurance." At the southern end of the line, the gap between Squamish and Vancouver was covered by barge or steamer. In the north, some seventy miles between Quesnel and Prince George remained untouched.

The railway had been started by private interests in 1912, and initial construction had been carried out in stages from various points. By 1915 a few miles of unballasted track ran from North Vancouver to Whytecliffe, a few more miles went north from Squamish, and another short stretch headed north from Quesnel. Rising costs due to the war, construction problems in difficult terrain, and a scandal, involving allegations of fraud, forced the builders to give up.

In 1918 the bits and pieces of the troubled mountain railway became the responsibility of the provincial government, which completed the stretch beween Squamish and Quesnel in 1921. There the line began and ended for many years to come. It neither touched Vancouver, seaport and metropolis of western Canada, nor did it link up with the tracks of the Grand Trunk Pacific, now operated by the CNR. There was no "eastern" connection, and there certainly was nothing very "great" about the PGE in those days, except for the spectacular views of towering mountains and gaping canyons it offered its passengers along the lonely Cariboo Trail. The railway's rolling stock was old and ramshackle. Leaky car roofs, overcrowded coaches with pot-bellied stoves, whimsical stops of the engine crew in the middle of nowhere, were part of a trip on the PGE. Avalanches, rock slides, washouts or a flock of sheep would often impede the progress of the train. With breakdowns a frequent occurrance, trains were simply never on time. The story is told of a salesman who was astonished to find the train come in on schedule for the first time in fifteen years of travelling over the line, only to be informed by the conductor that this was yesterday's train.

Despite its shortcomings, British Columbians were fond of their line. Periodically, someone turned up with an offer to buy the PGE, lock, stock and rails, but nothing ever came of it. The railway continued to operate, mostly in the red, until the economic boom following the Second World War brought about a change in affairs. The potential of the Pacific Great Eastern was at long last to be exploited.

Construction started in 1949 on the Quesnel-Prince George section, and the 82-mile stretch was completed in 1952. In 1955 track laying commenced between Squamish and Vancouver at the southern end of the line. Citizens of the suburb of West Vancouver were anything but pleased at the prospect of a railway running through their exclusive residential area. Rocks were thrown at trains, railway tools disappeared, and someone took a shot at a conductor, narrowly missing his head.

Nevertheless, the old "Please Go Easy" was finally going somewhere. From Prince George construction was being pushed farther north, via Summit Lake, down Crooked River Valley to McLeod Lake; thence in a northeasterly direction through Pine Pass to Dawson Creek, terminus of the Northern Alberta Railways and Mile O of the Alaska Highway. A branch line is now running from Chetwynd to Fort St. John, centre of the mineral-rich Peace River country. The Pacific Great Eastern has come into its own. Its freight trains are carrying British Columbia's wealth of lumber, coal, minerals and grain. Its sleek and modern pas-

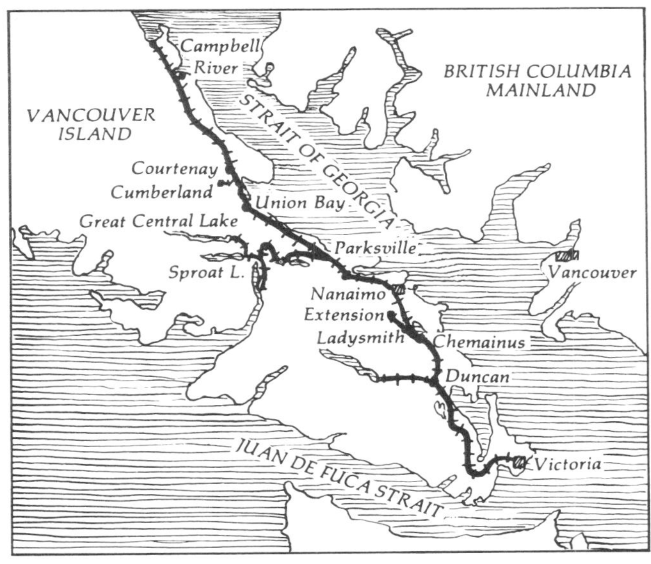

The Esquimalt and Nanaimo Railway

senger trains are taking travellers through some of the continent's most breathtaking wilderness areas. In the Fraser Canyon, above Moran, they can look straight down from their car windows at the river 2,200 feet below.

Work is under way to extend the Pacific Great Eastern through what is known as the "Rocky Mountain Trench," between the Rockies and the Coast Ranges, straight into the heart of Alaska.

ESQUIMALT AND NANAIMO RAILWAY

When the decision was announced in Ottawa that Vancouver was to become the western terminus of the Canadian Pacific Railway, outraged citizens of British Columbia's capital held a protest meeting at Victoria. In no uncertain terms they demanded a railway on Vancouver Island, threatening secession from Confederation if it were not built. Action followed swiftly. While Vancouver would remain the terminus of the CPR, the island would have its own railway, connecting Victoria with Nanaimo, seventy miles to the north. From Nanaimo, ferry service would provide the link with Vancouver and the east. As a result of this decision, the province surrendered nearly two million acres on the island to the dominion government, and Ottawa in turn promised a grant of 750,000 dollars. The deal also provided for the federal government to take over debts the province had incurred in connection with the building of a large dry dock at Esquimalt, which served as a Royal Navy base.

In 1883 an Act in the legislature of British Columbia authorized "the incorporation of persons, designated by Governor in Council, as Esquimalt and Nanaimo Railway Co. to build railway of same gauge as the Canadian Pacific Railway from Esquimalt Harbour to Nanaimo; Comox and to Victoria . . . "

Robert Dunsmuir, a Scotsman who had settled in Nanaimo and made his fortune in coal, undertook the construction of the Esquimalt and Nanaimo Railway in 1884. No stranger to railways, he had brought the first steam locomotive to Vancouver Island years earlier to haul coal over a narrow gauge colliery railway he had built to replace a wooden tramway operated by mule teams.

The last spike on the Esquimalt and Nanaimo was driven by Sir John A. Macdonald at Cliffside Station on August 18, 1886. When the first passenger train steamed into Victoria, the city fathers declared a half-holiday and completely forgot the fact that Robert Dunsmuir had never actually asked their permission to build a bridge across the

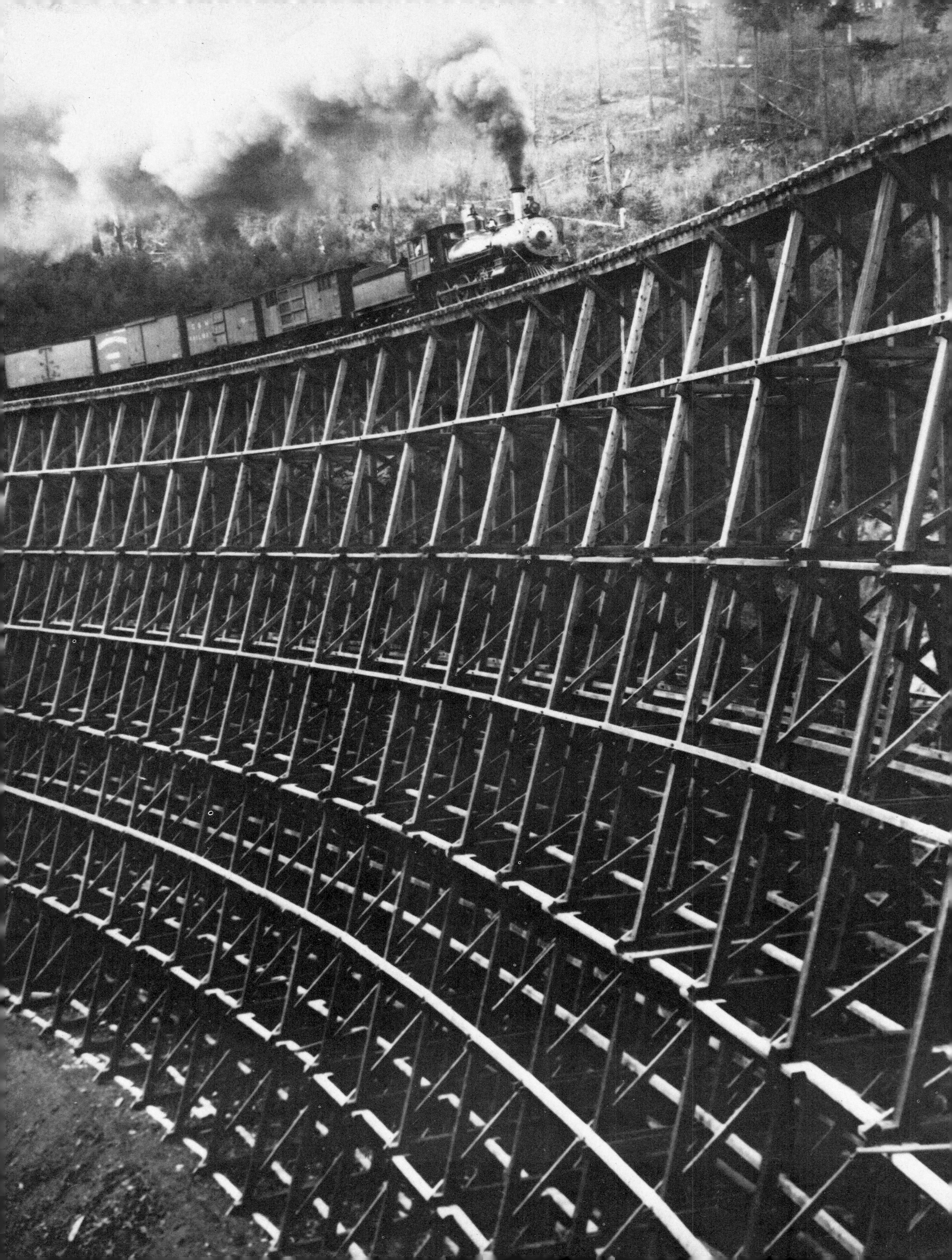

Gorge and bring his railway into Victoria. The Nanaimo industrialist usually did things his own way. To a minister who objected to trains running on Sunday, he pointed out that he was performing a public service by taking those who did not attend church from the city and removing them from "evil influences."

In the early days of Robert Dunsmuir's railway trains would occasionally make an unscheduled stop on the seventy-mile run to allow the engine crew to shoot grouse or chat with a prospector along the way. One of the engineers, Tony Silvene, who drove No. 4, allowed no one to touch his gleaming engine. When he went on holiday, she stayed in the roundhouse until he returned.

An extension of the line was opened to Port Alberni in 1911. A year later service was inaugurated to Lake Cowichan, northwest of Victoria. Unfortunately, the first engine to run over these rails went off the track and into the lake. In 1912 the Esquimalt and Nanaimo Railway was acquired by the CPR, along with vast tracts of land along the right of way.

An Esquimalt and Nanaimo Railway freight train crossing a trestle bridge (left)

Provincial Archives, Victoria, B.C.

The Esquimalt and Nanaimo Railway excursion train, drawn by Locomotive No. 5, on the trestle bridge 18 miles from Victoria, British Columbia, 1890

Steam in Newfoundland

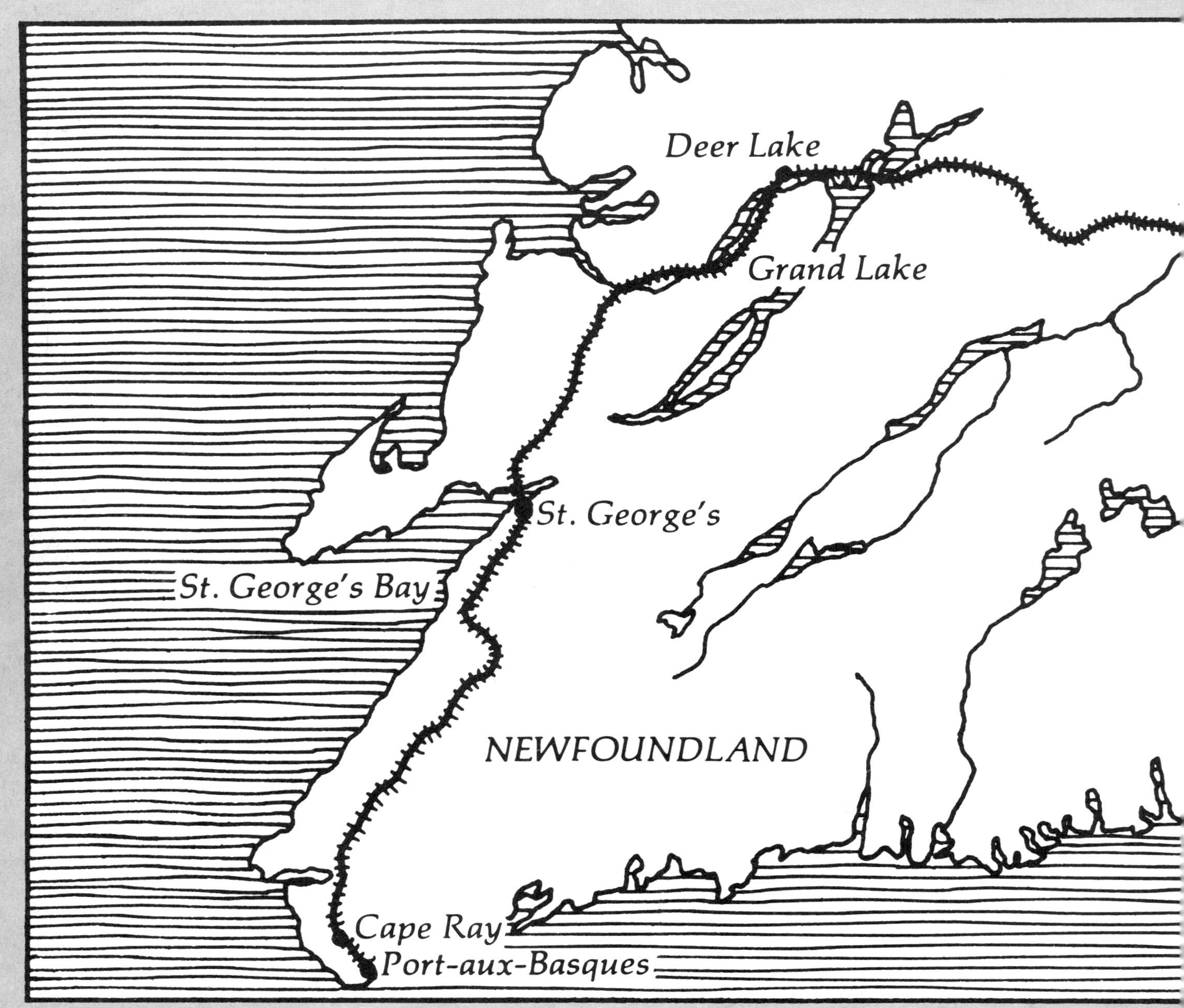

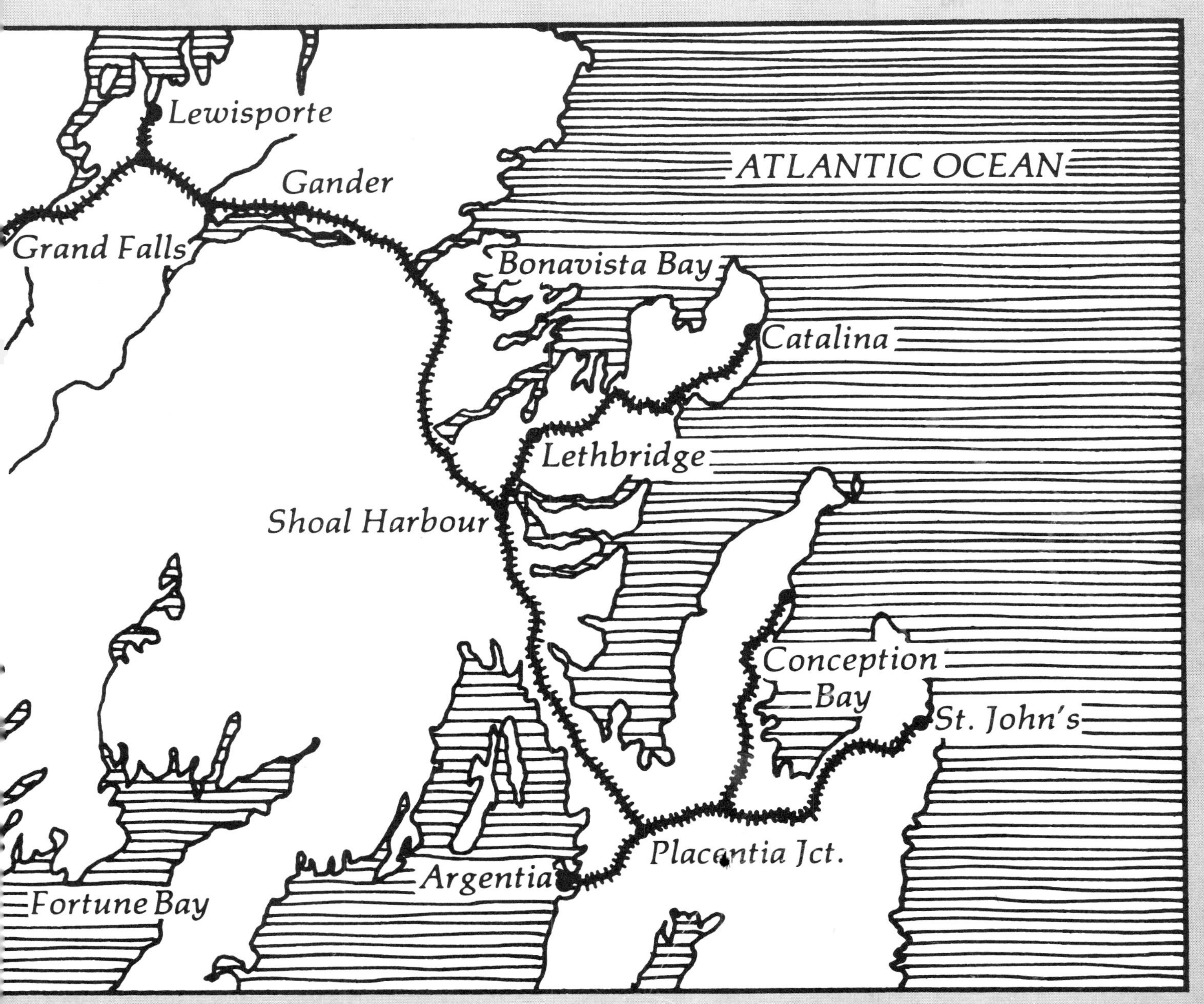
Lewisporte
Gander
ATLANTIC OCEAN
Grand Falls
Bonavista Bay
Catalina
Lethbridge
Shoal Harbour
Conception
Bay
St. John's
Placentia Jct.
Argentia
Fortune Bay

When the railway came to Newfoundland in 1881, most of the island's population lived near the sea in scattered fishing villages along the rugged coastline. Although talk of a railway had started in St. John's as far back as the 1840s, not everyone could visualize the value of a transportation system that would tap the wealth of the interior and help develop industries other than the ones connected with fishing. Inhabitants along the southern shore of Conception Bay, vigorously opposed to their government's railway policy of the 1880s, staged a riot (known as the "Battle of the Foxtrap") which temporarily put a halt to construction.

The narrow gauge line between St. John's and Harbour Grace had been started by the Newfoundland Railway Company, which was to receive five thousand acres of land a mile and a mail subsidy of $530 a mile per annum for thirty-five years. English money financed the American-based company's venture. Sixty miles of track had barely been completed, however, when funds ran out. The line reverted to the English bondholders who managed to complete the railway to Harbour Grace by 1884.

Under a new government, following that of Sir William Whiteway which ended in 1885, another section of the Newfoundland Railway was built and opened between Whitbourne and Placentia. The project, supervised by a railway commission and financed out of the colonial treasury, proved too costly to be continued, and in 1889, after the Whiteway administration had been returned to power, it was decided to leave the job to an independent contractor.

Among the tenders received was that of Robert Gillespie Reid of Montreal, well-known bridge builder and railway contractor, who in the past had successfully handled such difficult problems as the Jackfish Bay section of the CPR north of Lake Superior. The Newfoundland government accepted his $15,600 price per mile. Construction resumed, and by the time tracks reached Notre Dame Bay in 1893 it had been decided to extend the railway to Port-aux-Basques at the southwestern tip of the island, and R. G. Reid was awarded the contract for the western division. In the meantime, yet another agreement was being worked out with the builder, under the terms of which he was to operate the line for ten years in return for a five-thousand-acre land grant per mile.

The first cross-country train for Port-aux-Basques left St. John's on June 29, 1898. That same year the administration, then under Sir James Winter, signed a new contract with Mr. Reid concerning the railway's operation for the next fifty years. In Mr. Reid's favour the agreement provided further land grants, the right to purchase the colony's thousand-mile telegraph line, and the privilege of serving the entire seaboard of Newfoundland with a fleet of modern steamships he was to build. He thus became the owner of a large portion of the island's crown land and was assured the monopoly of its communication and transportation system. The contract, severely criticized in Newfoundland politics as the "Reid deal," was substantially modified a few years later, but it left the newly incorporated Reid Newfoundland Company committed to the development of the colony's prosperity.

Splendid ships were built to provide mail and passenger service in connection with the railway. In St. John's a large hotel was begun but later abandoned. Land was offered by the company to settlers for thirty cents an acre. Several branch lines were added to the railway system. By 1920, however, the Reid Company operated at an annual loss, and three years later the Newfoundland government took over all railway and steamship services for the price of two million dollars.

The first engines of the Newfoundland Railway were castoffs from Prince Edward Island. One of these locomotives toppled overboard from a schooner in transit at sea. The cross-country train, which came to be called the "Newfie Bullet," was once described as "the slowest crack passenger train in civilization." Early running orders limited its speed to five miles per hour, but even after the roadbed was improved and heavier steam engines were heading the train, passengers claimed they could hop off the first car and pick a cupful of berries by the time the caboose came in sight.

In the interior, where the main line passes over a barren high plateau with a group of peaks known as the Topsails, drifting snow in the winter could stall the train for days. In 1903, a train was stuck near Kitty's Brook for seventeen days. Stranded passengers were forced to tear up railway ties and use them for fuel in order to keep from freezing.

Newfoundland railroaders, like the early pioneers, had to be a stout-hearted breed, devoted, resourceful and hardworking. Curves on the line were sharp and numerous, grades were steep. And then there were the winds, which in Newfoundland can get fierce enough to blow a train right off the tracks. Not far from Port-aux-Basques lived Lauchie McDougall who used to gauge the wind velocity at his house and warn the train if it was unsafe to proceed. As a result, his home, came to be known as the "Wreck House."

The renowned "Bullet" no longer rides the rails of Newfoundland. Gone is a way of life, when people did not hurry and still thought a train trip was an adventure; when perfect strangers sang together or played cards on the train; when, on the way home after a summer at sea, they sat in the smoker and spun their yarns. Passenger trains in Newfoundland were discontinued recently and replaced by buses which cut the travelling time in half.

During the Second World War Newfoundland's railway was called upon to serve large military bases constructed on the island. Crews and equipment coped with a sudden staggering increase in passenger and freight traffic. The roadbed carried loads for which it had not been prepared. By the time Newfoundland joined Confederation in 1949 as Canada's tenth province, the antique rolling stock was badly in need of repair. A gigantic task—to bring the line up to date—faced the Canadian National Railways, which had taken over the Newfoundland system on Confederation. Equipment was overhauled, the roadbed rebuilt and steam engines were replaced by powerful diesel units. To this day, however, Newfoundland's railway remains a narrow gauge line, and mainland trains upon arrival at the dock have to be hoisted to have their wheels exchanged or else their cargo has to be transferred to narrow gauge cars.

Canadian National Railways

Tracklaying, Canadian National Railways

CN

CN

Clearing the right of way

In contrast to most other railways, Canada's largest transportation system was not born on the maps of surveyors and planners. It came into existence as the result of the amalgamation of privately and government-owned railways, and counts among its most prominent ancestors the Intercolonial, the National Transcontinental, the Canadian Northern, the Grand Trunk Pacific, and the Grand Trunk.

The collective name "Canadian National Railways" was authorized in 1918, but the organization of the company was not finalized until five years later. During the interim period some of Canada's financially troubled railways were operated by the Canadian Government Railways, a corporation established to hold title to railway properties owned by the dominion government.

One of the government holdings at the time was the Intercolonial, begun for the purpose of fostering closer ties between the provinces and completed as an instrument of Confederation. Another property was the National Transcontinental, built by the dominion government from Moncton, New Brunswick, to Winnipeg, Manitoba, as the eastern division of a new cross-Canada line. Upon completion in 1915 it was to have been operated by its western partner, the Grand Trunk Pacific. Saddled with heavy financial burdens and unable to generate sufficient traffic due to the lack of branch lines, the Grand Trunk Pacific, however, was incapable of living up to its obligations. Consequently, the National Transcontinental had become the dominion government's responsibility.

The first decade of the twentieth century had been a time of great general prosperity. During that period Canada's railway mileage nearly doubled, and by 1915 the country possessed three transcontinental lines competing with each other. Before World War I it was relatively easy to borrow money in Britain for railway ventures in Canada. Suddenly, however, it became difficult to float loans. Signs of economic recession had begun to appear. Money was needed to fight a war. The flow of immigration into Canada dwindled to a trickle. While the outbreak of the war brought increased traffic and earnings to railways in general, it also brought rising operating expenses and at the same time made further borrowing overseas all but impossible.

The Canadian Northern, Canada's third transcontinental railway, as well as the old Grand Trunk, parent of the Grand Trunk Pacific, were in dire financial straits. Loans granted from the federal treasury to keep the companies operating seemed a mere drop in the bucket.

CN

Canadian National Engine 1405, ten-wheeler, built in 1912

Eventually the government of the day had to make a decision regarding the country's railways. To withhold further urgently needed assistance and allow the companies to fall into the hands of the receivers would have meant not only disruption of rail transportation but a severe blow to Canada's future credit rating abroad. Immediate nationalization of the companies involved was not deemed advisable either. As an alternative, the government chose to appoint a Royal Commission to study Canada's railway problems and report to parliament. The commission consisted of Sir Henry Drayton, chairman of the Board of Railway Commissioners, W. C. Acworth, one of the most eminent financial experts of the United Kingdom, and A. H. Smith, president of the New York Central Railway. In May 1917, the report of the commission was tabled in the House of Commons. It was signed only by Sir Henry Drayton and Mr. Acworth. Mr. Smith filed a dissenting opinion.

The press of the day, calling the document "one of the most momentous ever presented to Parliament," informed the nation that the report recommended:

> *. . . immediate nationalization of all the railways of Canada, except the American lines and the Canadian Pacific Railway Co. The recommendation is that the Intercolonial (including the Prince Edward Island Railway), the National Transcontinental, the old Grand Trunk, the Grand Trunk Pacific and the Canadian Northern be brought into one system, to be owned by the people of Canada. This will create one of the greatest railway systems in the world, and build up a formidable competitor to the Canadian Pacific Railway Company.*

The report did not neglect to point out the many pitfalls of a government-owned and operated railway system of the proposed magnitude but, under the circumstances, it concluded that nationalization was the only feasible solution.

The privately owned railway empire of the CPR was deemed powerful enough to survive the resulting competition. A memorandum which had been presented to the government by Lord Shaughnessy, the CPR's president, analyzing the railway situation and proposing a plan under which the Canadian Pacific would operate the railways involved, was dismissed. Instead the government of Sir Robert Borden accepted the majority commission report. On August 1, 1917 the Minister of Finance, Sir Thomas White, electrified the House of Commons when he announced the immediate nationalization of the Canadian Northern system. Thus nearly ten thousand miles of tracks were added with a stroke of the pen to the government's railway holdings, and, in connection with the Intercolonial and the National Transcontinental, a national railway was established serving every province of the country. The acquisition of the Canadian Northern involved the control of a large number of subsidiary companies, including: Lake Superior terminals, with five elevators at Port Arthur with a capacity of ten million bushels; steamship lines on lake and ocean; the Canadian Northern and Great Northwestern Telegraph Companies, with 1,500 offices in Canada and direct connections with Western Union and other big cable companies; thirty underlying railway companies; stations and terminals in nearly every city of the dominion and a complete express service from coast to coast. The government was empowered to obtain possession of the railway by acquiring all shares of Canadian Northern stock which it did not already own. The value of the shares to be paid to the owners was to be determined by arbitration.

In his speech outlining future plans Sir Thomas intimated that the Grand Trunk Pacific, too, would soon become part of the publicly

Sorting the mail on the train

CN

A. Clegg

Outside Montreal, a steam train waits beneath the overhead electric wires of commuter trains

owned railway system, and possibly the Grand Trunk also. At any rate, the latter was not to be released from its commitments in regard to the Grand Trunk Pacific. For the present, however, a further loan was granted to the Grand Trunk Pacific to tide it over the coming year.

Early in 1919 the Grand Trunk advised the government that it could no longer operate the Grand Trunk Pacific. Consequently the Minister of Railways was appointed as receiver for the Grand Trunk Pacific. As for the old Grand Trunk, an agreement had yet to be worked out between the Canadian government and the company's directors. In October 1919 a bill was brought into the House authorizing acquisition of the capital stock of the Grand Trunk Railway Co. of Canada by the dominion government, and the appointment of a board of arbitration to determine the value of the stock.

Lord Shaughnessy once again offered his own practical remedy for the country's railway problems, submitting a revised memorandum outlining the terms under which the CPR would be prepared to administer a unified system and save taxpayers' money. Under the date of April 6, 1921, he wrote to Arthur Meighen, prime minister of Canada:

> *National railway affairs are, I am sure, to you a source of constant anxiety. To my mind the railway question, involving, as it does, such an enormous draft on the annual revenue of the country with no prospect of any improvement in the near future, is the most momentous problem before our country at this time.*
>
> *I fear very much that the Grand Trunk transaction will prove disappointing and expensive, and if it were my case I would go a long way to secure the consent of the Grand Trunk shareholders to the abrogation of the statutory contract.*
>
> *I am enclosing a memorandum giving in rough outline my opinion as to the only process through which the atmosphere can be cleared. Some people, whether they believe it or not, will find in my suggestions a selfish desire on the part of the Canadian Pacific to control the railway situation. The Canadian Pacific bogey has served its turn on every occasion in the past thirty-five years, when schemes were being promoted with disregard of the cost to the country.*
>
> *The Canadian Pacific has no fish to fry, and I am not sure that my plan would be viewed with favor by the executive, the*

Sir Henry W. Thornton, first president of Canadian National Railways

directors or the shareholders. Everybody connected with the company would prefer to see its status undisturbed, but it is impossible to accept with equanimity a situation which makes a demand on the public treasury of about $200,000 per day, wthout any compensating advantage, if there be any possibility of improving it.

My memorandum, as you will observe, merely brings up to date on very much the same lines a similar paper that I prepared about the end of 1917 and sent to Sir Robert Borden. He feared, I imagine, that as my plan would apparently create a Canadian Pacific monopoly in transportation it would not be acceptable to the country . . .

The Meighen government, too, rejected the CPR's suggestions. A board of arbitration, consisting of the Honourable W. H. Taft, Sir Thomas White and Sir Walter Cassels, was appointed in 1921 and entrusted with the task of placing a value on the stock of the Grand Trunk Railway. Mr. Taft valued the stock at $48 million, while the other two gentlemen held that there was no value in the common and preference shares. Needless to say, Grand Trunk stockholders were up in arms, but the majority decision stood.

On January 30, 1923 the Grand Trunk Railway of Canada, last of the country's major lines aside from the CPR, was formally amalgamated with the Canadian National Railways. Sir Henry Worth Thornton, a native of the United States and a man of wide-ranging experience in the field of railway transportation, became chairman and president of the new company. To him fell the gigantic task of coordinating the vast network of nearly twenty-two thousand miles of inherited tracks. Many of the lines had originally been built to compete with one another. Duplications of facilities and tracks running parallel for hundreds of miles were wasteful and had to be eliminated. Employees came from a variety of companies, each with its own traditions. Their loyalty had yet to be won. Sir Henry Thornton proved to be the man for the job. The speaker at a dinner given in his honour by the General Officers of the Canadian National System two years after he had taken office summed up the high esteem in which he was held by his staff:

Sir Henry Thornton has made us feel that every worker on the Canadian National Railways, from car sweeper to vice-president, is necessary to the good operation of the system and that every employee, provided he takes the necessary care with his work, can grow up from clerk, trainman, engineer to the highest posts in the service. He has proved to be one of the biggest men that has administered any railroad.

In the years to follow the system was unified, branch lines were built to round it out, services were improved and properties rehabilitated at a cost of some four hundred million dollars. Then came the great depression of the thirties which inevitably affected Canada's railways. Twice during this period Canadian National Railways' annual deficit rose to over sixty million dollars. Another Royal Commission was called upon to study the plight of the nation's railways. Proposals to amalgamate the CNR and the CPR were rejected. Instead, the commission advised coordination of services in an attempt to stop wasteful competition between the two rivals. The report of the commission led to the Canadian National-Canadian Pacific Act of 1933, which incorporated many of the commission's recommendations. World War II brought increased revenues, and the CNR managed to average a substantial annual surplus. In 1946, however, the company once again went into a deficit position. A staggering load of inherited debt in the amount

CN

No. 9000, Canadian National's first diesel engine, 1929

of nearly one and a half billion dollars had to be carried by the CNR until the Capital Revision Act of 1952 allowed one-half of it to be changed into preferred stock.

Meanwhile, the Canadian National system continued to grow. When Newfoundland entered Confederation as Canada's tenth province in 1949, its seven hundred miles of railway became part of the Canadian National Railways. The Hudson Bay Railway from The Pas to Churchill, Manitoba, was made a division of the CNR in 1951. Another railway added back then was the Temiscouata, a line built in 1889 between Rivière du Loup in Quebec and Edmundston, New Brunswick, and later extended along the Saint John River to Connors. It had never been a profitable venture, but its service to an agricultural and lumbering district was considered important enough for the CNR to rehabilitate and operate the line.

Under the leadership of Donald Gordon, who had been appointed president in 1950, the CNR began to change its image. Relics from the past, which had been pressed into service during wartime, disappeared from the road. With the emphasis on speed and comfort, millions were spent to lure travellers away from the family car and back to the rails. Streamlined diesels replaced the steam locomotives and the trains they headed took on a modern look. Canadian National's last regular steam engine No. 6043 was retired in Winnipeg on April 25, 1960 in a nostalgic farewell ceremony. After that only the occasional museum or special excursion train was to be hauled by steam.

Aware of the demand for rail transportation in the new frontier areas of Canada, CN pushed north, building branch lines to serve new mining, agricultural and manufacturing developments. The maple leaf in a slanted square, which once identified Canadian National engines and cars, has long since been replaced with the smooth-flowing CN symbol representing the flow of men, goods and ideas across this great land.

Canadian National Railways, headed by N. J. MacMillan, continues to serve the nation not only by rail, but also by road, water and sky. CN car ferries and steamers link the mainland with the islands off the east and west coasts of the country. The capital stock of Air Canada is owned by Canadian National. The company has a network of hotels from coast to coast and operates a vast telecommunication service which includes broadcasting.

Tracing its origin back to Canada's first pioneer railway of 1836 from Laprairie to St. Johns, Quebec, Canadian National has become the largest railway system on the North American continent and one of the most extensive in the world.

As a result of the amalgamation of individual railways, Canadian National Railways inherited among the rolling stock a most colourful assortment of over three thousand steam engines, ranging from the standard 4-4-0s of the early days to the powerful Santa Fe locomotives of later years. Many of the engines of the Canadian Northern and the Grand Trunk were integrated into the Canadian National's roster and continued to serve their new owner for years to come. Some of the older locomotives were scrapped while others were modernized or placed in yard service and on secondary lines.

One of the old-timers was Mogul engine No. 674 dating back to 1899. The Moguls owed their name to the fact that they were extremely powerful, and they remained on the CN roster as freight engines until the end of the steam era. Until more recent years, No. 674

A CNR passenger train in the Province of Quebec

CN

CN

Confederation Limited *in the Rockies*

used to haul the museum train of Canadian National Railways and thus became the oldest operating steam engine of its kind. After she was scrapped, she was replaced by another survivor of the Mogul family, locomotive No. 713.

There was another veteran which made the occasional journey into yesterday at the head of the museum train. It was engine No. 40, the first standard gauge woodburner built for the Grand Trunk in 1872. In her heyday she was considered a very fast locomotive. She worked for the Grand Trunk until 1903 at which time she was sold to a Quebec lumber company. In the early 1950s Canadian National purchased the engine. No longer operating, the historic relic can now be seen at the National Museum of Transportation in Ottawa.

The Canadian National acquired its first new steam locomotives in 1919 at a time when the company existed merely in name. They were 4-6-2 Pacific engines No. 5100 to 5124, built by the Montreal Locomotive Works. More Pacific-type locomotives were ordered from different builders during the years to follow. In 1929 CN bought the last three engines of this class. They were numbered 5632 to 5634. Among the earlier inherited Pacifics was No. 5093, built by the Montreal Locomotive Works, and eventually put on display in Regina, Saskatchewan.

The Canadian National motive power roster shows a considerable number of 2-8-2 Mikado engines. Some were taken over from their previous owners, others were built for CN. In 1923 the company ordered thirty-five Mikados, popular freight engines capable of relatively fast speed.

A special place in the hearts of many a railroader was occupied by the 4-8-2 Mountain-type locomotives of the Canadian National. The first of these, Nos. 6000 to 6003 and Nos. 6005 to 6015 were built in 1923 by the Canadian Locomotive Company of Kingston, Ontario. Number 6015, which served extensively in the Maritimes and in British Columbia, has since gone on display at the Railway Museum in Delson near Montreal. Mountain-type locomotive No. 6057 made history when the then Princess Elizabeth took the throttle between Yates and Piers, Alberta, during her 1951 visit to Canada.

The last twenty steam engines, numbered from 6060 to 6079, were all of the 4-8-2 type. Among the several new features incorporated in their design was a conical nose containing headlight and number lamps. Because of this they were called "Bulletnose Betties." Finished in

CN

CN Engine 5704, Hudson type 4-6-4. On April 2, 1933 this locomotive headed the first pool train in Canadian railway history

CN 6060, Mountain type 4-8-2, one of the "Bulletnose Betties." Now on display on the CN station grounds at Jasper, Alberta

No. 6405 belonged to the Grand Trunk Western. The 6400 class of 4-8-4's were capable of high speed

CN

CN

green and black with gold striping, their overall appearance was semi-streamlined.

Among the new engines purchased by the company in the early years were five Santa Fe locomotives, Nos. 4100 to 4104, built in Kingston during 1924. The 4100s were popular with their crews for more than one reason. Not only did the mighty engines have the highest tractive power of any steam locomotive built for Canadian roads and perform beautifully, but their appearance was impressive and their weight in excess of 409,000 pounds. The latter was important, since in the old days salaries of the crews were based on the weight of the locomotive which they operated. Number 4100, renumbered 4190 in 1957, is now part of the Delson exhibit. A large number of Santa Fe 2-10-2 type engines were operated by CN, mainly in transfer service between terminals.

Another workhorse of the Canadian National was the 2-8-0 Consolidation-type. CN possessed a family of over eight hundred of these engines, most of them dating back to the days prior to the company's birth. The Consols, particularly suited for mountain regions,

No. 6218, Northern type 4-8-4, Canadian National's last operating steam engine; retired July 4, 1971

CN

Canadian National's Confederation Museum *train on Trans-Canada tour, 1967*

CN

helped to open up the west. Number 2601, formerly a Grand Trunk freight engine, now represents this type of locomotive at Delson.

In 1927 a new 4-8-4 type engine of Canadian design was introduced by CN. It was No. 6100, and since she was born in the year of the sixtieth anniversary of Confederation, she became known as a Confederation-type. Later the name "Northern" was adopted for her successors which proved to be worthy opponents in the competition with the almighty diesels. Over two hundred Northerns were acquired by CN during the steam era. The 6200 class, outwardly resembling their predecessors but greatly improved, became a familiar sight on the Montreal-Toronto passenger run. Then there were the 6400s which gained fame because of their striking, streamlined appearance. They were capable of reaching a speed of one hundred miles an hour. Number 6400 pulled the royal train carrying King George and Queen Elizabeth over CN lines in 1939. Numbers 6401 and 6403 were selected to serve during the 1951 tour of Princess Elizabeth and her husband.

Among the five Hudson 4-6-4 locomotives built for Canadian National was No. 5704. This locomotive headed the first "pool train" in Canadian railway history from Montreal to Toronto on April 2, 1933. The Hudsons were extremely fast and featured the highest driving wheels of any CN locomotive. Measuring 80", they made an average-size man standing beside the wheel look rather short.

During the age of steam a total of 4,064 steam locomotives served on the roads of Canadian National Railways. Today the sights and sounds of an era have become but a memory. Canadian National no longer owns an operating steam engine.

Number 6218, a Northern-type engine built in 1942 by the Montreal Locomotive Works, was the last of five steam engines used to head special excursion trains which CN began to run for steam buffs in 1960, after it had retired its steam locomotives from regular service. In 1963 the engine had been rescued from storage, and after she was thoroughly overhauled, she embarked on a career that took her over CN lines in Ontario and Quebec, and over the company's United States subsidiaries, to the terminals of Chicago, Illinois, and Portland, Maine. To the enjoyment of thousands her whistle echoed through the countryside as a reminder of a glorious past. And then came a day when reality had to be faced. It had become too costly to keep the old engine running. Major boiler repairs were needed and few old-timers were left in the

CN

Experimental Turbo train tested on Canadian National Railways, Toronto to Montreal, 1968

employ of CN who could repair, service and operate the steam giant of yesteryear.

And so began what has become known in the annals of Canadian Railways as "Countdown 6218." Early in July 1971 the locomotive arrived in Belleville, Ontario. Polished to a gleam, she stood on the tracks at the station. Her last short excursion run was scheduled for the morning of July 4th. Early that afternoon, a crowd of several thousand gathered along the tracks to bid her farewell at a nostalgic retirement ceremony. Dignitaries of Canadian National Railways were welcomed by Mayor R. Scott of Belleville, a city which has been a railway centre since the days of the old Grand Trunk.

Omer Lavallée, well-known railway historian and member of the public relations staff of the Canadian Pacific Railway, had been invited by CN officials to give the valedictory for the occasion. He closed his address with the words of an old English poem:

My engine now is cold and still,
No water does my boiler fill.
My wood affords its flame no more,
My days of usefulness are o'er . . .

After the haunting music of a bagpipe band, Keith Hunt, vice-president of transportation, Canadian National Railways, gave the train order to engineer, L. Langabeer, for the last run-past. As No. 6218, under a magnificent plume of steam, went past the hushed crowd, many an old railway fan wiped a tear from his eye. Then the engine went to the roundhouse to have her fires drawn.

Acknowledgments

Railways of Canada is the result of several years of research. A great deal of our work was done at the Douglas Library of Queen's University in Kingston. Our sincere thanks go to Mr. D. A. Redmond, Chief Librarian, for permitting us to use the library's facilities. We are greatly indebted to Mr. William F. E. Morley, Curator of Special Collections at the Douglas Library, for his assistance and advice throughout the past years. To him and his competent staff at the library we say a sincere thank you.

We also wish to thank Miss H. Dechief and Mr. J. G. Coté of the CNR Headquarters Library in Montreal for their kindness and the help they have given us during the time the book was in preparation. To Mr. F. G. Feeny of the CPR Library in Montreal we wish to express our appreciation for the assistance he has given us in our research; many thanks also to Mr. H. T. Coleman and to Mr. Omer Lavallee for helping us in the selection of illustrations.

To Mr. J. Norman Lowe of Montreal we would like to say thank you for allowing us to use pictures in his personal collection. We are very grateful also to Mr. A. Clegg of Montreal for his kindness whenever we called on him. Miss Olive Delaney, Chief Librarian of the Belleville Public Library, and her staff who helped us find much of the material needed for our research also merit our grateful appreciation.

And last but not least, we are glad to thank our good friend, Mr. James Plomer, who is always ready to give us a helping hand.

Bibliography

Canadian Pacific Railway Co. *Confederation and the Canadian Pacific,* 1927.

Currie, A. W. *The Grand Trunk Railway of Canada.* Toronto: University of Toronto Press, 1971.

Dorman, Robert (compiler). *A Statutory History of the Steam and Electric Railways of Canada.* Ottawa: Petenaude, 1938.

Fleming, Sandford. *The Intercolonial.* Montreal: Dawson Brothers, 1876.

Gibbon, John Murray. *Steel of Empire.* Toronto: McClelland & Stewart, 1935.

Hedges, James B. *Building of the Canadian West.* New York: Macmillan, 1939.

Phillips, R. A. J. *Canada's Railways.* Toronto: McGraw-Hill, 1968.

Ramsey, Bruce. *PGE: Railway to the North.* Vancouver: Mitchell Press, 1962.

Skelton, O. D. *The Railway Builders.* Toronto and Glasgow: Brook & Co., 1916. Reprinted Toronto: University of Toronto Press, 1964.

Stevens, G. R. *Canadian National Railways.* Toronto and Vancouver: Clarke, Irwin, 1960.

Thompson, G. N. V. and J. H. Edgar. *Canadian Railway Development from the Earliest Times.* Toronto: Macmillan, 1933.

Trout, J. M. and Edward. *The Railways of Canada for 1870-71.* Toronto, 1871.

Bulletins and Newsletters, Upper Canada Railway Society

Canadian Geographical Journal

Canadian Illustrated News (Issues of 1870s and 1880s)

Canadian National Railways Magazine

Canadian Rail

CRHA News Reports

Encyclopedia Canadiana

Illustrated Historical Atlases, 1875-1901 (various Ontario counties)

Maclean's Magazine

Mika Collection of Newspaper Clippings, 1850-1930

Papers and Records, Ontario Historical Society

Railroad Magazine

The Globe Magazine

The Maritime Advocate and Busy East

Trains

Western Ontario History Nuggets, Lawson Memorial Library

Index

CP

The text material in this book is set in 10 point Palatino,
with Palatino headings.
The book was designed by Julian Cleva:
composition by HCM Group
lithography and binding by Ronalds Federated Limited.